VIRAL PATHOGENESIS AND IMMUNITY

VIRAL PATHOGENESIS AND IMMUNITY

Neal Nathanson

Co-authors
Rafi Ahmed
Margo A. Brinton
Louise T. Chow
Francisco Gonzalez-Scarano
Diane E. Griffin
Kathryn V. Holmes
Frederick A. Murphy
Julie Overbaugh
Harriet L. Robinson

LIPPINCOTT WILLIAMS & WILKINS
A **Wolters Kluwer** Company

Philadelphia • Baltimore • New York • London
Buenos Aires • Hong Kong • Sydney • Tokyo

Acquisitions Editor: Jonathan Pine
Developmental Editor: Delois Patterson
Production Editor: Frank Aversa
Manufacturing Manager: Ben Rivera
Cover Designer: Karen Quigley
Compositor: Maryland Composition
Printer: Courier-Westford

Library of Congress Cataloging-in-Publication Data

Viral pathogenesis and immunity / editor, Neal Nathanson; co-authors, Rafi Ahmed. . . [et al.].
    p. ; cm.
    Includes bibliographical references and index.
    ISBN 0-7817-1103-7
    1. Virus diseases—Pathogenesis. 2. Virus diseases—Immunological aspects. 3. Host—virus relationships. I. Nathanson, Neal. II. Ahmed, Rafi.
    [DNLM: 1. Viruses—pathogenicity. 2. Virus Diseases—etiology. 3. Virus Diseases—immunology. QZ 65 V8132 2001]
    QR201.V55 V533 2001
    616′.0194—dc21

                                                                2001038535

Care has been taken to confirm the accuracy of the information presented and to describe generally accepted practices. However, the authors, editor, and publisher are not responsible for errors or omissions or for any consequences from application of the information in this book and make no warranty, expressed or implied, with respect to the currency, completeness, or accuracy of the contents of the publication. Application of this information in a particular situation remains the professional responsibility of the practitioner.

The authors, editor, and publisher have exerted every effort to ensure that drug selection and dosage set forth in this text are in accordance with current recommendations and practice at the time of publication. However, in view of ongoing research, changes in government regulations, and the constant flow of information relating to drug therapy and drug reactions, the reader is urged to check the package insert for each drug for any change in indications and dosage and for added warnings and precautions. This is particularly important when the recommended agent is a new or infrequently employed drug.

Some drugs and medical devices presented in this publication have Food and Drug Administration (FDA) clearance for limited use in restricted research settings. It is the responsibility of the health care provider to ascertain the FDA status of each drug or device planned for use in their clinical practice.

10 9 8 7 6 5 4 3 2 1

Contents

IV. PREVENTION OF VIRAL INFECTIONS

viral pathogenesis and immune protection; comparison of vaccine modalities, including inactivated and attenuated virus vaccines, and newer vectors such as recombinant viruses and naked DNA; mechanisms of protection by established vaccines; roles of antibody and cellular immunity; the correlates of protection; sterilizing immunity vs partial protection; and challenges for an AIDS vaccine

Contributing Authors

Rafi Ahmed, M.D.
Institute for Vaccine Research and
Department of Microbiology and Immunology
Emory University
Atlanta, Georgia
(Chapters 5, 6)

Margo A. Brinton, M.D.
Department of Biology
Georgia State University
Atlanta, Georgia
(Chapter 12)

Louise T. Chow, M.D.
Department of Biochemistry
University of Alabama at Birmingham
Birmingham, Alabama
(Chapter 11)

Francisco Gonzalez-Scarano, M.D.
Departments of Neurology and Microbiology
University of Pennyslvania Medical Center
Philadelphia, Pennsylvania
(Chapters 8, 9)

Diane E. Griffin, M.D.
Department of Molecular Microbiology and
Immunology
School of Hygiene and Public Health
Johns Hopkins University
Baltimore, Maryland
(Chapters 4, 7)

Kathryn V. Holmes, M.D.
Department of Microbiology
University of Colorado Health Sciences Center
Denver, Colorado
(Chapter 3)

Frederick A. Murphy, M.D.
Medical and Veterinary Virology
School of Veterinary Medicine
University of California, Davis
Davis, California
(Chapters 1, 2)

Neal Nathanson, M.D.
Departments of Microbiology and Neurology
University of Pennyslvania Medical Center
Philadelphia, Pennsylvania

Julie Overbaugh, M.D.
Molecular Medicine
Fred Hutchinson Cancer Center
Seattle, Washington
(Chapter 13)

Harriet L. Robinson, M.D.
Yerkes Regional Primate Research Center and
Department of Microbiology and Immunology
Emory University
Atlanta, Georgia
(Chapters 10, 14)

Preface

To wrest from nature the secrets which have perplexed philosophers in all ages, to track to their sources the cause of disease . . . these are our ambitions.

Sir William Osler

This book is the direct offspring of *Viral Pathogenesis*, first published in 1997. Having read several drafts of all the chapters in *Viral Pathogenesis*, it was clear that the large book contained a wealth of information but that it was unsuited for readers who desired an introduction to the topic. From that observation sprang the plan for a short version that could be used as a textbook or for self-education.

An introductory text clearly benefits from the coherence provided by a single author, but suffers from the finite expertise of any single researcher. Therefore, a compromise was devised, in which each chapter was co-authored by an expert in the specific area under consideration. This strategy was facilitated by the successful collaborations that had been developed during the preparation of *Viral Pathogenesis.*

I acknowledge the contributions of the co-authors. Their advice and expertise have been essential to the planning and execution of this undertaking, and it has been a continual pleasure to work with them. They have provided extremely cogent suggestions that have given the book an enhanced level of authority that could not otherwise have been achieved.

Once again, the staff at Lippincott Williams & Wilkins has been an ongoing source of support because of their enthusiasm for this book and their highly professional expertise in all phases of the project. Delois Patterson has helped to maintain momentum through the inevitable distractions that plague authors, while Jonathan Pine has provided overall guidance. Wendy Jackelow provided the outstanding illustrations rendered from a wide variety of often primitive sketches.

Neal Nathanson
Philadelphia
July 2001

To the Reader

Viral Pathogenesis and Immunity explores all aspects of viral infection of the animal host, including the sequence of events from entry to shedding, the clearance or persistence of the virus, the immune response of the host, and the subsequent occurrence of disease. Particular attention is focused on mechanisms that explain the complex interaction between parasite and host.

This book is designed to provide an introductory overview of viral pathogenesis in a format that will be easy for the reader to absorb without recourse to additional information. Principles are emphasized, and no attempt is made to provide a virus-by-virus or disease-by-disease compendium, since these are already available in texts of microbiology and infectious diseases. Representative, not encyclopedic, examples are given to illustrate principles. By keeping to essentials, we hope to provide a coherent introduction in a brief compass, leaving the reader to acquire more detailed information about specific examples from well-documented comprehensive texts.

It is assumed that the reader knows the fundamentals of essential virology, including the structure of viruses, the organization of their genomes, the basic steps in viral replication, assembly, and release. In addition, a basic background in cell biology, immunology, and pathology will be useful. Students who have taken an introductory course in microbiology will have acquired this background and should be well equipped to use this book. For those who wish to review these essentials, many excellent texts are available and some outstanding ones are noted below. In addition, at the end of each chapter a few selected references are provided for readers who wish to delve more deeply into the subject matter or who wish to read a few of the classic original contributions to the field.

Viral Pathogenesis and Immunity is divided into four sections. Section I, *Essentials of pathogenesis*, acquaints the reader with the sequential events in viral infections, the dissemination of virus in the host, and the variety of cellular responses to infection. Section II, *Host response to viral infection*, describes the nonspecific and specific immune responses to infection, including the immunopathological and immunosuppressive consequences of infection. Section III, *Virus–host interactions*, deals with virus virulence, virus persistence, virus-induced oncogenesis, and the determinants of host susceptibility. Section IV, *Prevention of viral infections*, applies the principles of pathogenesis to methods of prevention or treatment of infection.

This organization permits readers to select those subjects of particular interest to them, depending upon their background, goals, and available time. Thus, it would be possible to base an abbreviated introduction to the subject upon Sections I and III alone, particularly for readers with some background in immunology.

BACKGROUND READING

Brooks GF, Butel JS, Morse SA. *Jawetz, Melnick, and Adelberg's Medical Microbiology*, 21st ed. Lange, Norwalk, 1998. Basic chapters on properties of viruses and on immunology will provide sufficient background for readers who have not taken a course in microbiology or immunology.

Abbas AK, Lichtman AH, Pober JS. *Cellular and Molecular Immunology,* 4th ed. WB Saunders, Philadelphia, 2000. Selected chapters will provide a thorough and clear introduction for readers who have not taken a course in immunology.

FURTHER READING (general references)

Nathanson N, Ahmed R, Gonzalez-Scarano F, et al,, editors. *Viral Pathogenesis*. Lippincott Williams & Wilkins, Philadelphia, 1997. This reference work provides much more detailed information and an extensive bibliography for readers who wish to acquire more in-depth information about a specific topic in viral pathogenesis.

Knipe DM, Howley PM, editors. *Fields' Virology,* 4th ed. Lippincott Williams & Wilkins, Philadelphia, 2001. This exhaustive reference work provides definitive information about basic virology and about individual viruses and virus families.

Mandell GL, Bennett JE, Dolin R. *Mandell, Douglas, and Bennett's Principles and Practice of Infectious Diseases,* 5th ed. Churchill Livingstone, New York, 2000. This authoritative reference describes specific viral infections of humans.

VIRAL PATHOGENESIS AND IMMUNITY

PART I
Essentials of Pathogenesis

Chapter 1
Historical Roots

HISTORY OF INFECTIOUS DISEASES AND MICROBIOLOGY

The history of viral pathogenesis is intertwined with the history of medicine. Ancient physicians recorded clinical illnesses, understanding that classification of diseases was a prerequisite for prescribing remedies, although treatments were often of questionable value. Viral diseases that were clearly recognized in ancient times were those that produced distinctive or unique signs and symptoms, such as poliomyelitis. In a few instances, such as rabies in which illness often followed the bite of a rabid dog or wolf, even the transmissible nature of the illness was clearly understood (Fig. 1.1).

From the time of early civilizations, relatively little progress was made until the Renaissance, when development of modern science was sparked by the prolific genius of Leonardo da Vinci (1452–1519). Girolamo Fracastoro, writing in the 16th century, proposed a theory of contagions caused by "small imperceptible particles" that were transmitted by contact, by fomites, or over distances. Fracastoro's theoretical treatise, though based on speculation, was a remarkably prescient vision that paved the way for the discovery of microbial organisms. The actual beginnings of microbiology are dated by some historians to the late 17th century when Antony van Leeuwenhoek described bacteria and other unicellular organisms seen through the microscopes that he built himself. Microorganisms could readily be observed in infusions or in putrefying materials, and one controversial question was whether they arose by spontaneous generation. In the late 18th century, Lazaro Spallanzani devised some simple but telling experiments showing that organisms in a flask could be killed by heating or boiling and did not reappear if the flask had been sealed to preclude reseeding from the air.

Nevertheless, the understanding of the nature of infection was relatively primitive in the late 18th century. For instance, the yellow fever epidemic in Philadelphia in 1793 engaged the best medical minds of 18th-century America, including Benjamin Rush, generally considered the leading physician of the colonies. Rush hypothesized that the disease arose from some effluvium deposited on the docks by ships recently arrived from the Caribbean, apparently not focusing on the human cases of yellow fever imported by the same ships. Furthermore, he prescribed a regimen of frequent "cupping" (therapeutic bleeding) that only served to debilitate the mortally affected sufferers.

Viral epidemics, in which most clinical cases were due to a single organism, indicated the transmissible nature of infections and demonstrated stages in the evolution of the infectious process. One instance is the epidemic of measles that occurred in the Faroe Islands in the North Atlantic in 1846, recorded by Peter Panum, a young Danish physician. The disease was introduced by a cabinet maker from Copenhagen who arrived on March 28 and developed measles in early April. Between April and October, more than 6,000 cases occurred among the population of almost 7,900, with more than 170 deaths.

Based on a number of simple clinical observations, several important inferences were drawn. First, the disease was clearly transmitted from person to person by direct contact, and it spread in this fashion to overtake almost the whole population. This strongly suggested that measles was caused by a specific agent, contradicting the vague miasma theory of febrile diseases that had been popular for

3

FIG. 1.1

Rabid dog biting a man. (Arabic painting by Abdallah ibn al-Fadl, Baghdad school, 1224. Courtesy of the Freer Gallery of Art, Washington, DC. After Baer G, ed., *The natural history of rabies,* 2nd ed. Boca Raton: CRC Press, 1991, with permission.)

centuries. Second, it appeared that most cases in the epidemic exhibited consistent signs and symptoms, such as the typical rash, suggesting that each transmissible disease might be attributable to a distinct agent. Third, the interval from exposure to onset was about 2 weeks, and the patient was contagious at the onset of illness, indicating a stereotyped natural history. For measles, this included a silent incubation period, followed by a febrile rash with virus shedding. Finally, the outcome of illness was influenced by age, with highest mortality among infants and the very elderly. This was one of the first documented instances of variable host responses to a single infectious agent.

In the 19th century, the theory of spontaneous generation was definitively refuted by a number of workers, particularly Schwann in 1837 and Cagniard-Latour in 1838. In 1857, Pasteur found that different fermentations were associated with different microbial agents, providing further evidence against spontaneous generation and setting the stage for the idea that each infection was caused by a specific agent. In 1850, Semmelweis inferred that physicians were spreading child-bed fever, a streptococcal infection, by failing to wash their hands, and in 1867 Lister showed that carbolic acid applied as an antiseptic could reduce postoperative infections. These advances strengthened belief in the microbial origin of infection and contributed practical applications of the concept.

In the second half of the 19th century, rapid advances were made (Table 1.1). One after another, the causal bacteria responsible for important infections were defined, beginning with isolation of the anthrax

TABLE 1.1

Age-specific differences in mortality from measles[a]

ANNUAL MORTALITY PER 100

Age group	1835–1845	1846	Excess in 1846
<1	10.8	30.0	19.2
1–9	0.5	0.5	—
10–19	0.5	0.5	—
20–29	0.5	0.7	0.2
30–39	0.8	2.1	1.3
40–49	1.1	2.7	1.6
50–59	0.9	4.4	3.5
60–69	2.0	7.7	5.7
70–79	6.5	13.1	6.6
80–100	16.8	26.0	9.2

[a] Data from the measles epidemic in the Faroe Islands, 1846, compared with average mortality for 1835–1845. The excess mortality for 1846 provides a crude estimate of measles-specific mortality during the epidemic, which involved at least 75% of the population. Data from Panum PL. *Observations made during the epidemic of measles on the Faroe Islands in the year 1846.* New York: American Public Health Association, 1940.

bacillus from the blood of infected animals by Davaine in 1865 and its transmission to mice by Koch in 1877. In 1881, Koch was able to grow bacteria on solid media, facilitating isolation of pure cultures of single organisms. In 1884, Koch, drawing on the ideas enunciated in 1840 by his teacher Jacob Henle, conceptualized the relationship between individual infectious agents and specific diseases as a series of axioms, commonly known as the Henle–Koch postulates (Sidebar 1.1).

Viruses were discovered as a direct outgrowth of these studies of bacterial agents. Between 1886 and 1892, Mayer and Beijerinck, working at the Agricultural Experimental Station in Wageningen, Holland, and Ivanovsky, working independently in Russia, demonstrated that mosaic disease of tobacco could be transmitted from plant to plant by extracts of infected vegetation. Furthermore, no bacterial agent could be grown from these extracts, and the infectivity could pass through Chamberland filters (porcelain filters with a pore size of 100–500 nm that excluded most bacteria).

We now consider these observations to represent the discovery of the first recognized virus, tobacco mosaic virus. However, at the time there was a controversy regarding whether the causal agent was in fact a bacterium capable of passing through the filters and incapable of growing on the medium used, or the first representative of a new class of agents. Beijerinck showed that the infectious agent multiplied in plant tissues but not in the sap and championed the latter view, naming the class of causal agents "contagium vivum fluidum" or "con-

tagious living fluid." Shortly after these studies, and informed by them, the first animal viruses were identified: foot-and-mouth virus (a picornavirus) and yellow fever virus (a flavivirus).

Foot-and-mouth disease was a highly contagious, sometimes fatal vesicular disease of cattle and swine that was a serious problem for farmers in Germany. Loeffler and Froesch from the Berlin Institute for Infectious Diseases, one of the foremost institutions for infectious disease research in the late 19th century, were commissioned to study the problem. Their published report is astoundingly modern. With impeccable logic, the investigators focused on fluids obtained by puncturing alcohol-sterilized early vesicles, the one source where the infectious agent could be obtained free of contaminating skin bacteria. Using rigorous techniques they excluded bacteria, although they scrupulously noted that they could not exclude bacteria incapable of growing on the media used and invisible in their microscopes.

Using filtration (controlled by samples to which known bacterial strains had been added to eliminate potentially undetected bacteria) they showed that the causal agent was present in high titer in filtered lymph from infected animals. Two explanations remained: either they were dealing with a toxin or with a subbacterial infectious agent. Careful calculations of the cumulative dilutions produced by serial pas-

■ ■ ■

SIDEBAR 1.1

The Henle–Koch Postulates, as Originally Framed

- The incriminated agent can be cultured from lesions of the disease.
- The agent can be grown in pure culture.
- The agent reproduces the disease when introduced into an appropriate host.
- The agent can be recultured from the diseased host.

Revised and expanded versions of these postulates have been developed because certain infectious agents—such as several hepatitis viruses—cannot be "cultured" or because there is no nonhuman host in which the disease can be reproduced. In such instances, the evidence contributed by virology, molecular genetics, immunology, epidemiology, and biostatistics are required to establish a causal relationship. In addition, it was emphasized by Evans that the postulated relationship between organism and disease "must make biological sense." ■

After Evans AS. Causation and disease: the Henle–Koch postulates revisited. *Yale J Biol Med* 1976;49:175–195.

sage indicated that either the toxin was even more virulent than tetanus toxin, the most potent bacterial toxin then known, or the disease was caused by a replicating agent. In the latter case, the organism was smaller than known bacteria and incapable of growth on bacterial media. In conclusion, the authors recognized that the foot-and-mouth disease agent might be the prototype of a new class of agents, and they nominated smallpox and vaccinia as potential members of this class. The combination of rigorous thinking, meticulous execution, and far-reaching insights marks their report as truly unique, a paper that is astonishing to read more than 100 years after its publication.

During the 17th, 18th, and 19th centuries, urban yellow fever was endemic in the major cities of South America and the Caribbean, and intermittently epidemic in many of the major ports of North America. In 1900, the U.S. Army sent Major Walter Reed to Cuba to head a commission to study yellow fever, which was causing devastating morbidity and mortality in troops stationed in the Caribbean theater. The commission arrived during a severe outbreak of disease and set to work to identify the causal agent. After eliminating a bacterial candidate, *Bacillus icteroides*, as a secondary invader, they decided to test the hypothesis that the agent was transmitted by mosquitoes, which had been proposed 20 years earlier by Carlos Finlay, a Cuban physician.

They devised a trial in which volunteer soldiers were divided into two groups. One group used bedding previously occupied by soldiers with acute yellow fever but were housed in barracks that were screened to exclude mosquitoes, whereas the other group occupied clean barracks that were unscreened. Only troops in the unscreened barracks developed the disease. Using colonized *Aedes aegypti* mosquitoes obtained from Dr. Finlay, Reed and his colleagues were able to transmit the disease by mosquitoes that had fed on acutely ill patients and then, about 2 weeks later, on human volunteers. Furthermore, they demonstrated that blood obtained from acutely ill patients transmitted the disease to volunteers. At the suggestion of William Welch, the famous pathologist from the Johns Hopkins University, who was aware of the work of Loeffler and Frosch, they injected three volunteers with serum from patients in the early phases of yellow fever, which had been diluted and passed through a Berkfeld bacteria-excluding filter; two of the volunteers developed the disease. This was the first demonstration that an infectious disease of humans was caused by a virus (Sidebar 1.2).

■ ■ ■

SIDEBAR 1.2

Origin of the Word "Virus"

The word *virus* is derived from the Latin for "poison" and was traditionally used to denote the cause of any transmissible disease. With the discovery of agents that could pass bacteria-retaining filters, the term "filterable virus" was introduced, and this was later shortened to "virus." Pioneering virologists crafted biologic definitions emphasizing that viruses were obligate intracellular parasites which, in their extracellular vegetative phase, formed particles smaller than bacteria (virus particles or virions range in size from about 15 to about 300 nm), and that these virions could—in some cases—be crystallized like chemical compounds. Subsequently, modern genetic and biochemical definitions of viruses were introduced, which emphasized that viral genomes consisted of RNA or DNA (or both nucleic acids) that encoded structural proteins, which were incorporated into the virus particle, and nonstructural proteins that were essential for replication, transcription, translation, and processing of the viral genome. Probably the most succinct description is that of Peter Medawar: "bad news wrapped in protein." ■

EARLY STUDIES OF VIRAL PATHOGENESIS: 1900–1950

Virology was severely constrained in the first half of the 20th century by several technical limitations, the most important of which was the lack of a cell culture system for growing and titrating viruses. In the absence of methods for detecting viruses in tissues, observations of experimental infections were limited to clinical signs and pathologic lesions that represented the end stage of disease. In spite of these adverse circumstances, extensive studies were undertaken of a few infections, such as poliomyelitis and Rous sarcoma.

Poliovirus was first isolated in 1908 by Landsteiner and Popper, who transmitted the infectious agent to monkeys by injection of a homogenate of the spinal cord from an acutely fatal human case. It was observed early on that the infection could not be transmitted to laboratory rodents; therefore, virus stocks were prepared by monkey-to-monkey passage, using the intracerebral route of inoculation. Investigators did not appreciate that this procedure neuroadapted the virus, changing its biological properties. Many experiments were performed using the mixed virus (MV) stock of poliovirus, later shown to be a type 2 strain that was an obligatory neurotropic virus. With the MV strain the only "natural" route by which rhesus monkeys could be infected was by intranasal instillation; it was later shown that the MV strain spread up the olfactory nerve to the brain stem, and thence to the spinal cord to destroy the lower motor neurons resulting in flaccid paralysis.

These experiments led to the conviction that all polioviruses were neurotropic (viruses that mainly replicate in neural tissues), a scheme of pathogenesis that had been widely accepted when it was summarized by Simon Flexner in 1931. This view of the pathogenesis of poliomyelitis led, in the summer of 1936, to a trial employing zinc sulfate as an astringent nasal spray; although the treatment produced some cases of anosmia, it failed to prevent poliomyelitis. The failure of this trial stimulated a reexamination of the pathogenesis of poliomyelitis, which was radically revised only after Enders's introduction in 1949 of cell culture methods that permitted the isolation and propagation of virus strains retaining the properties of wild virus during laboratory passage.

Peyton Rous's identification of the avian sarcoma virus that still bears his name is a remarkable example of pioneering work that earned a Nobel Prize in 1966. Experimental transplantation of tumors was first accomplished at the beginning of the 20th century by the immunologist Paul Ehrlich, who successfully adapted several mouse mammary carcinomas so that they could be transplanted to many strains of mice. These experiments demonstrated that transplantation was facilitated by the use of newborn or very young animals, by the intraperitoneal route of transfer, and by the use of cell suspensions rather than solid tumor masses.

Based on these observations, Rous began his studies of the sarcomas of domestic chickens. In the Plymouth Rock breed (a partially inbred line of chickens), tumors could be transferred by subcutaneous inoculation and became more aggressive on serial passage. Rous made the seminal observation that the tumors could be passed by cell-free extracts, which were still active after filtration through a bacteria-retaining filter. Furthermore, chickens could be immunized to resist tumor transplantation, and it was possible to experimentally differentiate immunity against whole tumor cells from immunity against the filtrable tumor-producing agent. These seminal studies were published between 1910 and 1913; however, due to the technical limitations of experimental virology little additional progress was made during the next 40 years. With the advent of new methods in cell biology and molecular genetics between 1955 and 1980, Rous sarcoma virus became a prototype system for the discovery of reverse transcriptase, the identification of retroviruses, and the discovery of oncogenes (Sidebar 1.3).

THE CLASSICAL ERA: 1950–1975

The study that ushered in the quantitative era of viral pathogenesis was Frank Fenner's classical investigation of mousepox (also called infectious ectromelia). Mousepox, a smallpox-like infection of mice, was shown to be caused by a transmissible virus in 1931, and in 1937 Burnet reported that the agent could be quantitatively assayed on the chorioallantoic membrane of embryonated chicken eggs; antibody could also be titrated by this method or by inhibition of its ability to agglutinate red blood cells under controlled conditions. These technical advances laid the way for Fenner to describe the sequential course of experimental infection, from entry by intradermal inoculation, to viremia, to spread to the liver and skin, and transmission by virus-contaminated skin shed from cutaneous pox. This information was summarized in 1948 in a classic diagram that conveys the dynamics of the infection (Fig. 1.2).

In the classical era, the most significant of the early breakthroughs was the development of methods for the culture of primary and continuous lines of mammalian cells. In 1952, exploiting cell cultures, Dulbecco demonstrated that viruses could be assayed by the plaque method, which was derived from the colony counts used to titrate bacteria and the plaque assays used for quantitating bacterial viruses (also called bacteriophages). A variant of this approach was used for tumor viruses that could be assayed in culture for their ability to produce foci of transformed cells.

A second significant advance was pioneered by Coons, who in 1953 introduced a procedure for the identification of viral antigens in cells. This made it possible to localize an agent to specific tissues and cell types in the infected host and to observe its progressive spread during the course of infection. This method depended on the development of techniques for the chemical labeling of antibody molecules with fluorescent "tags" so that the antibody could be visualized microscopically using an ultraviolet light source. Beginning about 1955, electron microscopy was introduced to permit morphologic observations at a subcellular level so that certain steps in the in-

■ ■ ■
SIDEBAR 1.3

The Development of Inactivated Poliovirus Vaccine

"In 1945, Professor Burnet of Melbourne (Macfarlane Burnet, subsequently to receive a Nobel Prize) wrote: `While I was in America recently I had good opportunity to meet with most of the men actively engaged on research in poliomyelitis. . . . The part played by acquired immunity to poliomyelitis is still completely uncertain, and the practical problem of preventing infantile paralysis has not been solved. It is even doubtful whether it ever will be solved.' Most of us doing research on poliomyelitis in 1945 were mainly motivated by curiosity, rather than by the hope of a practical solution in our lifetime" (Bodian, 1976). Yet, in 1954, less than 10 years later, a successful trial of inactivated poliovirus (Salk) vaccine was underway. What happened in the interval illustrates the importance of understanding pathogenesis for the development of practical methods for the control of viral diseases.

The chain of discoveries is readily followed. First, in 1949, Enders, Weller, and Robbins showed that it was possible to make cell cultures from a number of human tissues and that some of these cells would support the replication of poliovirus with a very obvious cytopathic effect. For the first time it was possible to readily isolate wild strains of poliovirus and show that the virus was excreted in the feces of patients undergoing acute poliomyelitis, strongly suggesting that the causal agent entered its host by ingestion and replicated in the gastrointestinal tract, a view that had been espoused by Swedish workers in the early 20th century but had been discarded by later investigators. Fresh field isolates grown in cell cultures were now available for experimental study in primates, and monkeys could be infected by feeding these isolates. Furthermore, and

critically, it was now possible to show that the virus produced a plasma viremia and traveled through the blood to reach the spinal cord where it attacked anterior horn cells to cause flaccid paralysis—its dreaded hallmark.

Tissue culture methods permitted the development of a simple and rapid method for the measurement of neutralizing antibodies, and a combination of studies, using cell culture assays and monkey challenges, showed that all poliovirus isolates could be grouped into three types, with neutralization and cross-protection within each type but not between types. Pooled sera from convalescent primates or from normal humans (gamma globulin) had substantial neutralizing titers and, administered prior to challenge with wild-type poliovirus, were shown to protect monkeys and chimpanzees against paralysis. It now remained to develop a vaccine to induce neutralizing antibodies. This was accomplished by Salk and his colleagues in the early 1950s using formalin to inactivate poliovirus purified from mass-produced batches of virus. His studies showed that infectivity could be ablated while antigenicity was maintained so as to induce the desired antibody response. Furthermore, a multivalent vaccine could be made, containing viruses of each antigenic type. In retrospect, understanding the role of viremia in infection, simple though it was, provided the logical basis for identifying neutralizing antibody as the immune correlate of protection, which established a rational basis for development of the vaccine. ■

This account (and the quotations) has been freely adapted from Bodian D. Poliomyelitis and the sources of useful knowledge. *Johns Hopkins Med J* 1976;138:130–136.

tracellular replication of viruses could be visualized together with the pathologic consequences in individual cells. These histologic methods complemented the quantitative assays of viral titers in tissue homogenates and body fluids.

A third important advance was the introduction of techniques for measurement of the immune response to viral infections. In the 1950s, methods were established for measurement of antiviral antibody, using neutralization, complement fixation, hemagglutination inhibition, and other assays. Primitive assays of cellular immunity, such as delayed hypersensitivity following intradermal injection of antigen, were introduced in the 1940s, but it was not

until the 1970s that more quantitative methods became available with the application of in vitro assays for cytolytic T lymphocytes by Doherty, Zinkernagel, and others.

Using these methods, and following Fenner's example, classic studies of a number of viral infections were conducted. Noteworthy examples are studies of poliomyelitis by Bodian, Howe, Morgan, and others (1940–1960), of arboviruses and ectromelia by Mims (1950s), of arboviruses and rabies by Johnson (1960s), of rabies by Baer and others (1970s), and of lymphocytic choriomeningitis by Armstrong, Rowe, Hotchin, Lehmann-Grube, and others (1945–1965).

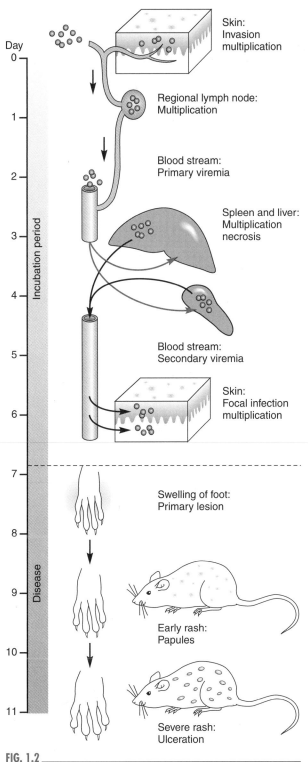

FIG. 1.2

The spread of ectromelia virus after intradermal infection of a mouse. (Redrawn from Fenner F. The pathogenesis of the acute exanthems: an interpretation based on experimental investigations with mousepox (infectious ectromelia of mice). *Lancet* 1948;2:915–920, with permission.)

THE ERA OF MOLECULAR BIOLOGY: 1975–PRESENT

With the advent of molecular biological methods, it became possible to sequence viral genomes and to modify them in order to determine the genetic basis of viral variation and virulence. Applied to in vivo studies, viral genomes and their transcripts could be visualized, using in situ hybridization and the in situ polymerase chain reaction. Starting in the 1990s, it became possible to manipulate the genomes of mammalian hosts, either by ablating a specific gene function ("knockout") or by inserting new or altered genes (transgenic animals), and these techniques have been used to tease apart the components of the host response, both those that protect and those that lead to disease.

Methods in immunology, particularly cellular immunology, have been radically upgraded in the 1990s, reflecting several important developments: first, an increased sophistication of flow cytometry that permits the separation and analysis of subpopulations of lymphocytes based on an array of surface markers; second, the discovery of an assortment of cytokines that transmit information among lymphocytes and monocytes; and third, the rapidly evolving field of molecular cell biology that has revealed a wide array of complex intracellular signaling pathways.

Currently, techniques in cell biology, immunology, molecular biology, and genetics, as well as virology, are being exploited to advance our understanding of specific problems in viral pathogenesis. A few selected examples will illustrate the advances made with these newer methods.

Live attenuated strains of poliovirus (oral poliovirus vaccine, OPV) were licensed for use as vaccines in the United States in 1961 and were widely used in the years that immediately followed. Epidemiologic surveillance soon documented that vaccine-associated cases of poliomyelitis were occurring both in vaccine recipients and in their immediate contacts. OPV is administered by feeding, replicates in the intestine, and is excreted in the stool. When virus isolates from recently vaccinated subjects were tested in cell culture or in monkeys for markers of attenuation, it was apparent that many isolates exhibited a phenotypic reversion to virulence.

The genetic sequences of the attenuated poliovirus vaccine strains differ at a number of sites from their virulent parent strains. Testing of chimeric viruses, constructed by substituting patches of avirulent genomes into the genetic "backbone" of virulent viruses, identified about 10 critical bases that were critical for attenuation, spread across the 7,000-base genome. As shown by Minor, Almond, and Racaniello, the reversion from attenuated to a more virulent phenotype after OPV feeding to humans was due to the selection of virus clones with mutations at several of these critical sites. Based on this information, it is now possible to explain the genetic basis for poliovirus attenuation and to construct variant strains that are less capable of reverting to a virulent phenotype after feeding to humans. Unfortunately, the complexities of licensing have made it impractical to utilize these "safer" variant viruses.

Genetic studies of Rous sarcoma and other retroviruses of chicken and mouse have elucidated the basis for the transforming viral phenotype which, in turn, has opened the new field of cellular oncogenes (see Chapter 10). Initial genetic studies showed that transforming retroviruses carry an open reading frame for an oncogene that confers the transforming activity on the virus. Furthermore, transforming viruses lack genetic sequences encoding the viral envelope. As a result, these viruses are replication-defective and can only be propagated in the presence of a closely related replication-competent "helper" retrovirus that supplies the envelope protein in trans. These findings clarified the role of the helper virus, a very enigmatic feature of transforming retroviruses that had perplexed a generation of investigators.

Bishop and Varmus, who pioneered the identification of the *src* oncogene (named after Rous sarcoma virus) in the 1970s, were surprised to find that it was similar to a host gene that encoded a cellular tyrosine kinase (Sidebar 1.4). This discovery led to the insight that viral oncogenes were derived from host genomic sequences by recombination. In the process of this genetic exchange, the majority of oncogenic retroviruses have lost part of the viral genome that encodes the viral envelope protein, explaining the need for a helper virus.

Further investigation showed that the expression and activity of the normal cellular src enzyme was controlled by a complex network of other cellular proteins involved in the cell cycle, whereas the viral variant escaped regulation and perturbed the cell cycle so that transformed cells were no longer subject to contact inhibition and other growth restraints. These findings initiated the discovery of oncogenes, which has revolutionized our understanding of the cell cycle and the multiple mecha-

■ ■ ■
SIDEBAR 1.4

The Discovery of a Cellular Homolog of the *src* Oncogene

"Infection of fibroblasts by avian sarcoma virus (ASV) leads to neoplastic transformation of the host cell. Genetic analyses have implicated specific viral genes in the transforming process, and recent results suggest that a single viral gene is responsible. . . . We demonstrate here that the DNA of normal chicken cells contains nucleotide sequences closely related to at least a portion of the transforming gene(s) of ASV; . . . Our data are relevant to current hypotheses of the origin of the genomes of RNA tumour viruses and the potential role of these genomes in oncogenesis."

"Our procedures to isolate $cDNA_{sarc}$ exploited the existence of deletion mutants of ASV which lack 10-20% of the viral genome (transformation defective or td viruses); results of genetic analysis indicate that the deleted nucleotide sequences include part or all of the genes responsible for oncogenesis and cellular transformation. . . . The preparation of $cDNA_{sarc}$ used was a virtually uniform transcript from about 16% of the Pr-C ASV genome, a region equivalent in size to the entire deletion in the strain of td virus used in our experiments. . . ."

"DNA from several avian species . . . contains nucleotide sequences which can anneal with $cDNA_{sarc}$. . . . In contrast, we detected no homology between $cDNA_{sarc}$ and DNA from mammals. . . . We suggest that part or all of the transforming gene(s) of ASV was derived from the chicken genome or a species closely related to chicken, either by a process akin to transduction or by other events, including recombination. . . . The sequences homologous to $cDNA_{sarc}$ in the genome of ASV are slightly diverged from the analogous sequences in chicken genome; . . . We anticipate that cellular DNA homologous to $cDNA_{sarc}$ serves some function which accounts for its conservation during avian speciation. The nucleotide sequences which anneal with $cDNA_{sarc}$ . . . could represent either structural or regulatory genes. . . . We are testing the possibilities that they are involved in the normal regulation of cell growth and development or in the transformation of cell behaviour by physical, chemical or viral agents." ■

Quoted extracts from Stehelin D, Varmus HE, Bishop JM, Vogt PK. DNA related to the transforming gene(s) of avian sarcoma viruses is present in normal avian DNA. *Nature* 1976; 260:170–173. This report was the first evidence that viral oncogenes were derived from cellular homologues and led to the discovery of a plethora of viral and cellular genes that could transform cells and played a causal role in many types of cancer.

nisms by which cells can be released from normal control mechanisms to assume the transformed phenotype.

PATHOGENESIS IN THE NEW MILLENNIUM

The recent sequencing of the human genome and the expected mapping of the genomes of a number of other mammalian species has begun a new era in biology. Fueled by techniques for mapping and manipulating animal genomes, the fields of virology and immunology are focused increasingly on experiments done in animals. This represents a radical change from the reductionist and chemical approach that once was advocated by leaders in biology. In this new era of "molecular medicine" viral pathogenesis is taking on greater prominence, reflected in the addition of sections on virus–cell interaction and pathogenesis in leading journals of virology.

It is also worthy of mention that, despite the advances in biomedical knowledge, there remain many challenging and significant unsolved problems in viral pathogenesis. For instance, it is sobering to reflect that in the year 2001, when we have just completed the eradication of type 2 poliovirus, there are still many fundamental aspects of poliovirus pathogenesis that are poorly understood, such as the initial site of enteric replication, the cellular sites of replication in lymphoid tissue, the mechanism of central nervous system invasion, the localization of virus in anterior horn cell neurons, the role of the virus receptor in tissue tropism, and the precise mechanism of cell killing. Human immunodeficiency virus (HIV) provides another example of pathogenesis and immunity that is incompletely understood, offering dozens of yet-to-be-solved questions (discussed in Chapter 13). Our knowledge of HIV is still insufficient to deal with problems of immense significance, such as the possible "cure" for a persistent infection or the induction of protective immunity.

As we enter a new millennium, advances in biology provide a plethora of opportunities for research in disease mechanisms, treatment, and prevention. At the same time, we are confronted with an array of fundamental and applied questions that offer numerous challenges. This juxtaposition of opportunities and challenges has provided a major impetus for summarizing our current knowledge of viral pathogenesis, in the hope that it will provide a foundation for future research and discoveries.

FURTHER READING

Reviews and Chapters

Brock TD, ed. *Milestones in microbiology*. Washington, DC: ASM Press, 1961.

Levine AJ. The origins of virology. In: Fields BN, Knipe DM, Howley PM, eds. *Virology*, 3rd ed. Philadelphia: Lippincott–Raven Publishers, 1996:1–14.

McNeill WH. *Plagues and peoples*. New York: Doubleday, 1977.

Mims CA. Aspects of the pathogenesis of viral diseases. *Bacteriol Rev* 1964;30:739–760.

Nathanson N. Introduction and history. In: Nathanson N, Ahmed R, Gonzalez-Scarano F, et al., eds. *Viral pathogenesis*. Philadelphia: Lippincott–Raven Publishers, 1997:3–12.

Tyler KL, Nathanson N. Pathogenesis of viral infections. In Knipe KM, Howley PM. *Fields virology,* 4th ed., Philadelphia: Lippincott Williams & Wilkins, 2001:199–243.

Methods

Haase AT. Methods in viral pathogenesis: tissues and organs. In: Nathanson N, Ahmed R, Gonzalez-Scarano F, et al., eds. *Viral pathogenesis*. Philadelphia: Lippincott–Raven Publishers, 1997:465–482.

Rall GF, Oldstone MBA. Methods in viral pathogenesis: transgenic and genetically deficient mice. In: Nathanson N, Ahmed R, Gonzalez-Scarano F, et al., eds. *Viral pathogenesis*. Philadelphia: Lippincott–Raven Publishers, 1997:507–531.

Original Contributions

Bodian D. Emerging concept of poliomyelitis infection. *Science* 1955;122:105–108.

Evans AS. Causation and disease: the Henle–Koch postulates revisited. *Yale J Biol Med* 1976;49:175–195.

Fenner F. The pathogenesis of the acute exanthems: an interpretation based on experimental investigations with mousepox (infectious ectromelia of mice). *Lancet* 1948;2:915–920.

Loeffler F, Frosch P. Report of the commission for research on the foot-and-mouth disease. Translated in Brock TD, ed. *Milestones in microbiology*. Washington, DC: ASM Press, 1961.

Panum PL. *Observations made during the epidemic of measles on the Faroe Islands in the year 1846*. New York: American Public Health Association, 1940.

Powell JH. *Bring out your dead: the great plague of yellow fever in Philadelphia in 1793*. Philadelphia: University of Pennsylvania Press, 1993.

Reed W. Recent researches concerning etiology, propagation and prevention of yellow fever, by the United States Army Commission. *J Hygiene* 1902;2:101–119.

Stehelin D, Varmus HE, Bishop JM, Vogt PK. DNA related to the transforming gene(s) of avian sarcoma viruses is present in normal avian DNA. *Nature* 1976;260:170–173.

Chapter 2
The Sequential Steps in Viral Infection

Infection of an animal host with many specialized organs and tissues is a complex multistep process. Viruses usually invade at a very specific site, which partly determines the subsequent route of spread both locally and systemically, and then spread by one of two routes: via the blood or via the peripheral nervous system. At each step in this process, the virus must overcome natural barriers to dissemination, such as the anatomic boundaries that separate organs and tissues. In addition, the restriction of viral replication to certain tissues and cells, a phenomenon also termed *tropism*, can influence the pattern of spread. Virus shedding can be either from the initial portal of entry or from distant sites that border on the external environment. Also, certain viruses are transmitted from the blood by transfusion, contaminated needles, or blood-sucking arthropods. The following account follows the teachings of Cedric Mims, one of the pioneers of viral pathogenesis.

During acute infection, viral replication is repeatedly checked by host defenses, both nonspecific and specific, such as immune response. In many acute viral infections, the host response succeeds in eliminating the invading virus completely within a few days to weeks. However, in a number of instances, the virus circumvents host defenses sufficiently to persist for varying periods of time. Although viruses vary widely in their patterns of dissemination, particular viruses tend to follow very stereotyped patterns based on properties encoded in the viral genome.

Some viruses are confined to the site of initial infection and spread only locally whereas others disseminate widely. Blood-borne viruses can invade almost any organ or cell type, whereas neurotropic viruses are usually confined to the peripheral and central nervous systems and replicate in relatively few peripheral tissues. The alternative patterns of entry, dissemination, and shedding that are used by a blood-borne and a neurotropic virus are shown in Figs. 2.1 and 2.2, respectively.

ENTRY

Skin and Mucous Membranes

The skin consists of the epidermis and underlying dermis. From the surface, the layers of the epidermis are the stratum corneum, a layer of dying cells covered by a superficial layer of keratin; the stratum granulosum; the stratum spinosum; and the stratum germinativum, a germinal layer of dividing cells that gives rise to the more superficial layers that are constantly being sloughed and replaced. Below the epidermis lies the dermis, a layer of highly vascularized connective tissue containing fibroblasts and dendritic cells (specialized macrophages).

Many viruses replicate in cells of the skin or mucous membranes (Table 2.1). It is unlikely that any virus can invade the intact skin since there are no viable cells directly on the surface; in fact, the exterior of the skin constitutes a relatively hostile environment due to its dryness, acidity, and bacterial flora. Rather, virus invades through a break in the barrier that allows contact with living cells. Skin invaders

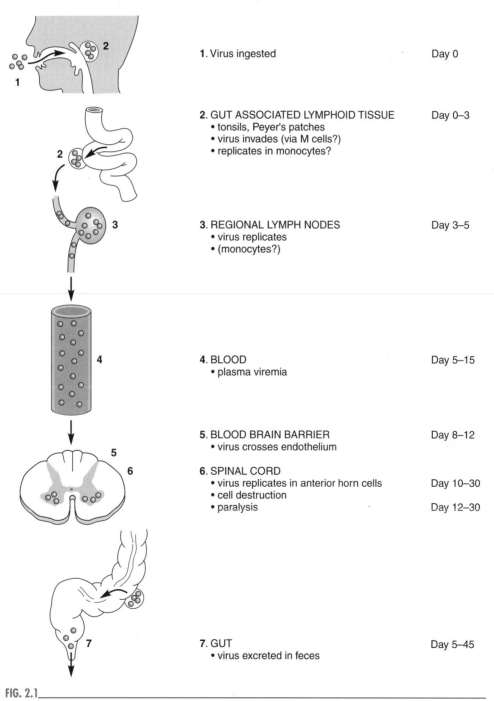

1. Virus ingested — Day 0

2. GUT ASSOCIATED LYMPHOID TISSUE — Day 0–3
• tonsils, Peyer's patches
• virus invades (via M cells?)
• replicates in monocytes?

3. REGIONAL LYMPH NODES — Day 3–5
• virus replicates
• (monocytes?)

4. BLOOD — Day 5–15
• plasma viremia

5. BLOOD BRAIN BARRIER — Day 8–12
• virus crosses endothelium

6. SPINAL CORD
• virus replicates in anterior horn cells — Day 10–30
• cell destruction
• paralysis — Day 12–30

7. GUT — Day 5–45
• virus excreted in feces

FIG. 2.1

The spread of representative viruses. Poliovirus is an example of a virus that disseminates via the blood.

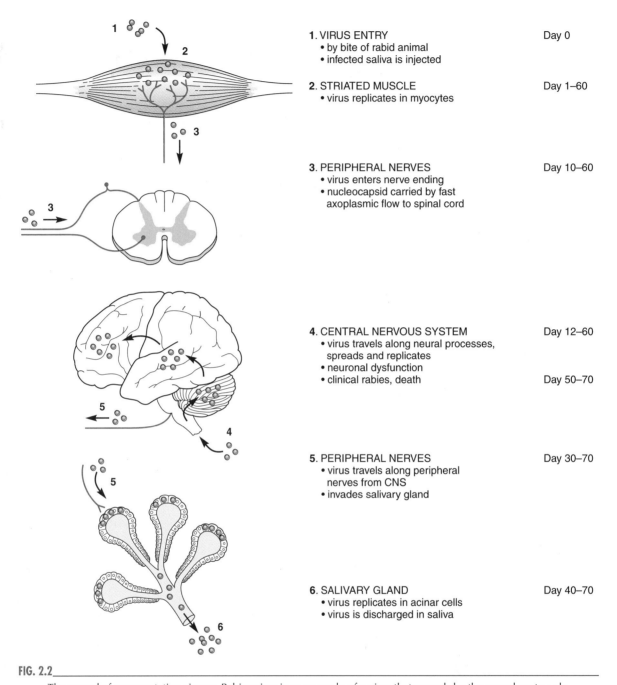

1. VIRUS ENTRY   Day 0
  • by bite of rabid animal
  • infected saliva is injected

2. STRIATED MUSCLE   Day 1–60
  • virus replicates in myocytes

3. PERIPHERAL NERVES   Day 10–60
  • virus enters nerve ending
  • nucleocapsid carried by fast
    axoplasmic flow to spinal cord

4. CENTRAL NERVOUS SYSTEM   Day 12–60
  • virus travels along neural processes,
    spreads and replicates
  • neuronal dysfunction
  • clinical rabies, death   Day 50–70

5. PERIPHERAL NERVES   Day 30–70
  • virus travels along peripheral
    nerves from CNS
  • invades salivary gland

6. SALIVARY GLAND   Day 40–70
  • virus replicates in acinar cells
  • virus is discharged in saliva

FIG. 2.2_____

The spread of representative viruses. Rabies virus is an example of a virus that spreads by the neural route only.

typically replicate in specific cells. For instance, both herpes simplex virus (HSV) and poxviruses replicate in germinal cells of the epidermis as well as macrophages and fibroblasts of the dermis. By contrast, papillomaviruses initially infect only the germinal cells of the epidermis; however, this group of viruses cannot mature in germinal cells and instead complete their replication cycle in the stratum granulosum.

Most superficial invaders can also infect epithelial cells on the surface of mucous membranes, although they must first penetrate a mucous barrier that may contain immunoglobin A (IgA) and other virocidal proteins. The conjunctiva of the eye, a specialized mucous membrane, is the primary site of entry of a few viruses, such as certain adenoviruses and enteroviruses (such as coxsackievirus A24 and enterovirus 70) that can cause conjunctivitis.

TABLE 2.1

Representative viruses that invade via skin and mucous membranes

Site of entry	Route	Virus family	Representative example
Skin	Minor breaks	Papillomaviridae	Papillomaviruses
		Herpesviridae	Herpes simplex virus 1
		Poxviridae	Ectromelia virus
Conjunctiva	Contact	Picornaviridae	Enterovirus 70
		Adenoviridae	Adenoviruses
Oropharynx	Contact	Herpesviridae	Epstein–Barr virus
Genital tract	Contact	Retroviridae	HIV
		Papillomaviridae	Papillomaviruses
		Herpesviridae	Herpes simplex virus 2
Rectum	Contact	Retroviridae	HIV

Modified after Mims CA, White DO. *Viral pathogenesis and immunology.* Oxford: Blackwell, 1984.

Transcutaneous Injection Some viruses breach the cutaneous barrier by injection. A wide variety of viruses are arthropod borne (arboviruses) and have a life cycle that alternates between an insect vector and a vertebrate host. These agents are injected by the infected insect when it takes a blood meal, which involves probing for a capillary with consequent injection of virus-contaminated saliva that is deposited mainly in the subcutaneous tissues but also in the circulation. It is estimated that during feeding a mosquito deposits about 10^{-4} µL of saliva, which would contain 100 plaque-forming units (PFU) of virus if the saliva titer is 10^9 PFU/mL.

Viruses may also be injected in other ways. Rabies virus and herpes B virus are often transmitted by bite of an infected animal; in this instance, infection is initiated by intramuscular inoculation of virus-contaminated saliva. Several medically important viruses, such as hepatitis B virus (HBV), hepatitis C virus (HCV), and human immunodeficiency virus (HIV), are frequently transmitted by blood or blood products, or by contaminated needles.

Urogenital Tract Viruses that are sexually transmitted fall into two entry types. Some, such as HSV-2 and papillomaviruses, replicate in mucous membranes of the genital tract following the pattern described above. Other sexually transmitted viruses, such as HBV and HIV, which do not replicate in epithelial cells, are associated with persistent viremia and may be transmitted via minute "injections" of blood during sexual contact. HBV may transit mu-

cous membranes to directly invade the circulation through surface capillaries. HIV infects CD4$^+$ T lymphocytes in the skin and submucosal tissues and is then carried to draining lymph nodes.

Oropharynx and Gastrointestinal Tract

The oropharynx and gastrointestinal tract are the portals of entry for many viruses; particular viruses may invade at specific sites ranging from the tonsils to the colon (Table 2.2). Some enteric invaders remain confined to the intestinal tract whereas others spread via the blood to produce systemic infection. Viruses that replicate in the gastrointestinal tract may or may not produce enteric disease.

The intimate details of entry are not well characterized for many enteric viruses but are quite well established for reoviruses (Fig. 2.3). In the alkaline environment of the small intestine, reovirions are converted to infectious subvirion particles that attach to M (microfold) cells, which form part of the specialized epithelium that overlies Peyer's patches (focal accumulations of lymphoid tissue in the wall of the intestine). Virions are endocytosed into M cells and appear to transit these cells within vesicles, to be released by exocytosis on the basal surface. From this point, virions may invade other intestinal epithelial cells through their basal surface, or may be taken up by macrophages or endings of the autonomic nervous system. Different reovirus types disseminate through the circulation or along peripheral

TABLE 2.2

Representative enteric viruses that do or do not cause gastroenteritis

Localization of disease	Replicate in the pharynx and/or gastroenteric tract	Virus family	Representative example
Gastroenteritis	Yes	Astroviridae	Astroviruses
	Yes	Caliciviridae	Norwalk virus
	Yes	Coronaviridae	Transmissible gastroenteritis virus of swine
	Yes	Rotaviridae	Rotaviruses
	Yes	Parvoviridae	Canine parvoviruses
	Yes	Adenoviridae	Adenoviruses 40, 41
No enteric illness (± systemic illness)	Yes	Picornaviridae	Poliovirus
	Yes	Picornaviridae	Coxsackieviruses
	Yes	Picornaviridae	Enteroviruses
	No	Picornaviridae	Hepatitis A virus
	Yes	Adenoviridae	Adenoviruses

Modified after Mims CA, White DO. *Viral pathogenesis and immunology.* Oxford: Blackwell, 1984.

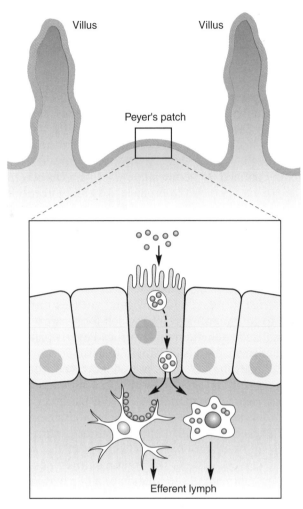

FIG. 2.3

Viral invasion of the intestine, showing the pathway taken by reovirus in the mouse. The virus binds to M cells, is carried by transcytosis to the basolateral surface, where it infects dendritic cells and macrophages in the lamina propria. This well-studied experimental model probably resembles many natural infections. (After Wolf JL, et al. Intestinal M cells: a pathway for entry of reovirus into the host. *Science* 1981;212:471–472.)

neural pathways, and dissemination phenotypes have been mapped to specific viral genes.

Although most enteric viruses replicate only in the intestinal tract, some, such as poliovirus, also infect the tonsils. By contrast, other viruses, such as HIV and HBV, can invade via the rectum or colon, as indicated by the importance of anal intercourse as a risk factor for infection.

Barriers to Infection There are many barriers to infection via the gastrointestinal tract. Invading virus may remain sequestered in the intestinal contents or

fail to penetrate surface mucus. Acidity of the stomach and alkalinity of the intestine, the proteolytic enzymes secreted by the pancreas, the lipolytic activity of bile, the neutralizing action of secreted IgA, and scavenging macrophages can all reduce viral infectivity. Thus, viruses that successfully utilize the gastrointestinal portal of entry must be resistant to this hostile environment, or actually exploit it by activation into an infectious particle, as in the case of reovirus. There are a few viruses, such as coronaviruses, that are susceptible to this hostile environment but when ingested in milk or food are sufficiently protected to initiate infection by the enteric route.

Respiratory Tract

Many viruses utilize the respiratory portal of entry (Table 2.3) and are acquired by aerosol inhalation or by mechanical transmission of infected nasopharyngeal secretions. Depending on their size, aerosolized droplets are deposited at various levels in the respiratory tract. Droplets larger than 10 µm in diameter lodge in the nose; those 5–10 µm lodge in the airways; and those smaller than 5 µm reach the alveoli of the lower respiratory tract. The respiratory tract offers several barriers to invading organisms, including the protective coating action of mucus, the ciliary action of the respiratory epithelium that sweeps particles from the airways, and the

TABLE 2.3		
Representative viruses that invade via the nasopharynx or respiratory tract, according to localization of disease		
Localization of disease	Virus family	Representative example
Upper respiratory	Picornaviridae	Rhinoviruses
	Adenoviridae	Adenoviruses
Lower respiratory	Coronaviridae	Bovine coronaviruses
	Orthomyxoviridae	Influenza viruses
	Paramyxoviridae	Respiratory syncytial virus
	Bunyaviridae	Sin Nombre virus
No respiratory illness (± systemic illness)	Togaviridae	Rubella virus
	Paramyxoviridae	Mumps virus
	Bunyaviridae	Hantaan virus
	Arenaviridae	Lassa fever virus
	Reoviridae	Reovirus
	Papovaviridae	Murine polyomavirus
	Herpesviridae	Varicella zoster virus
	Poxviridae	Variola virus

Modified after Mims CA, White DO. *Viral pathogenesis and immunology.* Oxford: Blackwell, 1984.

activity of immunoglobulins and macrophages that engulf foreign particles. In addition, there is a temperature gradient between the nasal passages (33°C) and the alveoli (37°C) that plays an important role in the localization of infection. Thus, rhinoviruses, which infect the nasopharynx and cause the common cold, replicate well at 33°C but grow poorly at 37°C, whereas influenza virus, which infects the lower respiratory tract, shows the inverse temperature preference. Temperature sensitivity has been used to select attenuated influenza vaccines because cold-adapted viruses are much less virulent but replicate sufficiently in the upper respiratory tract to induce immunity against wild-type influenza virus challenge.

The initial sites of infection have been characterized for some respiratory viruses. Rhinovirus has been shown to replicate in the epithelial lining of the nose, whereas poxviruses, some of which enter via aerosol transmission, replicate initially in macrophages free in the airways and then in the epithelial lining of small bronchioles. By contrast, those reoviruses that can enter via the respiratory route infect via M cells that overlie bronchus-associated lymphoid tissue.

SPREAD

Local Spread

Viruses can be divided into two groups: those that spread locally from their site of entry and those that disseminate widely (Fig. 2.1). Local spread occurs by infection of contiguous cells and can result in lesions such as the cold sores produced by HSV. Epithelial cells have "polarized" plasma membranes, and certain proteins are targeted almost exclusively to either the apical or the basolateral surface. When epithelial cells are infected, virus may be released through the apical surface, in which instance it tends to remain localized, or through the basolateral surface, in which case it may disseminate more widely. Released virus is often carried from epithelial surfaces via afferent lymphatic channels to regional lymph nodes. If the virus can replicate in one of the cell types found in the node, such as monocytes, or T and B lymphocytes, it is likely to disseminate via the thoracic duct into the blood.

Viremia

Viremia is the most important mode of viral dissemination within the host and can spread infec-

tion to any organ or tissue. Blood-borne virus either circulates free in the plasma or is cell-associated, and these two kinds of viremia have different characteristics and implications (Fig. 2.4). Most viremias are acute, lasting no more than 1–2 weeks, but certain viruses are able to evade immune defenses and persist in the blood for months or years.

During the course of viremia different sequential phases can be distinguished. In order to follow these events, experimental models have been used to elucidate natural infections (Fig. 2.5). When a virus is injected by intramuscular, intravenous, intracerebral, or other routes, a portion of the injected bolus enters the circulation without any intervening replication stage and produces a very short-lived passive viremia of a few hours' duration. If the virus replicates locally at the site of entry or in the draining regional lymph node, then a brief active primary viremia may occur that lasts for 1–2 days. The primary viremia serves to disseminate the virus systemically to permissive cells in various tissues; when virus is released from these secondary sites of replication an active secondary viremia occurs. This sequence is illustrated in Fig. 2.5, but it should be noted that it is often hard to document these different stages in viremia except in carefully studied experimental models.

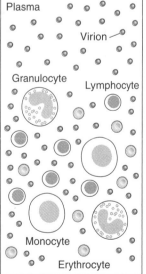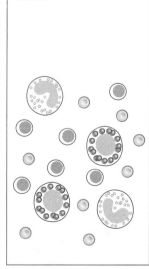

FIG. 2.4

Diagram of plasma or cell-associated viremia.

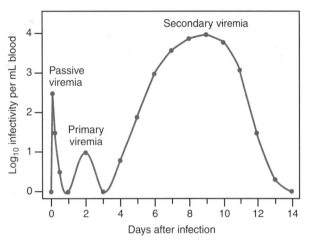

FIG. 2.5_____

Stages in acute viremia, a reconstruction from experimental observations. Although this reconstruction is based on intraperitoneal or footpad injection, it likely mimics the events that follow natural routes of infection. Passive viremia: unreplicated inoculum entering the circulation after intraperitoneal injection of La Crosse virus. Primary viremia: virus entering blood after local replication following footpad injection of a small inoculum of ectromelia virus. Secondary viremia: virus entering blood from widely dispersed sites of replication after footpad injection of ectromelia virus. (After Fenner F. The pathogenesis of the acute exanthems: an interpretation based on experimental investigations with mousepox [infectious ectromelia of mice]. *Lancet* 1948;2: 915–920, and Pekosz A, et al. Protection from La Crosse virus encephalitis with recombinant glycoproteins: role of neutralizing anti-G1 antibodies. *J Virol* 1995;69: 3475–3481.)

Sources of Viremia Most commonly, viremia is produced by virus that replicates in regional lymph nodes and is shed into efferent lymphatics where it is transported via the thoracic duct to the circulation. It is not clear whether virus that replicates in other tissues, such as striated muscle, can enter the vascular compartment directly or only via the lymphatic circulation. Some viruses replicate in the vascular endothelium and are released immediately into the circulation. A number of viruses replicate in monocytes, B cells, or T cells to create a cell-associated viremia; in some cases, virus may also be released from these cells to produce a concomitant plasma viremia.

Plasma Viremia In acute virus infections, plasma viremia often appears within a week after infection, lasts for 1–2 weeks, and comes to an abrupt termination concomitant with the appearance of circulating antibody, as shown in Fig. 2.6. Viremia may be prolonged if animals are immunosuppressed, demonstrating the role of the virus-specific immune response in terminating viremia.

Plasma viremia is dynamic: virus continually enters the circulation and is continually being removed. Viral clearance is mediated primarily by the sessile macrophages of the liver, spleen, and lung, which monitor the circulation for foreign particulates. The rate of clearance of virus can be expressed as the mean survival time of virus particles (t_m) within the vascular compartment (Sidebar 2.1) and, for many viruses, the t_m varies between 10 and 30 minutes. The titer of virus in the plasma is determined by the t_m and by the rate at which virus enters the circulation, and can vary from trace levels to $>10^6$ infectious units per milliliter of plasma.

Although plasma viremias are usually short lived, there are some exceptions due to two mechanisms. In some instances there is an antibody response: the antibody binds to circulating virus, but the immune complex retains its infectivity. Evidence for the circulation of infectious immune complexes is that the titer of plasma virus can be reduced by treatment with antisera directed against the host's immunoglobulins. Examples are lactic dehydrogenase virus infection of mice and Aleutian disease viremia of mink. Under special circumstances, the infected host may fail to recognize viral proteins as foreign (as state called "tolerance") and fail to induce serum-neutralizing antibodies. Tolerance is usually associated with infections acquired in utero or shortly after birth, prior to maturation of the immune system. Examples are HBV infection of humans and lymphocytic choriomeningitis virus (LCMV) infection of mice.

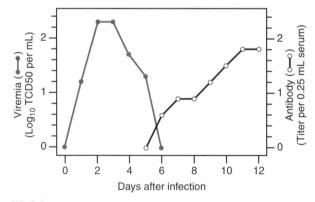

FIG. 2.6_____

The course of viremia in monkeys injected by the intramuscular route with wild-type poliovirus. (After Nathanson N, Bodian D. Experimental poliomyelitis following intramuscular virus injection. *Bull Johns Hopkins Hosp* 1961;108:320–333, with permission.)

■ ■ ■
SIDEBAR 2.1

Plasma Viremia: Measurement of Mean Transit Time (t_m)

In a plasma viremia, virus is constantly entering and being removed from the vascular compartment. The average duration of an infectious virion in the vascular compartment is the transit time (t_m), a parameter that differs for different viruses, and is important because it determines the titer of virus in the plasma. The t_m can be measured using the formula below.

If

V_i = virus entering the vascular compartment
V_o = virus leaving the vascular compartment
$[V]$ = concentration of virus in the vascular compartment
t_m = mean duration of virus particles in the vascular compartment

then, at steady state, the rate of removal equals the rate of entry:

$$dV_i/dt = dV_o/dt = [V] / t_m$$

and

$$t_m = [V]/dV_i/dt$$

t_m can be measured by intravenous infusion of an uninfected animal with a suspension of virus (dV_i/dt) and determining the level of viremia ($[V]$) that is reached after several hours of infusion when a steady state is achieved. ■

After Nathanson N, Harrington B. Experimental infection of monkeys with Langat virus. II. Turnover of circulating virus. *Am J Epidemiol* 1967;85:494–502.

Cell-Associated Viremia Some viruses replicate in cells found in the circulation, particularly B or T lymphocytes or monocytes, or (rarely) erythrocytes (Table 2.4), but usually each virus infects only a single cell type. Cell-associated viremias may persist for months to years, although the titers are often low so that isolation of virus requires cultivation of blood mononuclear cells, with highly susceptible indicator cells. Virus-infected cells in the blood are often shielded from attack by virus-specific cytolytic T cells or complement-fixing antibodies because the viral genome is latent or is so poorly expressed that infected cells carry few if any viral proteins on their plasma membranes. In persistent cell-associated viremias, infectivity is usually not found in the plasma because any virus released from the infected cells is rapidly neutralized by antibodies. However, there are exceptions, such as HIV, which produces concurrent cell-associated and plasma viremias.

Spread of Virus from Blood to Tissues The route by which viruses cross the vascular wall into tissues has not been well characterized, although several pathways are probably operative (Fig. 2.7). There are some localized regions where capillaries are fenestrated, offering the possibility for viral transit. One of these is the choroid plexus of the ventricles of the brain; certain blood-borne viruses, such as mumps, LCMV, and visna viruses, probably cross the blood into the cerebrospinal fluid by this pathway, which explains why they replicate in the epithelial lining of

---

TABLE 2.4
Representative viruses that replicate in blood cells

Cell type	Virus family	Representative example	Duration of viremia
Monocytes	Flaviviridae	Dengue viruses	Acute
	Togaviridae	Rubella virus	Acute
	Coronaviridae	Mouse hepatitis virus	Acute
	Orthomyxoviridae	Influenza viruses	Acute
	Paramyxoviridae	Measles virus	Acute
	Arenaviridae	LCMV	Persistent
	Retroviridae	HIV	Persistent
	Herpesviridae	Cytomegalovirus	Persistent
	Poxviridae	Ectromelia virus	Acute
B lymphocytes	Retroviridae	Murine leukemia virus	Persistent
	Herpesviridae	Epstein–Barr virus	Persistent
T lymphocytes	Retroviridae	HIV	Persistent
		HTLV-1	Persistent
	Herpesviridae	Human herpesviruses 6, 7	Acute
Erythroblasts	Reoviridae	Colorado tick fever virus	Acute

LCMV, lymphocytic choriomeningitis virus; HIV, human immunodeficiency virus; HTLV, human T-cell leukemia virus.

Modified after Nathanson N, Tyler KL. Entry, dissemination, shedding, and transmission of viruses. In: Nathanson N, et al, eds. *Viral pathogenesis*. Philadelphia: Lippincott–Raven Publishers, 1997.

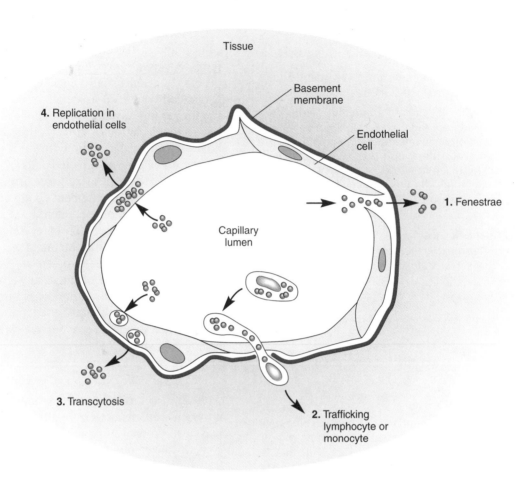

FIG. 2.7_____

Viral pathways from blood to tissues.

the choroid plexus or of the ventricles. Some viruses have been visualized to transit the endothelial cell lining of capillaries by a process of endocytosis, transcytosis, and exocytosis in order to be released from the basal surface of endothelial cells. Finally, a number of viruses can actually replicate in endothelial cells, so that they "grow" across the capillary wall.

A quite different pathway is used by viruses that infect lymphocytes or monocytes. Cells regularly traffic from the blood to the tissues, so that the virus is carried in the form of virus-infected cells—a route that has been called the "Trojan horse" mechanism. One example is HIV, which is carried to the central nervous system (CNS) by CD4$^+$ lymphocytes or monocytes with subsequent infection of the microglia, which are the resident macrophages of the brain.

Neural Spread

Neural spread is a process by which a virus is transmitted within the axoplasm of peripheral nerve fibers. The neural pathway plays an essential role in the dissemination of some viruses, although it is less common than viremia as a mode of spread. Rabies virus is the paramount example of a virus that is an obligatory neurotrope and is not viremogenic, whereas alphaherpesviruses (HSV-1 and 2) are often neurotropic in adults but viremogenic in newborn humans or animals. In some instances, a virus can use both pathways, but usually this involves different viral strains with diverse biological properties. For instance, type 1 reovirus is viremogenic, whereas type 3 reovirus uses the neural route. Also, it is possible to "neuroadapt" a viremogenic virus to select a strain that uses the neural route, exemplified by the MV strain of poliovirus (Table 2.5).

The classical evidence for neural spread is the demonstration that after viral injection into a peripheral site, a block of the innervating peripheral nerve will prevent virus from reaching the CNS and causing neurologic disease (Table 2.5). Neural spread involves axons or dendrites and not the

TABLE 2.5

Different tropism of two strains of poliovirus: neurotropic mixed virus and viremogenic Mahoney virus

	NEUROADAPTED MV STRAIN		VIREMOGENIC MAHONEY STRAIN	
	Control	Nerve Block	Control	Nerve Block
Paralysis	25/26	0/11	19/19	18/20
Site of initial paralysis:				
Injected leg	24	—	3	5
Other	1	—	16	13
Incubation to paralysis	5 days	—	7 days	7.5 days

After injection into the gastronemius muscle, the MV strain spreads only by the neural route, causes initial paralysis in the injected limb, and is impeded by a neural block, whereas the viremogenic Mahoney strain spreads by viremia, does not cause localized initial paralysis, and is not impeded by nerve block. Neural block was done just prior to virus injection by freezing the innervating sciatic nerve with dry ice proximal to the site of virus injection. MV, mixed virus.

After Nathanson N, Bodian D. Experimental poliomyelitis following intramuscular virus injection. *Bull Johns Hopkins Hosp* 1961; 108:308–319.

supporting cells, such as the Schwann cells or fibroblasts found in peripheral nerves. Presumably, viruses enter peripheral nerve endings by the same route used to enter other permissive cells. The viral nucleocapsid is probably transported by the machinery that mediates axoplasmic flow, since viruses move at a rate (>5 cm/d) similar to that of fast axoplasmic flow. Drugs that block fast axoplasmic flow, such as colchicine, will also interfere with the neural spread of viruses. Just as axoplasmic flow is bidirectional, both toward and away from the neural cell body, viruses can spread both from the periphery to the CNS, and from the CNS toward the periphery. Most RNA viruses can replicate within neural processes, but DNA viruses, such as HSV, must reach the nucleus within the neuronal cell body in order to replicate.

Although viremia and neural spread are classically considered as alternative modes of spread, some viruses may disseminate by both routes. For instance, varicella virus, an alphaherpesvirus, produces a viremia, invades peripheral nerve endings in the skin, and spreads along peripheral nerves to dorsal root ganglia where it becomes latent, occasionally emerging years later in the form of herpes zoster, also called shingles.

Viral localization and tissue tropism are described in the next chapter.

SHEDDING

Viruses may be discharged into respiratory aerosols, feces, or other body fluids or secretions, and each of these modes is important for selected agents. Viruses that cause acute infections are usually shed intensively over a short time period, often 1–4 weeks, and transmission tends to be relatively efficient. Viruses that cause persistent infections, such as HBV and HIV, can be shed at lower titers for months to years but will eventually be transmitted during the course of a long-lasting infection.

Oropharynx and Gastrointestinal Tract

Enteroviruses may be shed in pharyngeal fluids and feces (as shown for poliovirus in Fig. 2.8). In this case, the virus replicates in the lymphoid tissue of the tonsil and in Peyer's patches (lymphoid tissue accumulations in the wall of the small intestine) where it is discharged into the intestinal lumen. Other viruses may be excreted into feces from the epithelial cells of the intestinal tract (reoviruses and rotaviruses) or from the liver via the bile duct (hepatitis A virus).

Respiratory Tract

Viruses that multiply in the nasopharynx and respiratory tract may be shed by two distinct mechanisms, either as aerosols generated by sneezing or coughing, or in pharyngeal secretions that are spread from mouth to hand to hand to mouth. Typically,

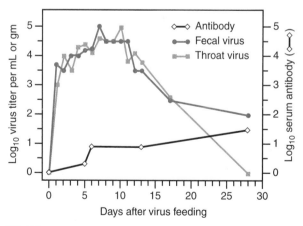

FIG. 2.8

Course of wild-type poliovirus excretion in the pharyngeal fluids and feces of chimpanzees after virus feeding. (After Bodian D, Nathanson N. Inhibitory effects of passive antibody on virulent poliovirus excretion and on immune response in chimpanzees. *Bull Johns Hopkins Hosp* 1960:107:143–162, with permission.)

viruses that cause acute respiratory illness, such as rhinoviruses and influenza viruses, are spread efficiently at high titer but for short periods, often not more than a week.

Skin

Relatively few viruses are shed from the skin, but there are exceptions. Papillomaviruses and certain poxviruses that cause warts or superficial tumors may be transmitted by mechanical contact. A few viruses that are present in skin lesions, such as variola virus (the cause of smallpox) and varicella virus (the cause of chickenpox) can be aerosolized and transmitted by the respiratory route. In fact, it is claimed that the earliest instance of deliberate "biological warfare" was the introduction into the villages of hostile Indian tribes of blankets containing desquamated skin from persons with smallpox.

Mucous Membranes, Oral and Genital Fluids

Viruses that replicate in mucous membranes and produce lesions of the oral cavity or genital tract are often shed in pharyngeal or genital fluids. An example is HSV (type 1 in the oral cavity and type 2 in genital fluids). A few viruses are excreted in saliva, such as Epstein–Barr virus (a herpesvirus that causes infectious mononucleosis, also called "kissing disease") and mumps virus. Probably the most notorious example is rabies virus, which replicates in the salivary gland and is transmitted by a bite that inoculates virus-contaminated saliva. Several important human viruses, such as HBV and HIV, may be present in the semen. It is estimated that in an HIV-infected male, an average ejaculate contains about 10^6 mononuclear cells of which 10^2–10^4 (0.01%–1%) may carry the viral genome.

Blood, Urine, Milk

Blood is an important potential source of virus infection in humans, wherever transfusions, injected blood products, and needle exposure are common (Table 2.4). In general, the viruses transmitted in this manner are those that produce persistent viremia, such as HBV, HCV, HIV, and cytomegalovirus (a herpesvirus). Occasionally, viruses that produce acute short-term high-titer viremias, such as parvovirus B19, may contaminate blood products. Although a number of viruses are shed in the urine, this is usually not an important source of transmission. One exception is certain animal viruses that are transmitted to humans; several

arenaviruses are transmitted via aerosols of dried urine. A few viruses are shed in milk and transmitted to newborns in that manner. The most prominent example is mouse mammary tumor virus, a retrovirus, and it appears that a few lentiviruses, such as HIV and human T-cell leukemia virus type I (HTLV I) of humans and visna maedi virus of sheep, can be transmitted in this manner.

Environmental Survival of Shed Virus

Transmission of a virus depends both on the amount and duration of shedding and on survival in the environment—a point often overlooked. For instance, viruses differ in their ability to survive in aerosols or after drying. Thus, poliovirus is sensitive to low humidity, and this is thought to account for its reduced transmission in the winter in temperate climates where humidity is low, whereas transmission continues year round in tropical climates. The gastrointestinal lumen constitutes a harsh environment that can inactivate all but the hardiest viruses. Thus, of the different hepatitis viruses, all of which are probably shed in the bile, only hepatitis A virus and hepatitis E virus behave as enteroviruses, presumably because the others, such as HBV and HCV, are inactivated before they can be transmitted by the fecal–oral route.

TRANSMISSION

Following shedding, a virus can be transmitted to a new host in several different ways, but individual viruses utilize only one or two of these potential modes. The most common mode of transmission of enteric and many respiratory viruses is probably by oral or fecal contamination of hands, with passage to the hands and thence the oral cavity of the next infected host. Inhalation of aerosolized virus is also an important mode of transmission for respiratory viruses. Another significant route is by direct host-to-host interfacing, including oral–oral, genital–genital, oral–genital, or skin–skin contacts. Transmission may involve less natural modes, such as blood transfusions or reused needles. In contrast to propagated infections are transmissions from a common source, such as food, water, biologicals, or other products of human society. Common source transmission is quite frequent and can produce explosive outbreaks that range in size depending on the number of recipients of the tainted product and the level of virus contamination.

PERPETUATION OF VIRUSES

Viruses That Cause Acute Infections

For viruses that can only cause acute infections, transmission must be accomplished during a relatively short time frame (frequently no more than one week of shedding). The efficiency of transmission can be measured by determining the number of new infections generated by each infected host (transmissibility or Ro); if Ro is greater than 1 the agent is spreading more widely, and if it is less than 1 the number of infections is declining. Although transmissibility may cycle above and below 1, overall it must be at least 1 if the agent is to be successfully perpetuated in the specified population. Acute viruses may fail to meet this criterion, in which case they "fade out" and disappear. Measles is probably the best documented example of this phenomenon because almost all cases of measles infection cause a readily recognized illness. Prior to measles immunization, in populations of under 500,000, measles periodically disappeared only to cause an outbreak when it was reintroduced. This was dramatically illustrated in islands such as Iceland with a population of approximately 200,000 (Fig. 2.9).

For acute viral infections, one indicator of transmissibility in a population is the age-specific prevalence of antibody, assuming that the initial infection

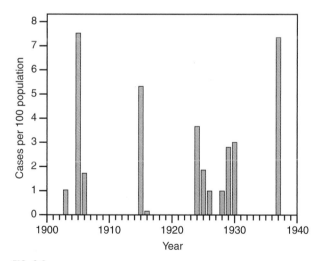

FIG. 2.9

Measles in Iceland showing its periodic disappearance and reintroduction during the period 1900–1940, prior to the use of measles vaccine. (After Tauxe, unpublished, 1979, and Nathanson N, Murphy FA. Evolution of viral diseases. In: Nathanson N, et al., eds. *Viral pathogenesis.* Philadelphia: Lippincott–Raven Publishers, 1997.)

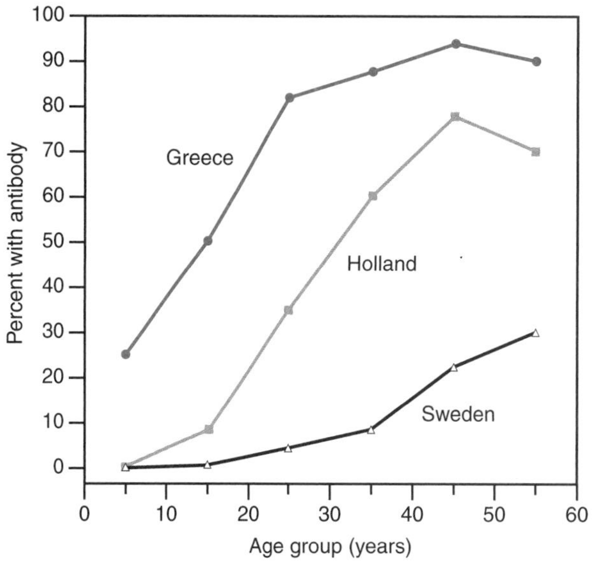

FIG. 2.10

Antibody against hepatitis A virus in selected countries to illustrate the differences in the transmissibility of a single virus in different populations. (After Frosner GG, et al. Antibody against hepatitis A in seven European countries. *Am J Epidemiol* 1979;110:63–69.)

confers lifelong detectable antibody as well as long duration immunity. Figure 2.10 shows the age-specific antibody profiles for hepatitis A in three countries, illustrating the difference in virus transmission in different populations. Likewise, if different viruses are studied in a single population, the differences in their transmissibility can be readily documented.

Viruses That Cause Persistent Infections

Viruses that cause persistent infections may be transmitted over a long period of time, in some cases for the lifetime of the infected host. In this instance, perpetuation of the virus still requires that each infection must generate at least one new infection, but this may take place over many years. Examples of human viruses that behave this way include HIV, HBV, and varicella zoster virus. Such persistent viruses can be perpetuated within very small populations. Studies of isolated primitive tribes have shown that most of the viruses that can be found are those that are capable of causing persistent infections in individual hosts, whereas acute viruses, when they appear, burn out very rapidly. One variant of this situation is viruses that are transmitted vertically, from mother to offspring, by perinatal or transplacental routes, or integrated into the host germline genes. Viruses such as LCMV of mice or

HTLV I or HTLV II of humans may persist with very limited horizontal transmission.

CONTROL AND ERADICATION OF HUMAN VIRUSES

The principles of virus shedding and transmissibility are relevant for the control and elimination of important human pathogens. Preexposure immunization can diminish the number of susceptible hosts in a population and reduce transmission to $\ll 1$, with the consequent disappearance of a virus from the immunized population, if virus perpetuation depends on acute infections. This principle has been successfully applied to the global eradication of variola virus, the cause of smallpox, and has led to the elimination of wild polioviruses from the Western Hemisphere. Conversely, although there is a highly efficacious vaccine for HBV, it can be calculated that it will take generations for this virus to disappear, due to persistent infections in a small number of unimmunized persons.

FURTHER READING

Reviews

Nathanson N, Tyler KL. Entry, dissemination, shedding, and transmission of viruses. In: Nathanson N, Ahmed R, Gonzalez-Scarano F, et al. *Viral pathogenesis*. Philadelphia: Lippincott–Raven Publishers, 1997:13–34.

Nathanson N, Tyler KL. Pathogenesis. In: Mahy B, ed. *Topley and Wilson's microbiology and microbial infections*. New York: John Wiley & Sons, 1998:400–425.

Original Contributions

Baer GM, Cleary WF. A model in mice for the pathogenesis and treatment of rabies. *J Infect Dis* 1972;125:520–529.

Card JP, Whealy ME, Robbins AK, Enquist LW. Two alphaherpesvirus strains are transported differentially in the rodent visual system. *Neuron* 1992;6:957–969.

Fenner F. The pathogenesis of the acute exanthems: an interpretation based on experimental investigations with mousepox (infectious ectromelia of mice). *Lancet* 1948;2:915–920.

Mims CA. Aspects of the pathogenesis of viral diseases. *Bacteriol Rev* 1964;30:739–760.

Nathanson N, Bodian D. Experimental poliomyelitis following intramuscular virus injection. I. The effect of neural block on a neurotropic and a pantropic strain. *Bull Johns Hopkins Hosp* 1961;108:308–319.

Nathanson N, Harrington B. Experimental infection of monkeys with Langat virus. II. Turnover of circulating virus. *Am J Epidemiol* 1967;85:494–502.

Notkins AL, Mahar S, Scheele C, Goffman J. Infectious virusantibody complexes in the blood of chronically infected mice. *J Exp Med* 1966;124:81–97.

Racaniello VR. Poliovirus neurovirulence. *Adv Virus Res* 1988;34:217–246.

Tyler KL, McPhee D, Fields BN. Distinct pathways of viral spread in the host determined by reovirus S1 gene segment. *Science* 1986;233:770–774.

Chapter 3
Viral Tropism and Cellular Receptors

WHAT IS VIRAL TROPISM AND WHY IS IT IMPORTANT?

Following viral infection there are many different patterns of localization and dissemination, as described in the previous chapter. Circumscribed replication is instrumental in determining the focused nature of the pathologic or physiologic changes caused by each virus, an attribute so characteristic that it accounts for the disease "signature" of many viruses. Thus, smallpox was known for the rash that left survivors pock-marked for life, poliomyelitis by the paralytic attack and permanent lameness, yellow fever by acute jaundice, and rhinoviruses by the common cold (rhinitis). *Tropism* is the traditional term used to refer to this anatomic localization of a viral infection.

The theme of this chapter is the mechanisms of tropism. The most important determinants of tropism are the cellular receptors, which in general are different for each virus group. Since receptors are unequally expressed on the cells in different tissues, they limit the possible cell types that can be infected by each virus. Following virus entry, viruses utilize a wide variety of cellular proteins for the transcription and translation of their proteins and the replication of their genomes. Again, some cells provide the cellular proteins required for the replication of a specific virus, and this further limits the range of differentiated cells in which a given virus can repli-

cate. Finally, there are other physiologic factors that constrain the replication or survival of specific viruses and can therefore influence tropism.

VIRAL ATTACHMENT AND ENTRY

Cellular Receptors for Viruses

Peter Medawar described viruses as "bad news wrapped in protein," a succinct summary of the structure of all viruses, in which the nucleic acid genome is internal to an outer protein structure that protects the genome from adverse environmental factors. The other important role of the protein coat is to deliver the viral genome across the plasma membrane to the cellular interior where replication occurs. Rapid and efficient transportation across the plasma membrane is a major engineering challenge that viruses have met by exploiting the presence of many diverse proteins, sugars, and lipids on the cell surface. Each virus can bind to one (or a very few) of this multitude of molecules, that are exploited to act as cellular receptors. It should be kept in mind that viral receptors are naturally occurring cellular molecules that subserve physiologic functions for the cell, functions that have nothing to do with infection.

How do viruses interact with their cognate receptors? The receptor activity is due to the ability of a viral surface protein, often called the viral attach-

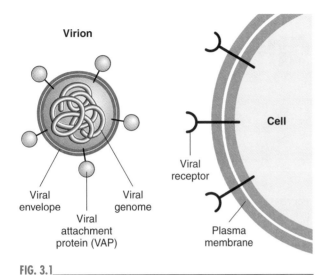

FIG. 3.1

Diagrammatic representation of a cellular receptor and its cognate viral attachment protein (VAP). Scale has been distorted to emphasize the interaction between the VAP and its receptor.

ment protein (VAP), to attach to the viral receptor. Figure 3.1 diagrams the interaction in a simplified illustration. In practice the interaction can be more complex, as exemplified by the entry of human immunodeficiency virus (HIV) shown in Fig. 3.2. Human immunodeficiency virus type 1 (HIV-1) binds to both a primary receptor (the CD4 molecule) and a coreceptor on the surface of susceptible cells. The virus will only enter and infect cells that bear both receptor and coreceptor, although there are a few special exceptions. Several different molecules, all of them chemokine receptors, can serve as coreceptors, and different CD4$^+$ cells express different coreceptors. Furthermore, some isolates of HIV-1 can utilize only one of the two coreceptors, producing a complex pattern (Fig. 3.2) of cellular susceptibility and viral host range.

Another example of a virus group that uses both a receptor and a coreceptor are the alphaherpesviruses, which includes herpes simplex viruses (HSV) types 1 and 2, the cause of "cold sores" and similar genital lesions of humans. In this instance, the virus undergoes primary binding to heparan sulfate (described further below), a substance found on the surface of many cells. However, for the virus to com-

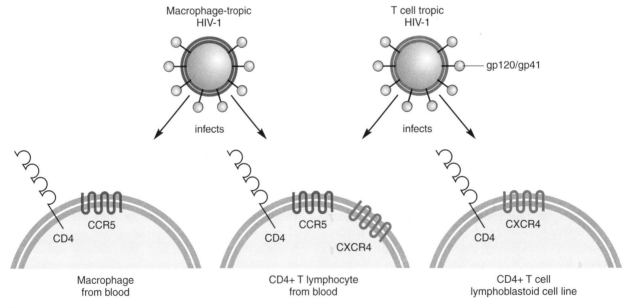

FIG. 3.2

Cells that are permissive for the entry and replication of human immunodeficiency virus type 1 (HIV-1) carry two receptors on their surface. The primary receptor is the CD4 molecule, a protein expressed on the surface of certain subsets of lymphocytes (so called CD4$^+$ cells). In addition, there is a coreceptor, which is either CCR5 or CXCR4; these proteins are members of a large family of molecules that serve as chemokine receptors on the surface of lymphoid cells. Some HIV-1 isolates are macrophage tropic because they use CCR5, a chemokine receptor that is expressed on the surface of macrophages, whereas other isolates are T-cell tropic because they utilize the CXCR4 molecule, another chemokine receptor that is expressed on the surface of T-cell lines. Both kinds of isolates can replicate on peripheral blood mononuclear cells, lymphocytes freshly cultured from peripheral blood, since these cells express both coreceptors. By the same token, some HIV-1 isolates (not shown) are "duotropic" because they can utilize both coreceptors. (Recent studies have shown that macrophages express low levels of CXCR4 at concentrations insufficient for entry of T-cell-tropic HIV-1.)

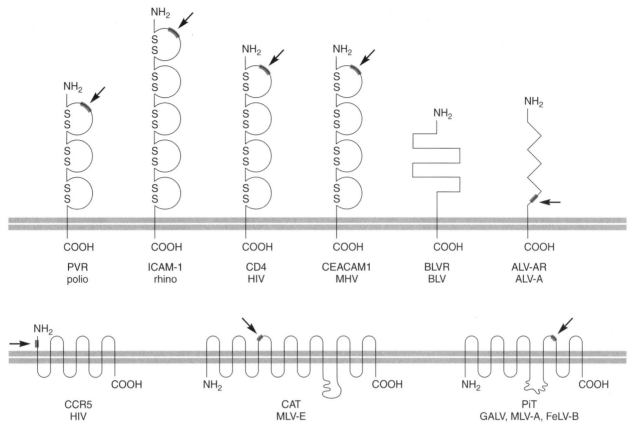

FIG. 3.3

Molecular backbone cartoons of some glycoprotein viral receptors. Receptors diverge widely in their structure and physiologic function. The amino and carboxy termini are shown, together with important disulfide bonds and the probable domains that bind virus. Abbreviations: PVR, poliovirus receptor; ICAM, intercellular adhesion molecule; BGP, biliary glycoprotein; BLVR, bovine leukemia virus receptor; ALV-AR, avian leukosis virus A receptor; CCR5, chemokine receptor 5; CAT, cationic amino acid transporter; PiT, inorganic phosphate transporter. Viruses: polio, poliovirus; rhino, rhinovirus, major group; HIV, human immunodeficiency virus; MHV, mouse hepatitis virus (a coronavirus); BLV, bovine leukemia virus; ALV-A, avian leukosis virus A; MLV-E, murine leukemia virus E; GALV, gibbon ape leukemia virus; MLV-A, murine leukemia virus A; FeLV-B; feline leukemia virus B. (After Holmes KV. Localization of virus infections. In: Nathanson N, et al., eds. *Viral pathogenesis.* Philadelphia: Lippincott–Raven Publishers, 1997; Wimmer E, ed. *Cellular receptors for animal viruses.* Cold Spring Harbor, NY: Cold Spring Harbor Laboratory Press, 1994; and Weiss RA, Tailor CS. Retrovirus receptors. *Cell* 1995;82:531–533.)

plete its entry process, it must first bind to a coreceptor. Recent studies have identified at least three coreceptors, which have been named HveA, HveB, and HveC. Each coreceptor is expressed on a different type of cell and partially accounts for the known tropism spectrum of HSV-1 and 2. Different glycoproteins on the surface of HSV bind to each of these receptors: glycoprotein C (gC) binds to heparan sulfate, and gD binds to Hve A, HveB, and HveC.

What cellular molecules can serve as viral receptors? Many viral receptors are glycoproteins because most proteins expressed on the cell surface have been glycosylated during their posttranslational maturation. In numerous instances, the physiologic role of the viral receptor has been determined, but there are some cases where the normal function

of the receptor is yet to be identified. Viruses have exploited cellular molecules that serve many different functions, as diagrammed in Fig. 3.3, which shows a representative group of glycoprotein viral receptors.

Glycoprotein Receptors The virus binds to a domain that represents a small part of the surface of the glycoprotein receptor molecule, and this domain may be either a polypeptide sequence or a carbohydrate side chain. CD4 is an example of a glycoprotein receptor whose binding domain is the amino acid backbone of the protein. CD4 is a member of the immunoglobulin (Ig) superfamily of molecules and has four globular domains linked together. Mutation of CD4 has shown that the VAP of HIV (called gp120) binds to

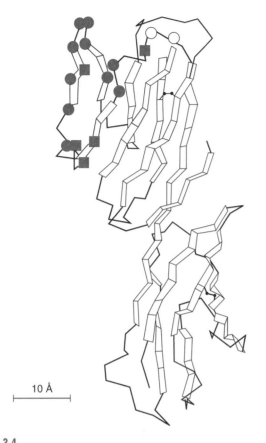

FIG. 3.4

Molecular structure of the two outermost domains of CD4 showing the site that binds gp120, the VAP of HIV. The filled circles and squares indicate amino acids that are part of the receptor domain. (After Wang J, Wan Y, Garrett TP, et al. Atomic structure of a fragment of human CD4 containing two immunoglobulin-like domains. *Nature* 1990;348:411–419.)

a small region within the outermost globular domain of CD4 (Fig. 3.4). Chemokine receptors that act as coreceptors for HIV are seven transmembrane proteins, and the binding domain of the HIV has been mapped to specific extracellular domains within these molecules.

An example of a carbohydrate side chain that acts as a receptor domain is provided by the influenza type A viruses. This receptor is sialic acid (or neuraminic acid), a modified sugar that is part of the branched carbohydrate side chains of glycosylated proteins. Different influenza viruses bind preferentially to terminal sialic acid residues, depending on the linkage of the sialic acid to a proximal galactose molecule in the carbohydrate chain. Thus, human type A influenza viruses bind most avidly to sialic acid α-2,3-galactose configurations, whereas equine type A influenza viruses bind best to sialic acid α-2,6-galactose. This subtle distinction illustrates the exquisite specificity of the interaction between the VAP and its cellular receptor. Influenza virus has a neuraminidase protein that can cleave the neuraminic (sialic) acid residue from the carbohydrate side chain of the receptor, destroying the ability of cells to bind the virus. Because of this property the neuraminidase is also called the "receptor-destroying enzyme." It may appear paradoxical that the virus can destroy its own cellular receptor, but this is thought to facilitate the release of newly budded virus from the cell surface.

Nonprotein Receptors In addition to proteins, glycolipids or glycosoaminoglycans can serve as receptors (Fig. 3.5). For instance, HIV-1 can infect some cell types (certain neural and intestinal cells) that do not express the CD4 molecule. In this case, it appears that galactosylceramide, a glycosphingolipid, serves as an alternative receptor. The virus appears to bind to the galactose moiety on galactosylceramide via the V3 loop on the VAP, a domain different from that which binds to the major CD4 receptor. Sialic acid, the receptor for influenza viruses, also occurs as part of some complex lipids on cell surfaces and, in addition to sialylated glycoproteins, sialic acid–containing lipids can also act as influenza virus receptors. Glycosoaminoglycans are sulfated carbohydrate polymers that are a component of proteoglycans, complex macromolecules composed of proteins and carbohydrates that coat the surface of cells and form the "ground substance" or intercellular matrix found between cells in many tissues. Heparan, one such glycosoaminoglycan, acts as a receptor for HSVs, although additional cell surface molecules, acting as coreceptors, are required for entry of these viruses. Some generalities about viral receptors are summarized in Sidebar 3.1.

Viral Attachment Proteins

As noted above, attachment of the virus particle to its cellular receptor is conferred by a virion surface protein, often called the VAP. As a rule, there is a single VAP, although other viral surface proteins often play an essential role in the steps that follow the initial attachment of virions to the cell surface. For enveloped viruses the VAP is almost always a surface glycoprotein that oligomerizes to form spikes that protrude from the viral envelope. Mutational mapping indicates that a restricted domain on the outermost region of the VAP is responsible for binding, as shown in Fig. 3.6 for the influenza virus hemagglutinin (HA). For nonenveloped naked viruses, the VAP is one of the surface proteins that

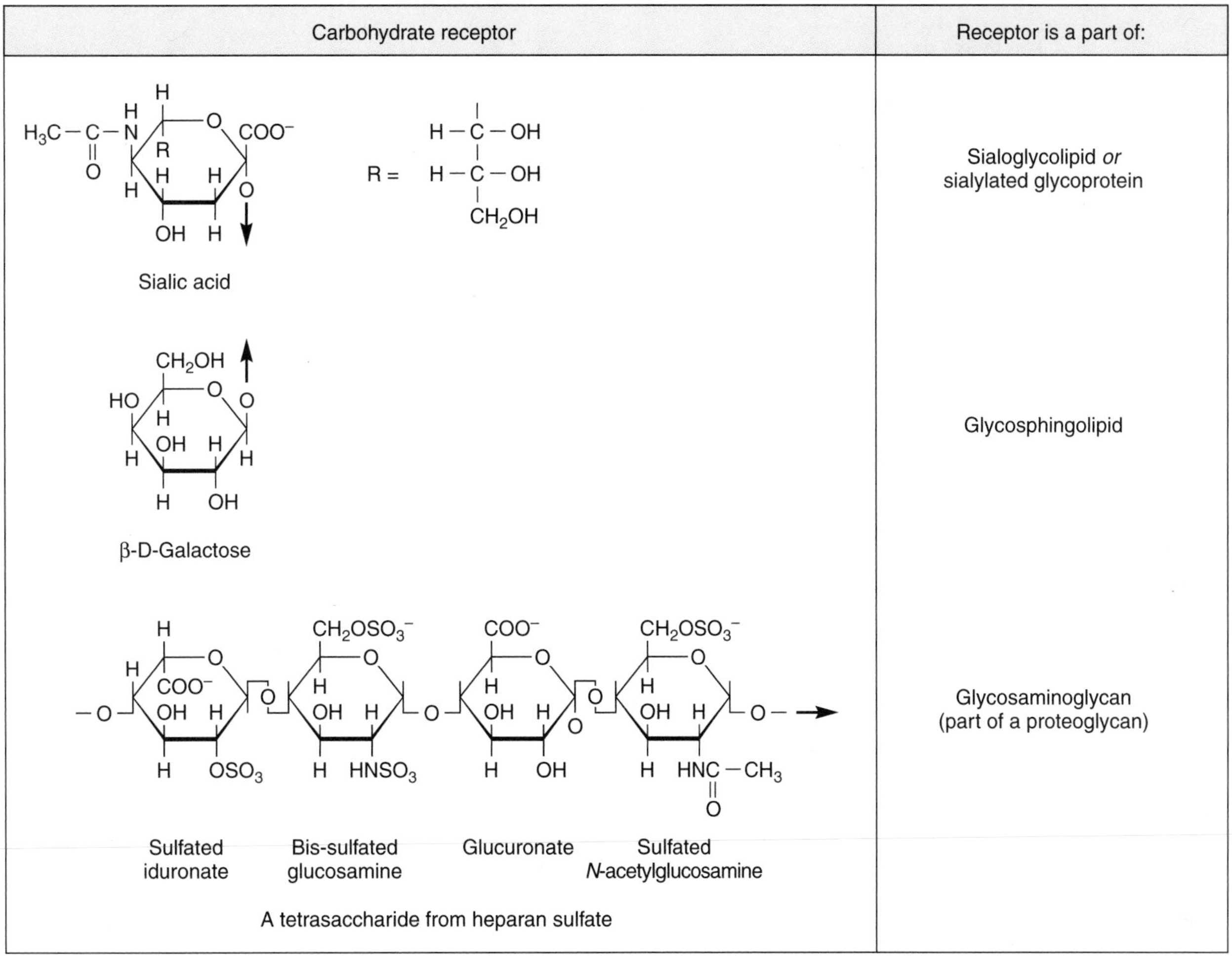

Carbohydrate receptor	Receptor is a part of:
Sialic acid	Sialoglycolipid *or* sialylated glycoprotein
β-D-Galactose	Glycosphingolipid
Sulfated iduronate Bis-sulfated glucosamine Glucuronate Sulfated *N*-acetylglucosamine A tetrasaccharide from heparan sulfate	Glycosaminoglycan (part of a proteoglycan)

FIG. 3.5

Some examples of nonprotein viral receptors. This diagram illustrates a sialic acid receptor for type A influenza viruses; a galactose receptor (part of a glycosphingolipid, galactosylceramide) that is an alternative receptor for HIV-1; and a glycosaminoglycan (heparan sulfate, part of a proteoglycan) that is a receptor for herpes simplex viruses. The diagrams show only the sugar residues and the arrows indicate where they are bound to the remainder of the molecules of which they are a part. Cognate viral attachment proteins are as follows: HIV-1, the V3 loop on gp120; influenza virus, the distal tip of the HA1 molecule; HSV, the gC glycoprotein (domain not mapped). In all instances, the sugar residue is responsible for binding the viral attachment protein, and this residue may be part of a glycolipid (galactosylceramide; sialic acid), a glycoprotein (sialic acid), or a complex proteoglycan (heparan sulfate). (After Stryer L. *Biochemistry*. New York: WH Freeman, 1988.)

forms the external ordered structure of the viral surface. Many cell surface proteins bind other cell surface proteins, thereby mediating cell–cell interactions, and it is possible that some viruses have "pirated" and adapted such cellular proteins to use as VAPs, but this is only a speculation.

Viral Entry

Viral entry is a multistep process that follows attachment of the virion to the cellular receptor and results in deposition of the viral genome (nucleocapsid) in the cytosol. The entry of enveloped viruses is exemplified by the influenza virus (Fig. 3.7). The sequential steps in entry include attachment of the HA spike (the VAP) to sialic acid receptors on the cellular surface; internalization of the virion into an endocytic vacuole; fusion of the endocytic vacuole with a lysosome, with marked lowering of the pH; a drastic alteration in the structure of the HA1 trimer, with reorientation of the HA2 peptide to insert its proximal hydrophobic domain into the vacuolar membrane; fusion of viral and vacuolar membranes; and release of the viral nucleocapsid into the cytosol. For some enveloped viruses, such as HIV, fusion occurs at the plasma membrane (without internalization of the virion and without the requirement

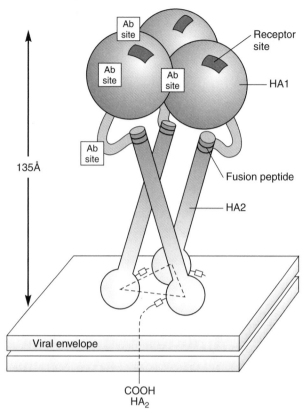

FIG. 3.6

Diagram of the structure of the hemagglutinin (HA) spike (a trimer of the HA1 and HA2 peptides) that acts as the viral attachment protein for influenza virus, showing the site at the distal end of the spike that binds to sialic acid, the cellular receptor. Ab site: viral epitopes that bind monoclonal antibodies that block viral infectivity. The C terminus of HA2 includes a transmembrane domain that anchors the HA in the lipid bilayer. (After Wilson IA, Skehel JJ, Wiley DC. Structure of the haemagglutinin membrane glycoprotein of influenza virus at 3 A resolution. *Nature* 1981;289:366–373.)

for an acidic pH), although there are similar steps involving conformational changes in the structure of the VAP (gp120 surface protein of HIV) leading to fusion of viral and plasma membranes (see Chapter 13). Some of these steps in entry are not yet entirely understood in molecular detail and are the subject of ongoing research.

For naked capsid nonenveloped viruses, viral entry also begins with attachment to the cellular receptor and ends with release of the viral nucleocapsid into the cytosol. However, there are fundamental differences in the process because membrane–membrane fusion is not involved. Apparently, the interaction between the cellular receptor and the VAP causes structural changes in the viral capsid, which releases internal components of the virion. However, it is not clear how this results in transport of the nucleocapsid across the plasma membrane into the cytosol.

Most DNA viruses, with the exception of poxviruses, replicate within the nucleus of the cell. To initiate replication, the nucleocapsid must be transported through the cytosol to the nuclear membrane, cross via nuclear pores, and enter the nucleus. The steps in transmission of the nucleocapsid into the nucleus, which probably involve specific transport mechanisms, are poorly understood.

TROPISM

Tropism Determined by Cellular Receptors

As indicated at the beginning of this chapter, cellular receptors are unequally expressed on different cell types, and this distribution limits the cell types that are permissive for a given virus group. One classic example is the poliovirus receptor.

Poliovirus Poliovirus was one of the first animal viruses to be isolated (1908). It was early recognized that the virus would only infect humans and other primates, a contrast to many other viruses of humans (such as influenza virus and HSV) that would readily infect rodents and other experimental animals. Two findings focused attention on the viral receptor: homogenates of primate tissues would bind virus

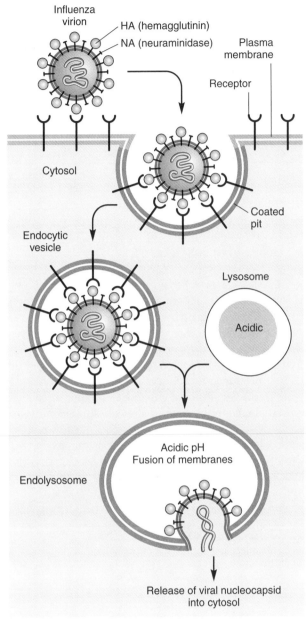

FIG. 3.7

Diagram of the stepwise entry of influenza virus at a cellular level. Key events are attachment of the virion; internalization of the virion by endocytosis; lowering the pH of the endocytic vacuole leading to drastic reconfiguration of the viral attachment protein (hemagglutinin, HA1 and HA2); insertion of a hydrophobic domain of HA2 into the vacuolar membrane; fusion of the viral and vacuolar membranes; release of the viral nucleocapsid into the cytosol.

much more avidly than similar homogenates from nonprimate tissues, and RNA extracted from virions would produce one round of infection in nonprimate cells. Taken together these observations suggested that the species specificity (or tropism) of poliovirus was determined by the viral receptor. Furthermore,

cytogenetic experiments indicated that the putative poliovirus receptor was encoded by human chromosome 19.

In 1989, using DNA transfection, the gene encoding the poliovirus receptor was isolated and shown to encode a previously unknown surface protein of the Ig superfamily. The species restriction of poliovirus is apparently due to the close similarities between the receptor expressed by all primates, which differs from the homologous gene expressed in nonprimate species. Furthermore, there are a few poliovirus strains that have been adapted to replicate in mice, and adaptation has been shown to be associated with a small domain in the VAP VP1. Molecular exchange of this domain confers mouse tropism on polioviruses otherwise unable to infect mice, presumably because the mouse-adapted virus can bind to a murine receptor. It is not known whether this putative receptor is the mouse homolog of the human poliovirus receptor or another molecule.

Poliovirus is an enterovirus that replicates in the gut and is excreted in the feces, being transmitted by the fecal–oral route. It produces a viremia, invades the central nervous system, and then infects and destroys a very specific set of lower motor neurons in the medulla and spinal cord, thereby producing a characteristic flaccid paralysis. A survey of primate tissues indicated that the poliovirus receptor was expressed at differing levels in many organs. A high level of expression in the anterior horn cells—lower motor neurons of the spinal cord—was consistent with their susceptibility to poliovirus infection. However, there were many discrepancies between receptor expression and viral replication in that certain tissues or cells that expressed the receptor robustly were not infected whereas other cells that expressed only modest levels of receptor were infected (Table 3.1). In this instance, it appears that receptor expression is necessary for susceptibility to infection, but is not sufficient to explain the restricted tissue and cellular replication of poliovirus.

Parvovirus B19 Parvovirus B19 causes a mild exanthem (known as "fifth" disease or exanthem subitum) in children but is mainly of clinical importance because of its predilection for erythrocyte precursors in the bone marrow. This ability to replicate in and destroy red cells, which can cause a severe anemia, is an unusual tropism exhibited by very few viruses. Autonomous parvoviruses, such as B19, are restricted in their replication to dividing cells because some of the cellular enzymes required for replication of parvoviral DNA are only expressed during the S phase of cell division. The receptor for B19 virus has been identified as globoside, a neutral glycosphingolipid, and the receptor domain is probably the terminal galactose residue. Tissue surveys show that globoside is present on many tissues of mesodermal origin but constitutes a higher proportion of the glycosphingolipids on erythrocytes than on other tissues. In this instance, tropism is apparently associated with robust expression of the viral receptor plus the restriction of replication to rapidly dividing cell populations. Again, receptor expression is necessary but is not the only determinant of tropism.

Lymphocytic Choriomeningitis Virus Different strains of lymphocytic choriomeningitis virus (LCMV) exhibit differences in their tropism in vivo; Clone 13 replicates preferentially in the liver and spleen whereas Armstrong replicates best in the brain. Furthermore, these differences are determined by a single amino acid at position 260 in the viral glycoprotein (the VAP), implying that tropism may be correlated with ability to bind to the viral receptor. Recent identification of the receptor as α-dystroglycan has led to the recognition that differences in the degree of affinity to this single receptor appear to correlate with differential tropism. Thus, LCMV strains with high affinity for the receptor replicate preferentially in the white pulp and low-affinity strains in the red pulp of the spleen.

> ■ ■ ■
> ### SIDEBAR 3.2
> #### Determinants of Viral Tropism: A Selected List
> - Cellular receptors
> - Activation of viral attachment or fusion proteins by cellular proteases
> - Temperature sensitivity
> - pH lability
> - Cellular transcriptases ■

Other Determinants of Tropism

In addition to receptor specificity, there are a number of other determinants of tropism, some of which are summarized in Sidebar 3.2.

Cellular Protease Requirement A number of enveloped viruses are not infectious when they bud from cells as mature virions because one of the viral surface proteins (either the viral HA or the viral fusion protein) requires proteolytic cleavage to be activated (Fig. 3.8). In such instances, infectious virus is only produced by replication in cell types that secrete a cellular protease capable of mediating the required cleavage. In many instances, susceptibility to proteolytic cleavage is determined by a few amino acids (such as one versus several arginines) at the site of cleavage, so that mutations in one or two critical amino acids can alter the tissue tropism. As a

TABLE 3.1				
Expression of the human poliovirus receptor in transgenic mice, and expression of poliovirus RNA following infection in the same				
Tissue	Cells	PVR expression	Poliovirus replication	Cellular destruction
Central nervous system	Neurons			
	Anterior horn spinal cord	High	High	Severe
	Posterior horn spinal cord	High	High	Minimal
	Medulla	High	High	Moderate
	Cerebellum	High	Moderate	Minimal
	Midbrain	High	Moderate	Minimal
	Forebrain	High	Moderate	Minimal
Thymus	T lymphocytes	High	None	None
Kidney	Epithelial, tubule cells	High	None	None
Lung	Alveolar cells	High	None	None
Adrenal	Endocrine cells	High	None	None
Intestine	Many	Low	None	None
Spleen	Lymphocytes	Low	None	None
Skeletal muscle	Myocytes	Low	Moderate	None

Receptor expression is necessary but not sufficient to explain viral tropism. Virus replication is confined to cells where the PVR is expressed at a high level but there are many tissues that are not infected in spite of PVR expression. PVR, poliovirus receptor.

After Ren R, Racaniello VR. Human poliovirus receptor gene expression and poliovirus tissue tropism in transgenic mice. *J Virol* 1992;66:296–304.

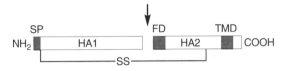

FIG. 3.8

Stick diagram of two viral proteins that require proteolytic cleavage for activation. Cleavage (*arrow*) can be mediated either during intracellular maturation by furin, a protease found in the Golgi apparatus, or by extracellular secreted proteases. The sequence of amino acids—one to three basic arginines—immediately adjacent to the cleavage site determines whether the viral protein is susceptible to furin, a more fastidious enzyme, or whether it can only be cleaved by extracellular proteases, which have a broader range of substrates. NH_2 and COOH, the amino and carboxy termini of the peptides; FD, fusion domain; TMD, transmembrane domain; SP, signal peptides. (After Nagai Y. Virus activation by host proteases: a pivotal role in the spread of infection, tissue tropism and pathogenicity. *Microbiol Immunol* 1995;39:1–9.)

general rule, viruses with an extended cellular host range spread more widely in vivo and cause more serious disease.

One example is the paramyxoviruses. These enveloped viruses carry two surface glycoproteins: the HA, which acts as the VAP, and the fusion protein, which mediates fusion between the viral envelope and the plasma membrane. In many cases, the fusion protein exists on newly budded virus as an F_0 polypeptide that is unable to mediate fusion. Conversion to the active F_1 form requires proteolytic cleavage of the F_0 protein, and the virus depends for its activation on the presence of secreted cellular proteases in the intra- or extracellular milieu. Not all cells secrete the activating proteases, and this limits virus replication to selected cell types and explains why some paramyxoviruses do not spread beyond the respiratory tract.

An example is Newcastle disease virus, a paramyxovirus of birds. Virulent isolates of Newcastle disease virus encode a fusion protein that is readily cleaved by furin, a proteolytic enzyme present in the Golgi apparatus, so that the protein is activated during maturation prior to reaching the cell surface and before budding of nascent virions. This makes it possible for virulent strains of the virus to infect avian cells of many different types, thereby increasing its tissue host range and causing systemic infections that are often lethal. By contrast, avirulent strains of Newcastle virus encode a variant fusion protein that is not cleaved during maturation in the Golgi, so that nascent virus requires activation by an extracellular protease. The required protease is found only in the respiratory or enteric tracts, thereby limiting tropism to surface tissues and conferring an attenuated phenotype on the virus.

Temperature of Replication Most human viruses replicate optimally at 37°C, which is the core body temperature of humans. However, some mucosal surfaces, such as the upper respiratory tract, have a lower temperature, about 33°C. Certain viruses that replicate in the epithelial cells of the nose and throat, such as rhinoviruses, have been selected to grow optimally at 33°C. Rhinoviruses replicate poorly at 37°C and were originally isolated only when cell cultures were maintained at 33°C. Such viruses are restricted in their tissue distribution by their relative inability to replicate at 37°C, which limits their spread beyond the upper respiratory tract.

Acid Lability A number of viruses are enteric, initiate infection by ingestion, and are excreted in the feces to be transmitted by the fecal–oral route. The gastrointestinal tract presents a harsh environment—first the acidic pH of the stomach, followed by the alkaline pH of the intestine and the destructive effects of pancreatic digestive enzymes. In general, enterotropism is limited to viruses that can survive these adverse conditions, although there are some exceptions. Thus, most respiratory viruses are swallowed to some extent, but few are enterotropic; exceptions include adenovirus, which can survive and replicate in the gastrointestinal tract. Some enterotropic viruses, such as reoviruses, have exploited the adverse conditions of the enteric tract, so that they undergo conversion to an infectious subvirion particle by digestive enzymes (Fig. 3.9). Enzymatic processing leads to a conformational change in the VAP $\sigma 1$, permitting the virions to bind to a receptor on microfold (M) cells, the first step in reoviral entry in the gut.

Transcriptional Control of Tropism Following uncoating and entry of the viral genome, there is a complex program of replication, in which the virus utilizes a number of host proteins to transcribe its genome, translate virus-encoded messages, and assemble and release new infectious virions. Since cells vary in their ability to support each of these multiple steps,

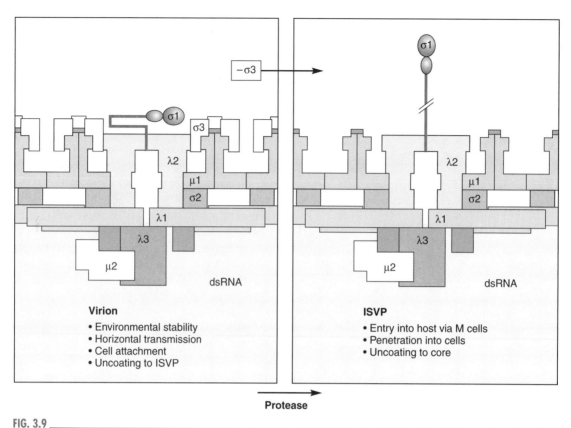

FIG. 3.9

Activation of reovirion to an infectious subvirion particle (ISVP) by action of protease on the viral coat. Digestion of the σ3 protein and of the μ1 protein leads to conformational changes in the σ1 protein. The σ1 protein is the viral attachment protein that mediates binding of virus to cellular receptors, but only after protease has converted the virion into the infectious subvirion particle (ISVP). (Redrawn after image provided by ML Nibert.)

tropism can be determined by cell-specific differences in replication.

One example is the papillomaviruses, DNA viruses that replicate in skin and may cause tumors ranging from benign warts to malignant cancer of the cervix (Fig. 3.10). The skin is a somewhat unique tissue in which new cells are continuously produced at the innermost basal layer of the epidermis, where germinal cells divide. These cells gradually move outward, differentiating into keratinocytes that are constantly being sloughed from the superficial layers of the skin. Papillomaviruses commence their replication in germinal cells that are permissive for replication of the viral genomes. However, germinal cells produce proteins that block the transcription of late structural genes of the virus. As the infected basal cells move outward and begin to differentiate, they become permissive for transcription and translation of the papillomavirus structural genes, so that complete infectious virus is only formed in cells that are about to be sloughed. Release of virus from the superficial layers of dead cells promotes the transmission of infection to new

sites on the infected person as well as to new uninfected hosts. Not surprisingly, only the skin provides the complex cascade of transcription factors that can support the complete program of viral replication, explaining the very specialized tropism of papillomaviruses. Papillomavirus receptors are widely expressed on many nondermal cell types and appear to play no role in tropism.

Retroviruses Retroviruses are RNA viruses that have a complex life cycle involving transcription of their genome into a DNA transcript; this transcript is imported to the nucleus where it dictates both the replication of nascent RNA genomic transcripts and the transcription of viral messenger RNA (discussed in Chapter 10). Promoters and enhancers that regulate these transcriptional events are located in a "long terminal repeat" (LTR) at the ends of the viral genome. Some avian retroviruses (avian leukosis sarcoma viruses) are oncogenic, but various members of the group differ in the location and nature of the tumors that they induce, reflecting variation in tissue tropism. Using molecular methods to con-

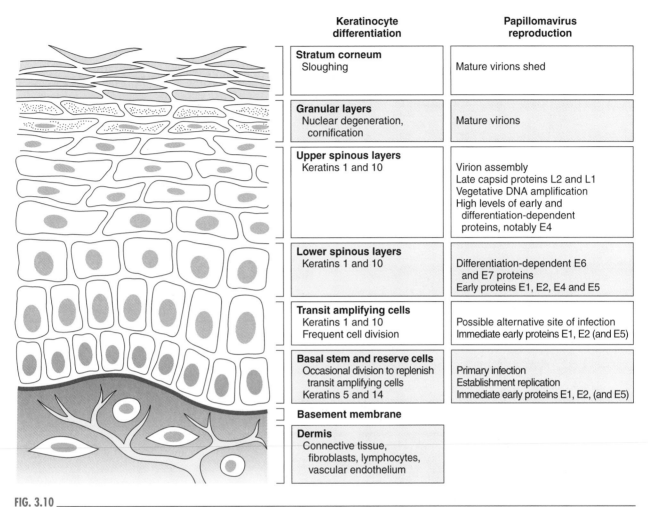

Keratinocyte differentiation	Papillomavirus reproduction
Stratum corneum Sloughing	Mature virions shed
Granular layers Nuclear degeneration, cornification	Mature virions
Upper spinous layers Keratins 1 and 10	Virion assembly Late capsid proteins L2 and L1 Vegetative DNA amplification High levels of early and differentiation-dependent proteins, notably E4
Lower spinous layers Keratins 1 and 10	Differentiation-dependent E6 and E7 proteins Early proteins E1, E2, E4 and E5
Transit amplifying cells Keratins 1 and 10 Frequent cell division	Possible alternative site of infection Immediate early proteins E1, E2 (and E5)
Basal stem and reserve cells Occasional division to replenish transit amplifying cells Keratins 5 and 14	Primary infection Establishment replication Immediate early proteins E1, E2, (and E5)
Basement membrane	
Dermis Connective tissue, fibroblasts, lymphocytes, vascular endothelium	

FIG. 3.10 _____

Sequence of transcriptional events in the synthesis of papillomavirus in different layers of the epidermis. This diagram shows the steps in generation and maturation of keratinocytes and the corresponding steps in viral replication. (After Chow LT, Broker TR. Small DNA tumor viruses. In: Nathanson N, et al. *Viral pathogenesis*. Philadelphia: Lippincott–Raven Publishers, 1997:267–301.)

struct genetic hybrids between viruses with different patterns of tropism, these differences have been mapped in part to the LTR (Table 3.2). In this instance, viral tropism is partially determined by cellular transcription factors.

Tropism and Viral Variation

Variation in the pathogenicity or virulence of different isolates of a single virus is an important topic that is the subject of Chapter 8. However, it is appropriate to mention a few examples in which pathogenicity can be associated with variation in tropism.

Polyoma virus is a DNA virus of mice that causes both acute death in newborn animals and tumors at a later age. Sialic acid residues, in one specific linkage to proximal oligosaccharides, act as the cellular receptor and the cognate VAP is VP1, a

TABLE 3.2
Genetic chimeras of avian leukosis viruses with different tropisms

		REPLICATION	
Env	LTR	Muscle	Bursa
A	RAV-1	++++	++++
A	RAV-0	–	++++
E	RAV-1	–	++++
E	RAV-0	–	++++

Tropism of avian leukosis virus for muscle and bursal tissues of chickens is determined by the viral envelope and by the viral LTR (long-terminal repeat). Replication in muscle appears to require the A envelope and also the RAV-1 LTR, illustrating the multifactorial effect of viral genes on tropism. Env, genotype of the envelope protein that serves as the viral attachment protein for avian leukosis virus; LTR, genotype of the long term repeat was derived from either of two viruses, RAV-1, Rous-associated virus, or RAV-0.

After Brown DW, Robinson HL. Influence of env and long terminal repeat sequences on the tissue tropism of avian leukosis viruses. *J Virol* 1988; 62:4824–4831.

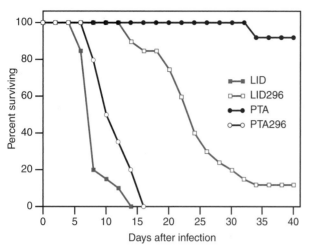

FIG. 3.11

Alteration of binding domain of the viral attachment protein can influence binding to the cellular receptor and also alter the pathogenicity of the virus. Comparison of the pathogenicity of different polyoma virus strains in newborn mice as reflected in survival curves. Change of a single amino acid, at position 296 on the viral attachment protein, VP1, markedly alters pathogenicity. Virulence is associated with a valine and avirulence with an alanine at position 296. LID, virulent strain; LID296, LID with a valine to alanine mutation at position 296; PTA, avirulent strain; PTA296, PTA with an alanine to valine mutation at position 296. (After Bauer PH, Bronson RT, Fung SC, et al. Genetic and structural analysis of a virulence determinant in polyomavirus VP1. *J Virol* 1995;69:7925–7931.)

component of the viral capsid. The interaction between VP1 and sialic acid has been determined at high resolution by x-ray crystallography. Two strains of polyoma virus, one virulent and the other attenuated, differ by only two amino acids in VP1. Reciprocal exchange of one of these amino acids interconverts the virulent and attenuated phenotypes (Fig. 3.11). The same mutation markedly alters the ability of polyoma virus to bind to erythrocytes (a surrogate for binding to susceptible cells, since erythrocytes carry the viral receptor) and significantly changes the hydrophobic interaction between sialic acid and the binding pocket on VP1. Somewhat surprisingly, virulence is associated with an amino acid substitution that is predicted to reduce binding to the receptor. It is speculated that less avid binding enhances the ability of virus to spread in vivo.

Another striking example is avian influenza virus. A particular serotype (known by its HA and neuraminidase as H5N2) circulates in domestic fowl as a relatively innocuous cause of respiratory infection. However, on two occasions a virulent mutant of this virus appeared and caused pandemic fatal disease in commercial poultry flocks (first in Pennsylvania in 1983 and subsequently in Mexico in 1995). In both epidemics, the virus acquired a mutation—

deleting an O-linked glycosylation site close to the HA1/HA2 cleavage site—that renders the viral more susceptible to proteolytic activation. As a consequence, the mutated virus can multiply in a wider range of host tissues and replicate to higher titers, converting a mild sublethal infection to a rapidly fatal disease.

A final example is the differences in pathogenesis between two similar strains of SHIV (chimeric viruses bearing the envelope of HIV inserted into the genetic backbone of simian immunodeficiency virus, SIV) that use different coreceptors (Fig. 3.12). As described above, HIV can utilize either of two cellular chemokine molecules, CCR5 or CXCR4, as coreceptors for viral entry; HIV strains using CCR5 can infect macrophages and CD4$^+$ T lymphocytes in peripheral blood, whereas strains using CXCR4 can only infect

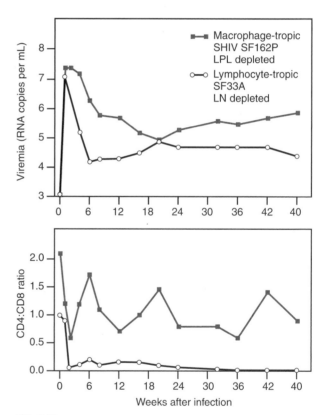

FIG. 3.12

Different pathogenesis in rhesus monkeys infected with two strains of SHIV, one macrophage tropic (SHIV SF162P) and the other T-lymphocyte tropic (SHIV SF33A). *Upper panel:* Both viruses produced similar levels of viremia. *Lower panel:* The T-cell-tropic SHIV (SF33A) depleted circulating CD4 T lymphocytes in the lymph nodes and the blood, whereas the macrophage-tropic SHIV (SF162P) depleted CD4 T lymphocytes in the lamina propria of the gut but not in the blood. (After Harouse JM, Gettie A, Tan RCH, et al. Distinct pathogenic sequelae in rhesus macaques infected with CCR5 or CXCR4-tropic strains of SHIV. *Science* 1999;284:816–820.)

T lymphocytes. When SHIVs are constructed bearing the two different types of HIV envelope genes, there is a striking difference in pathogenesis (Fig. 3.12). The CCR5-using SHIV (macrophage-tropic virus) preferentially infect and destroy T cells in the lamina propria of the intestine with little reduction of $CD4^+$ T lymphocytes in the blood, whereas the CXCR4-using SHIV (T-lymphocyte-tropic virus) preferentially destroy T cells in the regional lymph nodes and cause a marked depletion of $CD4^+$ T lymphocytes in the blood, a depletion associated with AIDS.

FURTHER READING

Reviews and Chapters

Chow LT, Broker TR. Small DNA tumor viruses. In: Nathanson N, Ahmed R, Murphy FA, et al. *Viral pathogenesis*. Philadelphia: Lippincott–Raven Publishers, 1997:267–301.

Dimitrov DS. Cell biology of virus entry. *Cell* 2000;101:697–702.

Holmes KV. Localization of viral infections. In: Nathanson N, Ahmed R, Murphy FA, et al. *Viral pathogenesis*. Philadelphia: Lippincott–Raven Publishers, 1997:35–53.

Racaniello VR, Ren R. Poliovirus biology and pathogenesis. *Curr Top Microbiol Immunol* 1999;206:305–325.

Virgin HW III, Tyler KL, Dermody TS. Reovirus. In: Nathanson N, Ahmed R, Murphy RA, et al. *Viral pathogenesis*. Philadelphia: Lippincott–Raven Publishers, 1997:669–699.

Weiss RA, Tailor CS. Retrovirus receptors. *Cell* 1995;82:531–533.

Wimmer E. *Cellular receptors for animal viruses*. Cold Spring Harbor, NY: Cold Spring Harbor Laboratory Press, 1994.

Classic Papers

Kawaoka Y, Naeve CW, Webster RG. Is virulence of H5N2 in chickens associated with loss of carbohydrate from the hemagglutinin? *Virology* 1984;139:303–316.

Mendelsohn C, Wimmer E, Racaniello VR. Cellular receptor for poliovirus: molecular cloning, nucleotide sequence, and expression of a new member of the immunoglobulin superfamily. *Cell* 1989;56:855–865.

Wilson IA, Skehel JJ, Wiley DC. Structure of the haemagglutinin membrane glycoprotein of influenza virus at 3 A resolution. *Nature* 1981;289:366–373.

Original Contributions

Amerongen HM, Wilson GAR, Fields BN, Neutra MR. Proteolytic processing of reovirus is required for adherence to intestinal M cells. *J Virol* 1994;68:8428–8432.

Bauer PH, Bronson RT, Fung SC, et al. Genetic and structural analysis of a virulence determinant in polyomavirus VP1. *J Virol* 1995;69:7925–7931.

Brown DW, Robinson HL. Influence of env and long terminal repeat sequences on the tissue tropism of avian leukosis viruses. *J Virol* 1988;62:4828–4831.

Cao W, Henry MD, Borrow P, et al. Identification of alpha-dystroglycan as a receptor for lymphocytic choriomeningitis virus and Lassa fever virus. *Science* 1998;282:2079–2081.

Dockter J, Evans CF, Tishon A, Oldstone MB. Competitive selection in vivo by a cell for one variant over another: implications for RNA virus quasispecies in vivo. *J Virol* 1996;70:1799–1803.

Geraghty RJ, Krummenacher C, Cohen GH, et al. Entry of alphaherpesviruses mediated by poliovirus receptor–related protein 1 and poliovirus receptor. *Science* 1998;280:1618–1620.

Harouse JM, Gettle A, Tan RCH, Blanchard J, et al. Distinct pathogenic sequelae in rhesus macaques infected with CCR5 or CXCR4 tropic strains of SHIV. *Science* 1999;284:816–820.

Klenk HD, Rott R. The molecular biology of influenza virus pathogenicity. *Adv Virus Res* 1988;34:247–281.

Mothes, W, Boerger AL, Narayan S, et al. Retroviral entry mediated by receptor priming and low pH triggering of an envelope glycoprotein. *Cell* 2000;103:679–689.

Nagai Y. Virus activation by host proteinases: a pivotal role in the spread of infection, tissue tropism, and pathogenicity. *Microbiol Immunol* 1995;39:1–9.

Ren R, Racaniello VR. Human poliovirus receptor gene expression and poliovirus tissue tropism in transgenic mice. *J Virol* 1992;66:296–304.

Wang J, Yan Y, Garrett TP, et al. Atomic structure of a fragment of human CD4 containing two immunoglobulin-like domains. *Nature* 1990;348:411–418.

Webster RG, Kawaoka Y, Bean WJ, Jr. Molecular changes in A/chicken/Pennsylvania/83 (H5N2) influenza virus associated with acquisition of virulence. *Virology* 1986;149:165–173.

Wu Dunn D, Spear PG. Initial interaction of herpes simplex virus with cells is binding to heparan sulfate. *J Virol* 1989;63:52–58.

Chapter 4
Virus—Cell Interactions

HOW DO VIRUSES AFFECT CELLS?

Viral infection of the cell, with subsequent production of infectious virus, often results in cell death. However, there are many variations on this theme. Some viruses can cause highly productive infections without death of the cell, and release of progeny virus may occur without cytolysis. Conversely, viruses can induce marked effects by binding to the cell surface without entering the cytosol or as the result of an incomplete cycle of replication that fails to produce new infectious virions. Alterations in the host cell can be mediated by interfering with a wide variety of normal cell functions or by inducing a new cascade of cellular processes that leads to apoptosis or programmed cell death. This chapter describes this plethora of virus–cell interactions, many of which can best be studied in cell culture systems.

PRODUCTIVE VIRAL REPLICATION MAY NOT DESTROY HOST CELLS

Does viral replication necessarily compromise normal cellular function? Most viruses utilize only a small fraction of the total protein synthetic machinery of the host cell, often estimated as 1%. During a highly productive infection, a cell may produce approximately 1,000 virions during a 24-hour period, of which perhaps 100 are infectious. If each virion contains fewer than 5,000 molecules of protein, then fewer than 5 million molecules of viral proteins will be produced daily. This is dwarfed by the daily cellular production of about 1 billion protein molecules, implying that productive viral infection need not compromise the normal synthetic activity of the cell.

Can nascent virus be released without lysis of the host cell? Nonenveloped viruses assemble intracellularly, and efficient release often requires lysis of the host cell to free individual virions. However, enveloped viruses, particularly enveloped RNA viruses, normally mature by budding across a cellular membrane, either the plasma membrane or an internal membrane such as the endoplasmic reticulum or Golgi apparatus. Virions are then released from the plasma membrane or by exocytosis of the content of intracytoplasmic vesicles, in a manner similar to the release of secreted proteins. Under these

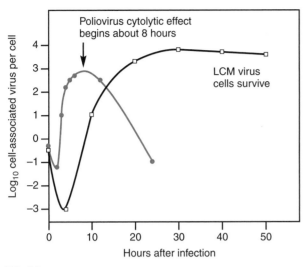

FIG. 4.1 _____

Comparison of replication of a cytolytic and a noncytolytic virus. Cytolytic poliovirus replicates rapidly, cells begin to die about 8 hours after infection, terminating virus replication and releasing cell-associated virus. Noncytolytic lymphocytic choriomeningitis virus (LCMV) replicates more slowly than poliovirus but does not kill most infected cells, which continue to produce cell-associated virus, much of which is released by budding through the plasma membrane. (Poliovirus: After Darnell JE, Levintow L, Thoren MM, Hooper JL. The time course of synthesis of poliovirus RNA. *Virology* 1961;13:271–279. LCMV: Lehmann-Grube F, Popescu M, Schafer H, Gschwinder HH. LCM virus replication in vitro. *Bull WHO* 1975;52:443–456.)

circumstances, virus dissemination does not depend on cytolysis and the infected cell may continue to produce infectious virions indefinitely, unless the immune response intervenes to destroy infected cells. Figure 4.1 compares intracellular virus accumulation for a typical cytolytic nonenveloped and a noncytolytic enveloped virus.

PRODUCTIVE VIRAL REPLICATION IS NOT REQUIRED FOR CYTOPATHOLOGY

Conversely, a virus may initiate pathologic processes in potential host cells without completing a productive cycle of infection. From the first contact, a virus may impact the potential host cell. Fusion initiated by enveloped viruses is a most dramatic example. As described in Chapter 3, fusion is a critical step in the cellular entry of enveloped viruses and leads to deposition of the viral nucleocapsid complex in the cytosol. When applied to uninfected cells, a viral inoculum may produce massive fusion of host cells, which are converted to a multinucleated syncytium within minutes. This "fusion from without" phenomenon does not require replication

because it can be induced with a virus stock whose genomes have been inactivated by radiation. In fact, the fusion activity of certain viruses is so striking that it is used by cytogeneticists as a method for production of heterokaryons, involving the fusion of two cell types to make a cell that contains two different nuclei. Sendai virus, a paramyxovirus, has frequently been used for this purpose. Figure 4.2 shows a typical virus-induced syncytium.

Many viruses cause abortive infections in certain cell types that fail to provide the required cellular proteins necessary to complete transcription, translation, and assembly of new infectious virions. However, such abortive infections may lead to cell death via some of the pathways described later in this chapter. Again, productive virus replication and cellular alterations are not necessarily linked.

CELLULAR RESPONSES TO VIRUS ATTACHMENT

Fusion induced by binding of enveloped viruses to host cells has just been described. A number of other effects can also follow binding of a virus to its cellular receptor. Apoptosis, or programmed cell death, is described later in this chapter. Reovirus can initi-

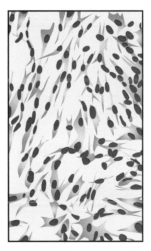

FIG. 4.2 _____

Syncytia produced by an enveloped RNA virus. Both panels show cell cultures exposed to La Crosse virus, a bunyavirus. In this instance, the culture on the right was then treated with a low-pH buffer for a few minutes, to complete the fusion process, while the control culture on the left was exposed to a neutral buffer (fusion did not occur). Acid-dependent fusion is characteristic of many RNA viruses that only produce fusion after they have undergone endocytosis and the pH has dropped in the endocytic vacuole following fusion with a lysosomal vacuole. (After Gonzalez-Scarano F, Pobjecky N, Nathanson N. La Crosse bunyavirus can mediate pH-dependent fusion from without. *Virology* 1984;132:222–225.)

TABLE 4.1										
Capacity of reovirus strains to induce apoptosis										
	GENOME SEGMENT									
Percent apoptosis	L1	L2	L3	M1	M2	M3	M4	S1	S2	S3	Virus strain
3	1	1	1	1	1	1	1	1	1	1	T1L
6	3	3	3	3	3	1	3	1	3	3	EB145
2	3	3	1	3	1	3	3	1	3	3	EB121
97	1	1	1	1	1	1	1	3	1	1	1.HA3
48	3	3	3	3	3	3	3	3 ↑	3	3	T3D

Differences in the capacity of reovirus strains to induce apoptosis are determined by the viral attachment protein σ1, which is encoded by the genome segment S1. Mouse fibroblasts (L929 cells) were infected with type 1 Lang (T1L) and type 3 Dearing (T3D) strains of reovirus and reassortants between them, and the degree of apoptosis was determined by the proportion of cells that showed characteristic chromatin changes (condensed or fragmented) when stained with acridine orange.

After Tyler KL, Squier MKT, Rodgers SE, et al. Differences in the capacity of reovirus strains to induce apoptosis are determined by the viral attachment protein σ1. *J Virol* 1995;69:6972–6979.

ate apoptosis simply by binding to certain cell types, and genetic studies show that the σ1 viral attachment protein of reovirus is responsible for this effect (Table 4.1). Likewise, the viral attachment protein of HIV-1 (the su or surface envelope protein, often called gp120) can induce apoptosis by binding to its cognate receptor, the CD4 molecule. Experimental evidence indicates that this effect is due to cross-linking of individual CD4 molecules on the cell surface, and the effect can be mimicked by anti-CD4 antibodies.

Many viral receptors are cell surface proteins whose function is to initiate transmembrane signaling, beginning a cascade of intracellular biochemical steps. In some instances, attachment of the viral surface protein will mimic the effects of the "normal" ligand for the receptor and initiate the signaling cascade. For instance, the gp120 envelope protein of HIV-1 uses members of the chemokine receptor family, such as CCR5 and CXCR4, as a viral coreceptor. In some instances, binding of gp120 to CCR5 or CXCR4 initiates transmembrane signaling events, as evidenced by activation of intracellular pathways or biological effects such as chemotaxis or apoptosis.

VIRAL MODULATION OF HOST CELL TRANSCRIPTION

Viral infection can affect RNA transcription of host cells in different ways, depending on the virus–cell combination. There may be a widespread down-regulation of transcription, or a mixed up- and down-regulation of different genes. These activities can lead to death of the infected cell, alterations in the cellular phenotype, or transformation and potential tumorigenesis.

Many viruses initiate a general reduction of host cell transcriptional activity that often begins soon after infection. Although long known and much studied, the molecular mechanisms of this phenomenon are not completely understood. A well-documented example is infection with poliovirus, which rapidly induces a reduction in transcription by the three major RNA polymerases, Pol I, Pol II, and Pol III, that catalyze the synthesis of ribosomal, messenger, and transfer RNA, respectively. Poliovirus induces a reduction in the availability of transcription factors, as well as a decrease in the amount of polymerase associated with the DNA templates, leading to an acute drop in the rate of synthesis of mRNAs for many cellular proteins.

In some instances, viral infection decreases the transcription of a specific gene in a differential manner. One of the most intriguing examples is the ability of certain strains of lymphocytic choriomeningitis virus (LCMV) to cause growth retardation in suckling mice (Fig. 4.3). This effect was traced to a reduced production of growth hormone (GH) by the pituitary gland, and it was shown that LCMV produced a nonlytic persistent infection of the GH-producing cells of the pituitary. LCMV infection was associated with a reduced level of GH mRNA and, surprisingly, this effect was messenger specific since the mRNA levels for another hormone, prolactin, and for "housekeeping" genes, such as actin, were reduced much less (Table 4.2). Genetic manipulation showed that the viral effect was associated with a domain in the GH promoter that bound a specific transactivator, although the precise molecular mechanism has yet to be determined.

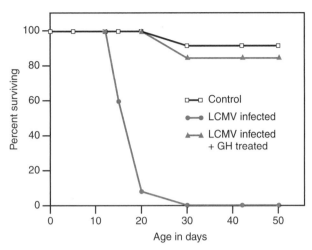

FIG. 4.3

The effect of infection with lymphocytic choriomeningitis virus (LCMV) on the survival of infant mice is due to a reduction in the production of growth hormone (GH). GH has two major functions: to enhance normal growth and to regulate glucose metabolism. LCMV-infected mice exhibit retarded growth (not shown in this graph) and hypoglycemia, which leads to death by age 1 month. (After Oldstone MBA, Rodriguez M, Daughaday WH, Lampert PW. Viral perturbation of endocrine function: disordered cell function leads to disturbed homeostasis and disease. *Nature* 1984;307:278–281.)

Some viruses have a differential effect on the expression of selected host genes. Using a general method to detect changes in levels of mRNA, it was shown that human cytomegalovirus infection was followed by changes in transcription of about 5% of cellular genes; approximately an equal number of genes exhibited enhanced or reduced transcription. These changes produced many different effects, most of which have yet to be studied. For instance, human cytomegalovirus infection drives host lymphocytes into cell division, and some genes associated with G_0 (cells in the vegetative state) are down-regulated whereas other genes associated with the cell cycle, such as the cyclins, are up-regulated.

Certain viruses increase transcription of cellular genes, often in a selective manner. For instance, human T-cell lymphotrophic virus type I (HTLV-I) can transform CD4$^+$ lymphocytes, and this activity has been traced to the tax protein, a nonstructural viral protein. The tax protein by itself can transform the cellular phenotype when transfected into T lymphocytes. The tax protein apparently acts indirectly, by binding to cellular transcription factors and increasing their ability to up-regulate host cell transcription. It has a similar effect on the viral LTR (long terminal repeat, the viral promoter), which contains an element called CREB (CRE binding) that binds a transactivator, the cAMP-responsive element (CRE). When tax binds to CRE it increases its ability to bind CREB and enhances the transcription of viral genes.

VIRAL MODIFICATION OF HOST CELL PROTEIN SYNTHESIS

As noted above, many viral infections are associated with the down-regulation of host cell protein synthesis. This is accomplished through a wide variety of biochemical mechanisms, most of which interfere with the initiation of translation of mRNA. Yet translation of viral messages is maintained to ensure the production of large amounts of structural proteins for assembly into nascent virions. How is this accomplished? Poliovirus, one of the best studied examples, illustrates the process.

TABLE 4.2				
Transcription of specific mRNAs from nuclei of pituitary cells of normal LCMV-infected mice				
		RELATIVE AMOUNT OF RNA		
DNA probe for:	Function of protein	Uninfected	LCMV	Ratio LCMV/uninfected
Actin	Housekeeping	0.11	0.10	0.9
Collagen	Housekeeping	0.96	0.46	0.5
TSH	Hormone	1.72	0.69	0.4
GH	Hormone	3.80	0.24	0.08

Transcription of specific mRNAs from the nuclei of pituitary cells isolated from normal and LCMV-infected mice reveals a differential reduction of the expression of the message for GH. The table shows the relative amounts of hybridization of P^{32}-labeled nuclear runoff products from pituitary nuclei isolated from mice infected 15 days previously with LCMV compared with uninfected control animals. LCMV, lymphocytic choriomeningitis virus; TSH, thyroid-stimulating hormone; GH, growth hormone.

After Klavinskis LS, Oldstone MBA. Lymphocytic choriomeningitis virus selectively alters differentiated but not housekeeping functions: block in expression of growth hormone gene is at the level of transcriptional initiation. *J Virol* 1989;168:232–235.

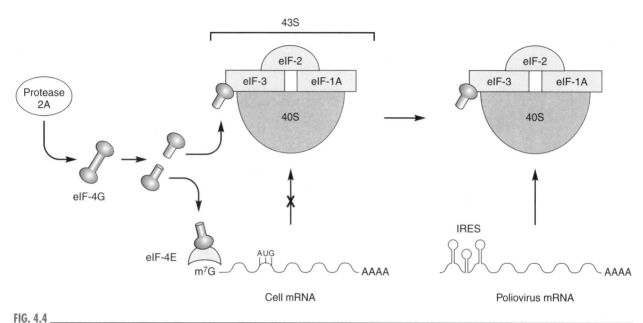

FIG. 4.4

Picornaviruses inhibit host cell protein synthesis by the action of the viral protease protein 2A, which cleaves a cellular protein, eIF-4G (eukaryotic initiation factor 4G). Cellular mRNA molecules are modified (capped) at their 5′ end and ribosomes (43S complex) bind to these capped ends to initiate translation. In the absence of intact eIF-4G ribosomes cannot efficiently bind to the 5′ end of mRNA molecules and are very poorly translated. Polioviral messages are not capped but contain an internal ribosomal entry site (IRES) that is recognized by the ribosome in the absence of eIF-4G, leading to efficient translation of viral mRNA. EIF, eukaryotic initiation factor; IRES, internal ribosomal entry site; 40S, 43S, ribosomal complexes. (After Hardwick JM, Griffin DE. Viral effects on cellular function. In: Nathanson N, et al., eds. *Viral pathogenesis.* Philadelphia: Lippincott–Raven Publishers, 1997.)

Poliovirus and many other picornaviruses contain a viral protease that markedly reduces synthesis of host proteins. The protease cleaves a cellular protein essential for the binding of the ribosome to mRNA, a step required for the initiation of translation. Viral messages are translated by a mechanism different from host RNA messages, involving an internal ribosomal entry site (or IRES element), and this process does not require the cellular protein cleaved by the viral protease (Fig. 4.4). This ingenious arrangement permits the virus to interfere with host protein synthesis while ensuring the continued synthesis of viral proteins.

VIRUS-INDUCED ALTERATIONS IN CELLULAR MEMBRANES

Cell fusion, described above, is probably the most striking effect of viruses on cellular membranes. There are a number of other important effects that viruses can have on the membranes of host cells.

Blocking or Down-regulation of the Cellular Receptor for Virus

Animals or cells infected with avian leukosis-sarcoma virus (ALSV) are known to resist superinfec-

tion by other closely related strains of retroviruses, and this phenomenon is used to classify these viruses into groups that exhibit reciprocal resistance. An infected resistant cell line sheds high titers of virus as well as many copies of its envelope protein, sufficient to saturate the specific viral receptor on the cell surface. This renders the cells resistant to superinfection by viruses that utilize the same receptor, while retroviruses that use different receptors can still infect such cells (Fig. 4.5).

A related phenomenon has been described for HIV, a member of the lenti (slow) subgroup of retroviruses. Quantitation of the expression of CD4—the primary viral receptor—on the surface of HIV-infected cells shows that CD4 expression is reduced at least 10-fold in comparison with uninfected cells. Three viral gene products contribute to this effect; early in infection the *nef* (negative factor) gene and late in infection the *env* (envelope) and the *vpu* genes down-regulate surface expression of CD4. The three viral proteins probably act by quite different mechanisms (Fig. 4.6). The envelope glycoprotein complexes with nascent CD4 in the endoplasmic reticulum; vpu increases the degradation of CD4 trapped in such complexes, whereas nef induces endocytosis of CD4 that has already reached the plasma membrane.

Modulation of Major Histocompatibility Complex Expression

Some viruses persist in vivo in spite of an immune response that includes virus-specific cytolytic T lymphocytes. As described in Chapter 5, virus-infected cells digest viral proteins and transfer oligopeptides to the endoplasmic reticulum where they complex with class I major histocompatibility complex (MHC) molecules. The MHC–peptide complexes migrate to the cell surface where they are

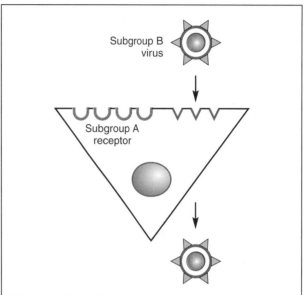

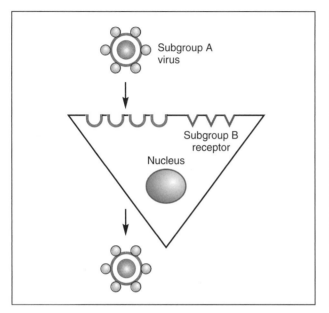

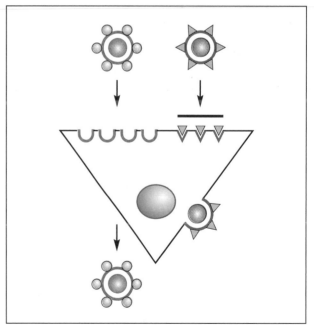

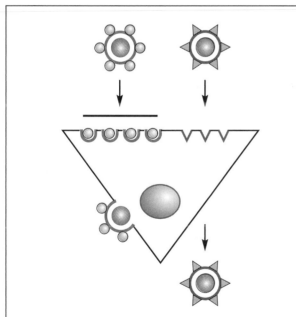

FIG. 4.5 _____

Blocking of infection among different retroviruses within the family of the avian sarcoma/leukosis virus (ALSV) families. The four panels show the identical chicken cell bearing two sets of receptors for subgroup A and subgroup B ALSVs, respectively. (*Top*) Uninfected cells are susceptible to either a subgroup A or a subgroup B virus. (*Bottom left*) Cells previously infected with a subgroup A virus are resistant to superinfection with a second subgroup A virus but retain susceptibility to a subgroup B virus. (*Bottom right*) Cells previously infected with a subgroup B virus are resistant to superinfection with a second subgroup B virus but retain susceptibility to a subgroup A virus. (After Coffin JM, Hughes SH, Varmus HE, eds. *Retroviruses*. Cold Spring Harbor, NY: Cold Spring Harbor Press, 1997.)

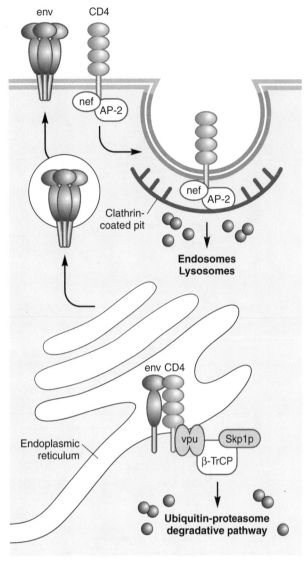

FIG. 4.6 _____

An example of viral proteins that down-regulate the expression of a cell surface protein. CD4 is a cellular protein expressed on the surface of a subset of T lymphocytes that have a "helper" function during immune induction; CD4 also acts as the primary receptor for HIV (see Chapter 3). Three HIV proteins—nef, env (gp160), and vpu—combine to down-regulate the expression of CD4 on the surface of HIV-infected lymphocytes. The nef protein binds to the cytoplasmic tail of CD4 and, through interactions with the AP-2 adaptor complex, directs CD4 to clathrin-coated pits, where it is endocytosed and targeted to degradative lysosomes. The env protein binds to nascent CD4 in the endoplasmic reticulum, and the vpu protein recognizes the complex and (in conjunction with B-TrCP and Skp1p) directs it to the ubiquitin-proteosome degradative pathway. (After Emerman M, Malim MH. HIV-1 regulatory/accessory genes: keys to unraveling viral and host cell biology. *Science* 1998;280:1880–1884.)

recognized by the antigen-specific receptor on cytolytic T lymphocytes, which then destroy the virus-infected cells. Therefore, it is surprising that virus-infected cells can persist in vivo. In some instances, it has been found that the viral infection down-modulates expression of class I proteins on the cell surface, thereby reducing the amount of viral peptides that can be displayed on the cell surface. Presumably, this enables infected cells to resist attack by the antiviral cellular immune response.

One interesting example of this phenomenon has emerged from investigation of the nef protein of HIV-1. Nef is a small nonstructural protein the function of which is somewhat enigmatic, since mutant strains of HIV-1 and SIV (simian immunodeficiency virus, the natural "HIV" of monkeys), engineered to delete the *nef* gene, were able to replicate in cell culture with the same kinetics as cognate nef-expressing viruses. However, when monkeys were infected with SIV strains containing a point mutation in nef, they quickly were replaced with viruses with a functional *nef* gene, implying that nef had an important function in vivo. If a large mutation was made in *nef* such that repair was impossible, the nef-deficient virus exhibited an attenuated phenotype with reduced viremia and delayed or absent development of immunodeficiency. The nef protein has several functions, one of which is the ability to down-regulate the expression of MHC class I molecules on the surface of infected cells. It is postulated that this effect protects the virus against immune surveillance, thereby increasing its ability to replicate and, hence, its virulence. Conversely, the nef-deficient virus is relatively avirulent, presumably because virus-infected cells are more susceptible to immune destruction.

Another important example of down-regulation of class I MHC is adenoviruses. Some adenovirus serotypes produce tumors in normal mice whereas others do not (Table 4.3). The ability to produce tumors is related to immune surveillance because strains of adenovirus that do not produce tumors in normal mice will produce tumors in immunocompromised mice, such as mice that have been thymectomized shortly after birth. Furthermore, direct measurements show that low levels of MHC class I proteins are expressed on cells transformed by oncogenic adenovirus strains, whereas high levels are expressed on cells transformed by nononcogenic strains of virus. Together these observations suggest that oncogenic strains of adenovirus down-regulate MHC class I expression, permitting adenovirus-transformed cells to escape immune surveillance.

TABLE 4.3			
Tumorigenic activity of adenovirus serotypes			
	TUMORS IN		
Adenovirus serotype *Tumorigenicity*	Normal mice	Immuno-compromised mice	MHC class I expression on tumors
Adenovirus 5 *Nontumorigenic*	No	Yes	High
Adenovirus 12 *Tumorigenic*	Yes	Yes	Low

Tumorigenic activity of different adenovirus serotypes correlates with down-regulation of MHC class I expression on adenovirus-transformed cells.

MHC, major histocompatibility complex.

After Ricciardi R. Adenovirus transformation and tumorigenicity. In: Prem S, ed. *Adenoviruses: basic biology to gene therapy.* Austin, TX: RG Landes, 1999.

How does adenovirus regulate the expression of class I proteins, proteins that are a product of the cell not the virus? This trait is related to a single adenovirus protein, the E1A (early) protein, which has a dominant negative effect on class I expression. Ordinarily, cellular proteins, such as NFκB, bind to enhancer sequences just upstream from the class I promoter and increase transcription of class I mRNA. The E1A protein interferes with this activity by modifying (possibly by dephosphorylating) NFκB, resulting in reduced expression of class I proteins on the surface of transformed cells. The E1A protein of nontumorigenic adenoviruses lacks the ability to down-regulate class I expression. As a result, nononcogenic adenovirus serotype 5 will transform cells, but these cells are destroyed by immune surveillance mechanisms (immune surveillance is discussed in Chapter 5).

Viral Proteins That Serve as Ion Channels Another unusual effect of viruses on cellular membranes is the alteration of membrane permeability by the introduction of viral proteins that serve as ion channels. The best studied example is the M2 protein of influenza virus type A, which is incorporated in the viral envelope. When influenza virus is endocytosed during the process of viral entry (see Fig. 3.7), the drop in pH leads to a conformational change that converts the M2 protein to an ion channel, thereby conducting H^+ ions into the interior of the virion. In turn, this internal drop in pH leads to a dissociation of the M1 (matrix) protein from the viral ribonucleoprotein complex. This dissociation is essential to infection because it frees the ribonucleoprotein complex for transport to the nucleus—a step required for the initiation of transcription of the viral genome.

VIRUSES AND THE CELL CYCLE

Many viruses replicate preferentially in cells that are in a specific phase of the cell cycle. The smallest viruses encode very few proteins and are dependent on the cell to provide a large number of the proteins required for replication of the viral genome and for the transcription and translation of viral messages. Often these required proteins are expressed only at a specific point in the cell cycle, limiting virus replication to cells at the required phase of the cycle. In some instances, the virus can control cycling of the cells that it has infected, in order to maximize its ability to replicate. One common strategy exhibited by a number of DNA viruses, whose genome replicates solely or preferentially in cells in the S phase (DNA-replicating phase) of the cycle (e.g., adenoviruses, papillomaviruses), is to induce infected cells to begin cycling, favoring replication of the viral genome.

One example is human herpesvirus 8 (HHV-8), a virus associated with Kaposi's sarcoma, an unusual vascular tumor that occurs most frequently in patients with AIDS. HHV-8 has a number of "pirated" cellular genes that the virus uses to control the cell cycle. HHV-8 expresses a G-protein-coupled receptor (GPCR) that is active in the absence of any ligand and can transform cultured fibroblasts by activating one or several intracellular signaling pathways. In addition, the viral GPCR initiates the synthesis and secretion of vascular endothelial growth factor, a powerful angiogenic agent that is expressed by the spindle cells that characterize Kaposi's sarcoma. Other examples of viruses that express modified forms of cellular proteins with transforming activity are discussed in Chapters 10 and 11.

Other viruses require (or prefer) cells that are at another point in the cell cycle. HIV preferentially infects $CD4^+$ T lymphocytes that are dividing, but it arrests the cell cycle in G_2, probably because the cellular environment is then optimal for transcription from the viral LTR. This effect is mediated by the vpr protein of HIV, a small regulatory protein expressed soon after infection, but the mechanism has yet to be explained.

VIRUS-INDUCED CELL DEATH: APOPTOSIS AND NECROSIS

In recent years it has been recognized that there are two pathways that lead to cell death: necrosis and apoptosis. Necrosis is the long established term for cell death and can be induced by any severe insult to cells, such as heating, cooling, nonphysiologic pH, other physical or chemical trauma, as well as by the action of specific viral proteins. Necrotic cells can be identified by loss of membrane integrity with spilling of cellular contents and random degradation of DNA. The residual cellular debris is then engulfed by local phagocytes, producing a transient inflammatory scar.

Apoptosis, or programmed cell death, has been recognized in the last decade as a major alternative pathway to cell death. Apoptosis is accomplished by a cascade of biochemical steps that produce several morphologic stigmata, including blebbing of the plasma membrane (which remains intact in contrast to the membranes of necrotic cells) and condensation of chromatin around the periphery of the nuclear membrane. Terminally, the cell shrinks, condenses, and breaks up into membrane-bound apoptotic bodies that contain cytoplasm, nucleoplasm, or both. Importantly, the plasma membrane is altered in such a way as to initiate phagocytosis and removal without the inflammation that characterizes necrosis. The cytologic changes that are characteristic of apoptosis can be detected by histologic tests developed for this purpose, such as the TUNEL assay. In addition, cellular DNA is enzymatically cut into nucleosome length fragments of 180–200 base pairs, or multiples of this size, which produce a ladder when the extracted DNA is electrophoresed on agarose gel.

Originally, apoptosis was thought to be confined to programmed growth and development of organs and tissues. In addition to cell multiplication, cell migration, and cellular differentiation, programs of normal development include the death of cells. Examples are formation of phalanges, development of the central nervous system, and elimination of lymphoreticular cells with antiself or other unwanted immune activities. More recently, it has been shown that viruses often kill cells by inducing apoptosis rather than via a necrotic pathway (Sidebars 4.1 and 4.2).

Initiation of Necrosis by Viral Proteins

One example of a viral protein that initiates cellular necrosis is Ebola virus, a filovirus (positive-stranded enveloped RNA virus) that is maintained in a yet-to-be-identified animal host in Africa. Rarely, Ebola virus crosses the species barrier to produce propagated human outbreaks that are associated with mortality as high as 50% of diagnosed cases. Ebola disease is an acute hemorrhagic shock syndrome, and pathologically there is severe damage to the endothelial cells that constitute the walls of capillaries. The envelope glycoprotein (gp) of Ebola virus is the viral attachment protein that binds to the cellular receptor

---

■ ■ ■
SIDEBAR 4.1

Viral Effects on Host Cells: Some Generalizations

- Viral infection may kill host cells but cytopathic effects are not a necessary consequence of viral replication. Conversely, cytopathic effects can be induced by viruses without completing a productive cycle of replication.
- Binding of a viral attachment protein to its cognate cellular receptor, in itself, can initiate dramatic effects such as cell-to-cell fusion, apoptosis, or signalling cascades.
- During infection, some viruses induce a global down-regulation of transcription of host genes, while viral genes are actively transcribed. Infection with other viruses may have a differential effect upon transcription of genes of the host cell, some of which are upregulated, others down-regulated, and some unchanged.
- Virus infection can reduce translation of host mRNA while viral messages are translated at a high rate, due to differential mechanisms of translation initiation.
- Viral infection can alter the expression of certain cell surface proteins, such as the cellular receptor for the virus or MHC (major histocompatibility complex) Class I proteins.   ■

---

and determines its tropism for endothelial cells and monocytes. Quite separate from its attachment function, intracellular expression of the gp appears to be responsible for the cytotoxic effect of the virus. Ebola virus gp contains a serine- and threonine-rich mucin-like internal domain that appears to confer its toxic activity because constructs in which this domain has been deleted lack the toxic activity while maintaining other functions of the protein. It has yet to be determined if (and how) gp initiates necrosis.

Initiation of Apoptosis by Virus Infection

It is now known that apoptosis can be induced at various steps in the virus life cycle, depending on the specific virus. For instance, reovirus can induce apoptosis via binding of the viral attachment protein, $\sigma 1$, to its cognate cellular receptor, and HIV can induce apoptosis via binding to one of its coreceptors, CCR5 or CXCR4. Adenovirus initiates apoptosis early in its program of transcription, via protein E1A, and the tat protein of HIV, another early protein that acts as a transactivator of transcription of the viral genome, can also initiate apoptosis. The structural proteins of

■ ■ ■
SIDEBAR 4.2

Apoptosis (Programmed Cell Death) in Viral Infections

- Cells may die through two pathways, necrosis or apoptosis. These two routes to cell death can be differentiated by several cytological and biochemical criteria.
- Many viruses kill infected cells by inducing apoptosis rather than necrosis. Apoptosis can be initiated at different points in the infectious cycle, by specific viral proteins.
- Some viral proteins act directly to induce apoptosis while others act indirectly, by neutralizing the effect of a host cell protein that blocks apoptosis. In some instances, the difference between a virulent and an avirulent strain of virus is mediated by variation in the ability of a viral protein to initiate apoptosis or to overcome the effect of a host cell protein that prevents apoptosis.
- Some virus proteins can block the apoptosis pathway, leading to prolonged cell life and increasing the yield of progeny virions. Blocking of apoptosis is also a characteristic of some transforming viruses. ■

certain viruses, made later in the replication cycle, may also induce apoptosis; examples are the envelope E2 protein of Sindbis virus, an alphavirus, and the VP3 protein of chicken anemia virus, a DNA virus.

In addition to direct induction of apoptosis, some viral proteins may induce apoptosis indirectly by inactivating a host protein that blocks a constitutive cellular apoptosis pathway. A neurovirulent strain of Sindbis virus, an alphavirus, produces encephalitis and neuronal death in mice, and the neurons are positive in the TUNEL assay indicating that neuronal death is mediated via the apoptosis pathway. By contrast, an avirulent strain of Sindbis virus fails to kill infected neurons. The *bcl-2* oncogene encodes a cellular protein that is known to block the apoptosis pathway in many normal cells, and cell lines that express *bcl-2* resist the lethal effects of Sindbis virus whereas those that do not express bcl-2 are often killed. A genetic recombinant strain of neurovirulent Sindbis virus, which expresses the *bcl-2* gene (in addition to viral genes), shows the phenotype of avirulent Sindbis virus when used to infect mice. These correlations suggest that the ability of virulent Sindbis virus to kill neurons may be

due to its ability to overcome the anti-apoptosis effect of the host cell genes such as *bcl-2*.

Poliovirus is another virus that initiates apoptosis via an indirect mechanism. The poliovirus 2A protein is a protease that has several activities required for preferential translation of poliovirus uncapped mRNA relative to cellular capped mRNAs. In addition, the 2A protease induces cellular apoptosis, presumably by cleaving selected cellular proteins, although the exact mechanism is yet to be determined.

Anti-apoptotic Actions of Viral Proteins

Although a number of viruses can induce apoptosis in infected cells, some viral proteins block the apoptosis pathway, thereby prolonging the production of virus by infected cells. Several herpes-, pox-, and adenoviruses have been shown to produce apoptosis-blocking proteins. The adenovirus example is interesting because one viral protein initiates apoptosis while another protein blocks the apoptosis pathway. As illustrated in Fig. 4.7, E1A, one of the transforming proteins of adenovirus, drives resting cells into cell cycle; at the commencement of cellular S phase, the intracellular environment becomes permissive for the replication of this small DNA virus. E1A, by itself, induces the transformed phenotype, manifested by the appearance of foci of continually dividing cells. However, these foci wither and die because E1A also induces apoptosis. However, E1B, another adenovirus protein, is capable of blocking the apoptosis pathway, and cells cotransfected with E1A and E1B are stably transformed.

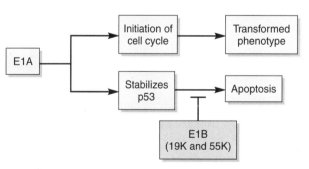

FIG. 4.7 _____

Some viruses produce proteins that both induce apoptosis and block apoptosis. The adenovirus E1A gene drives resting cells into cell cycle and thence into a transformed phenotype. The same gene product also stabilizes p53, an anti-oncoprotein, and p53 initiates apoptosis. Thus cells transformed with E1A alone undergo transient transformation but the transformed foci quickly die. Adenovirus E1B encodes a p19 and a p55 protein, and both of these proteins block the apoptosis pathway. Cells cotransformed with both EIA and E1B genes undergo stable transformation.

IMMUNE-MEDIATED EFFECTS ON VIRUS-INFECTED CELLS

As described above, many viruses replicate without causing death of the host cell. However, all infected cells express virus-encoded proteins that can act as neoantigens that make the infected cell a potential target for antiviral immune responses directed against the viral proteins. Infected cells can present viral antigens in two ways. The surface proteins of enveloped viruses are expressed on the cell surface, often in patches that constitute sites for budding of mature virus. Viral envelope proteins can be recognized by antiviral antibodies. Small peptides derived from viral proteins by either an endogenous or an exogenous pathway can be complexed with class I or class II MHC proteins on the cell surface where they are recognized as individual epitopes by T lymphocytes via their T-cell

receptors. (This process is described in more detail in Chapter 5.)

There are a multitude of specific immune responses that can attack virus-infected target cells (Fig. 4.8). These include binding of antibody that triggers a complement-mediated attack on the target cell, lysis by natural killer cells that are "armed" via their Fc receptors that bind to antiviral antibodies on the infected cell surface, and lysis by cytolytic CD8$^+$ effector T lymphocytes that recognize viral peptides bound to class I MHC molecules.

In addition to the destruction of individual host cells, antiviral immune responses can cause a variety of inflammatory processes in vivo. Virus-induced immunopathology is the subject of Chapter 6.

REPRISE

Virus infection can produce a wide array of effects on the host cell, from modulation of the expression of cellular genes to cell death. Viral replication does not necessarily cause cell death and, conversely, cells can be killed by viral interactions—such as attachment—that may not result in viral replication. The viral effects on cells can be associated with any stage in the viral life cycle, from attachment to release of new viral progeny, and viruses may modulate every aspect of normal cellular function, including transcription, translation, and cell division. Virus-induced cell death can be produced by necrosis or apoptosis, through a variety of different pathways. In summary, viruses use a plethora of mechanisms to enhance their ability to replicate in host cells, and as a result the virus–cell interaction ranges from commensal to lethal.

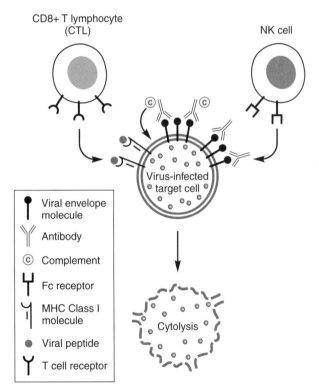

FIG. 4.8

Immune attack on virus-infected cells: selected mechanisms. (*Center*) Antibody binds to viral envelope proteins and initiates a complement cascade. (*Upper right*) Antibody binds to viral envelope protein and natural killer cells, via their Fc receptors, bind to antibody, triggering the release of toxic cytokines. (*Upper left*) CD8$^+$ T lymphocyte, via its T-cell receptor recognizes a viral peptide carried in the binding site of a class I molecule, leading to activation of a lytic cascade of proteins secreted by the T lymphocyte.

FURTHER READING

Reviews and Chapters

Coffin JM, Hughes SH, Varmus HE, ed. *Retroviruses*. Cold Spring Harbor, NY: Cold Spring Harbor Press, 1997.

Heemels MT, Dhand R. Apoptosis. *Nature* 2000;407:769–816.

Ricciardi R. Adenovirus transformation and tumorigenicity. In: Prem S, ed. *Adenoviruses: basic biology to gene therapy*. Austin, TX: RG Landes Co, 1999.

Original Contributions

Bais C, Santomasso B, Coso O, et al. G-Protein-coupled receptor of Kaposi's sarcoma-associated herpesvirus is a viral oncogene and angiogenesis activator. *Nature* 1998;391:86–89.

Chen BK, Gandhi RT, Baltimore D. CD4-down-modulation during infection of human T cells with human immunodeficiency virus type 1 involves independent activities of *vpu*, *env*, and *nef*. *J Virol* 1996;70:6044–6053.

Darnell JE, Levintow L, Thoren MM, Hooper JL. The time course of synthesis of poliovirus RNA. *Virology* 1961;13: 271–279.

Davis CB, Kikic I, Unutmaz D, et al. Signal transduction due to HIV-1 envelope interaction with chemokine receptors CXCR4 or CCR5. *J Exp Med* 1997;186:1793–1798.

Debbas M, White E. Wild-type p53 mediates apoptosis by E1A, which is inhibited by EIB. *Genes Dev* 1993;7:546–554.

Emerman M, Malim MH. HIV-1 regulatory/accessory genes: keys to unraveling viral and host cell biology. *Science* 1998; 280:1880–1884.

Goldstaub D, Gradi A, Bercovitch Z, et al. Poliovirus 2A protease induces apoptotic cell death. *Mol Cell Biol* 2000;20: 1271–1277.

Gonzalez-Scarano F, Pobjecky N, Nathanson N. La Crosse bunyavirus can mediate pH-dependent fusion from without. *Virology* 1984;132:222–225.

Gummuluru S, Emerman M. Cell cycle and vpr-mediated regulation of HIV-1 expression in primary and transformed T cell lines. *J Virol* 1999;73:5422–5430.

Hesselgesser J, Taub D, Baskar P, et al. Neuronal apoptosis induced by HIV-1 gp120 and the chemokine SDF-1α is mediated by the chemokine receptor CXCR4. *Curr Biol* 1998;8: 595–598.

Klavinskis LS, Oldstone MBA. Lymphocytic choriomeningitis virus selectively alters differentiated but not housekeeping functions: block in expression of growth hormone gene is at the level of transcriptional initiation. *Virology* 1989;168:232–235.

Lehmann-Grube F, Popescu M, Schafer H, Gschwinder HH. LCM virus replication in vitro. *Bull WHO* 1975;52:443–456.

Levine B, Goldman JE, Jiang HH, et al. Bcl-2 protects mice against fatal alphavirus encephalitis. *Proc Natl Acad Sci USA* 1996;93:4810–4815.

Oldstone MBA, Rodriguez M, Daughaday WH, Lampert PW. Viral perturbation of endocrine function: disordered cell function leads to disturbed homeostasis and disease. *Nature* 1984; 307:278–281.

Pinto LH, Holsinger LJ, Lamb RA. Influenza virus M2 protein has ion channel activity. *Cell* 1992;69:517–528.

Tyler KL, Squier MKT, Rodgers SE, et al. Differences in the capacity of reovirus strains to induce apoptosis are determined by the viral attachment protein σ1. *J Virol* 1995;69:6972–6979.

Yang Z-Y, Duckers HJ, Sullivan NJ, et al. Identification of the Ebola virus glycoprotein as the main viral determinant of vascular cell cytotoxicity and injury. *Nature Med* 2000;6:886–889.

Zhu H, Cong J-P, Mamtora, G, et al. Cellular gene expression altered by human cytomegalovirus: global monitoring with oligonucleotide arrays. *Proc Natl Acad Sci USA* 1998;95: 14470–14475.

PART II
Host Response to Viral Infection

Chapter 5

Immune Responses to Viral Infection

NONSPECIFIC AND SPECIFIC IMMUNE RESPONSES

Following viral invasion, the infected host mounts a number of responses to infection. Many of these responses involve the induction by the adaptive immune system of antibodies or cells that are specific for antigenic determinants expressed by the foreign pathogen, and most of this chapter is devoted to a description of these specific immune responses. Infected hosts also possess a number of nonspecific defenses against an invading virus. Some of these nonspecific defenses, such as interferon (IFN), are induced by infection, whereas others, such as complement and natural antibodies, exist prior to infection. In general, nonspecific defenses come into play at the time of infection or shortly thereafter, filling an important gap in protection against viral invasion, until the appearance of specific immune responses that characteristically require days to weeks for induction.

NONSPECIFIC DEFENSES

Interferon

The IFN system is probably the most important of the nonspecific host defenses against viral infection. IFNs are a group of proteins that are synthesized in response to viral infections and other stimuli. IFN induction occurs mainly at the level of transcription, and the inducing molecules, directly or indirectly, act on promoters upstream from the IFN coding sequences. The IFN response is characteristically transient, even in the continued presence of the inducing stimulus.

IFNs do not produce their effects directly; they are secreted and spread either locally or through the circulation to other cells where they induce the production of a characteristic set of proteins that are responsible for the multiple effects of the IFN system (Fig. 5.1). Of these effects, the most important are inhibition of the replication of viruses and some other parasites, inhibition of cell growth including that of certain tumor cells, and stimulation of the immune system by activating macrophages, lymphocytes, and natural killer (NK) cells and up-regulating the expression of class I major histocompatibility complex (MHC) molecules.

IFNs are classified as type 1 or type 2 (Table 5.1). Type 1 IFNs (most importantly, IFN-α and IFN-β) can be produced in most cell types, and they are typically induced by double-stranded RNA, which is synthesized in the course of infection with many but not all viruses. Type 1 IFNs are produced early in infection and therefore serve as a first line of

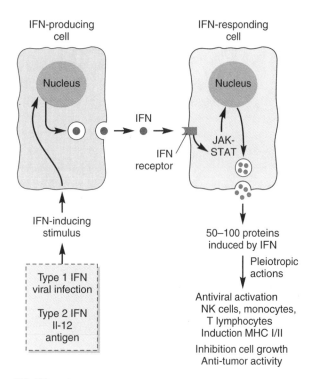

FIG. 5.1 _____

An overview of the interferon (IFN) system. A number of different stimuli can induce cells to synthesize and secrete IFNs; the stimulus for type 1 (α/β) IFNs is usually virus infection or double-stranded RNA, whereas the stimulus for type 2 (γ) IFN is IL-12 (IL-12) secreted by macrophages and dendritic cells that are activated by antigens presented in the context of major histocompatibility complex class I and II molecules. Many cells can produce type 1 IFNs, whereas type 2 IFNs are produced mainly by T lymphocytes and natural killer cells. The secreted IFNs bind to specific receptors—present on most cells—and induce the production of a characteristic range of proteins that mediate the pleiotropic actions of the IFNs. See text for details.

defense prior to immune induction. Type 1 IFNs can also be induced during infections with bacteria and parasites and by a variety of stimuli, including certain cytokines and growth factors, such as interleukins 1 and 2 (IL-1 and IL-2) and tumor necrosis factor (TNF).

Type 2 IFN (IFN-γ), or "immune" IFN, is not a "nonspecific" defense but rather one of the mediators of the antigen-specific immune response. Type 2 IFN is synthesized by a restricted group of cells, particularly T lymphocytes and NK cells, following their activation. Activation is mediated mainly by IL-12 which, in turn, is secreted by macrophages and dendritic cells as a result of antigen recognition in the context of class I and class II MHC molecules. Also, a number of cytokines, such as IL-2, and other intercellular mediators, such as estrogens, may activate IFN-γ-producing cells.

IFN receptors are found on most cell types, and are distinct for type 1 and type 2 IFNs. Most IFN receptors are heterodimers and include an extracellular ligand domain, a transmembrane domain, and an intracellular signaling domain. Characteristically, the receptors are specific for the IFN molecules produced by the same species, explaining the early observation that IFN produced by one species will not act on cells of another species. When IFNs bind to their cognate receptors, an intracellular signaling cascade is initiated that involves the JAK-STAT pathway (Fig. 5.1). Briefly, the intracellular signaling domain activates the Janus tyrosine kinase (JAK), which phosphorylates the signal transducer and activator of transcription (STAT). The phospho-

TABLE 5.1						
The interferons and some of their properties						
Type *Cell source*	Stimulus	*Class* No. of amino acids	No. of subtypes	Chromosome	Active form	Glycosylated
Type 1 *All cells*	Virus infection, dsRNA	α 165	14	9	Monomer	No
		β 166	1	9	Monomer	Yes
		ω 172	1	9		
		τ 172	1	9		
Type 2 *T lymphocytes NK cells*	IL-12	γ 146	1	12	Dimer	Yes

Type 1 interferons are structurally unrelated to type 2 interferon. Type 1 interferons are typically secreted in response to viral infection, triggered by intracellular appearance of double-stranded RNA (dsRNA). Type 2 interferon is typically secreted by T lymphocytes and NK cells following stimulation by IL-12, which in turn is secreted by macrophages and dendritic cells in response to antigen presentation and activation.

IL, interleukin; NK, natural killer.

rylated STAT complex moves to the nucleus where it enhances the transcription of selected genes. IFN induces the transcription of gene products within minutes of binding to the IFN receptor.

Viruses vary in their ability to induce IFN and in their sensitivity to IFN. IFN induces between 50 and 100 different proteins, and a number of these proteins interfere with virus replication in different ways, many of which have not been elucidated. As might be predicted from the large number of induced proteins, different viruses are inhibited by different mechanisms. Inhibition can occur at the level of genome replication, transcription, mRNA stability, or translation. One example is the Mx protein, a cellular protein induced by IFN that is known to be highly specific in its ability to depress the replication of influenza viruses. IFN may also have indirect effects on viral infection, such as the up-regulation of class I MHC, which renders virus-infected cells more sensitive to immune surveillance by antigen-specific T lymphocytes.

IFN is important as a host defense particularly against primary virus infection, whereas specific immune responses—such as antibody—play a major role in early defenses against reinfection. The role of IFN can be demonstrated in animal models in which the IFN response is reduced by treatment with anti-IFN antibodies or in mice whose IFN genes (or IFN receptor genes) have been "knocked out." In both instances, such animals, compared with controls, exhibit a reduced ability to contain viral infections and often show an increased incidence of illness or death. When the type 1 (IFN-α/β) response is abrogated, there is a global increase in susceptibility to most viruses, whereas a knockout of the type 2 response (IFN-γ) has a more modest effect, which is seen with some viruses. These observations suggest that type 1 IFN plays a crucial role as a nonspecific antiviral defense, whereas the type 2 IFN response is less critical because it is only one element in a multicomponent immune effector system.

A number of viruses have developed specific mechanisms to elude the effects of IFN, and some of these are described in Chapter 8 on virus virulence. The evolution and conservation of anti-IFN viral genes testifies to the importance of IFN as a host defense.

Natural Killer Cells

NK cells are large lymphocytes that are defined by their cytoplasmic granules. These granules contain perforin (which can produce pores in plasma membranes) and granzymes (proteins that can initiate apoptosis). The cytoplasmic granules released by NK cells can kill adjacent cells. NK cells do not express the T-cell receptor (TCR) and therefore cannot produce an antigen-specific attack on cells bearing peptides—derived from foreign proteins—on their class I molecules.

Instead, NK cells have activating receptors, not yet well defined, that may include integrins and other cell surface molecules. In addition, NK cells bear a large class of "killer-inhibiting receptors" that, when occupied, initiate intracellular signals that inhibit the activation pathways. Whether or not NK cells "attack" is determined by the balance between the activating and inhibitory pathways. A large class of killer-inhibiting receptors recognize MHC class I molecules and prevent NK cells from lysing most normal cells that express "self" antigens. In contrast, cells that have reduced levels of MHC class I expression fail to activate killer-inhibiting receptors and are targets for lysis. Thus, one important role of NK cells appears to be surveillance against foreign cells (cells that do not express self antigens).

The mechanism by which NK cells protect against viral infection is not thoroughly understood, but the following tentative reconstruction can be ventured. NK cell activity is dramatically increased early in viral infection, and this activation is mediated at least in part by type 1 IFNs (α/β) that are induced in the cells first infected. Many viruses downregulate the expression of class I MHC (see Chapter 4) by a variety of mechanisms, and such cells become nonspecific targets for NK cell attack, thereby synergizing the effect of type 1 IFNs. Later in infection, IFN induces an increase in MHC class I expression on infected cells, and this renders them simultaneously resistant to attack by NK cells but more sensitive to lysis by antigen-specific cytotoxic CD8-positive T lymphocytes (CTLs). The sequence of NK cell activity and CTL activity is shown in Fig. 5.2 for a model virus infection.

In addition to their nonspecific activity, NK cells can also mediate antibody-dependent cell-mediated cytotoxicity, an antigen-specific host defense that is described later in this chapter.

Complement and "Natural" Antibodies

The serum of a number of mammalian species exhibits nonspecific antiviral activity that may be mediated either by complement or by "natural" antibodies. The complement system consists of a group of proteins that circulate in the plasma. As shown in Fig. 5.3, these proteins can participate in a biochemical cascade that causes the production of channels

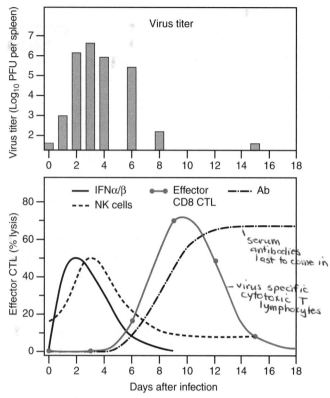

FIG. 5.2

The nonspecific responses to an acute viral infection precede the induction of antigen-specific responses. This diagram shows the sequence of interferon (IFN) α/β levels, natural killer (NK) cell activity, virus-specific cytotoxic T lymphocytes, and serum antibody (Ab) responses during acute infection with lymphocytic choriomeningitis virus. The IFN, NK, and Ab curves are drawn to an arbitrary scale. (After Welsh RM. Natural killer cells in virus infections. *Curr Top Microbiol Immunol* 1981;92:83–106; Biron CA. Cytokines in the generation of immune responses to, and resolution of, viral infections. *Curr Opin Immunol* 1994;6:530–538; and Lau LL, Jamieson BD, Somasundarma T, Ahmed R. Cytotoxic T cell memory without antigen. *Nature* 1994;369:648–652.)

in a lipid bilayer, resulting in lysis of viruses or other foreign pathogens. Complement acts as part of the effector mechanism for an induced immune response (described later in this chapter). However, complement may also act as a nonspecific host defense against certain viruses. For instance, a number of animal retroviruses are lysed by human complement in the absence of antiviral antibody. Apparently, p15e, one of the virus-encoded proteins on the surface of some groups of retroviruses, binds C1q, one of the complement proteins, thereby activating the complement cascade. Complement-initiated lysis has also been reported for some members of other classes of viruses.

Natural antibodies are found in sera of many animal species and often are directed against foreign antigens to which humans or animals may be exposed. One example is antibodies against a disaccharide consisting of two galactose molecules bound in an α1–3 linkage. This digalactose is not synthesized by humans or other old-world primates who lack a galactosyltransferase specific for the α1–3 bond. Since old-world primates are exposed to this disaccharide present on ingested proteins, they develop antigalactose (α1–3) antibodies. Viruses grown in cells that express digalactose incorporate this molecule into the carbohydrate side chains of their surface proteins. Such virions are neutralized by sera from old-world primates that have never been exposed to the virus itself.

OVERVIEW OF THE IMMUNE SYSTEM

The following cursory description is designed to refresh the reader's memory about the organization and function of the immune system. Students who are not familiar with basic immunology may wish to consult one of the many excellent introductory texts (see Further Reading).

Lymphocytes are responsible for both the induction and expression of cellular immunity. There are two major classes of lymphocytes: B cells and T cells. T cells are named after the thymus, and B cells are named after the bursa of Fabricius, an organ found in birds (but not in mammals) where B lymphocytes undergo early maturation. In mammals, the same steps in B-cell maturation occur in the bone marrow.

All types of lymphocytes are constantly produced from precursor stem cells residing in the bone marrow. The T lymphocytes migrate first to the thymus where they undergo maturation that prepares them to respond to an immune stimulus; they emerge from the thymus as "naïve" T cells that migrate to spleen, lymph nodes, and selected tissues such as the skin and submucosal sites. B lymphocytes emerge from the bone marrow and migrate directly to lymphoid tissues.

The other important cell involved in immune responses is the monocyte/macrophage, a cell that is also produced from stem cell precursors in the bone marrow and leaves the bone marrow to circulate as the blood monocyte, following which it can enter tissues where it resides as a macrophage or dendritic cell. Dendritic cells are found in skin and mucosal

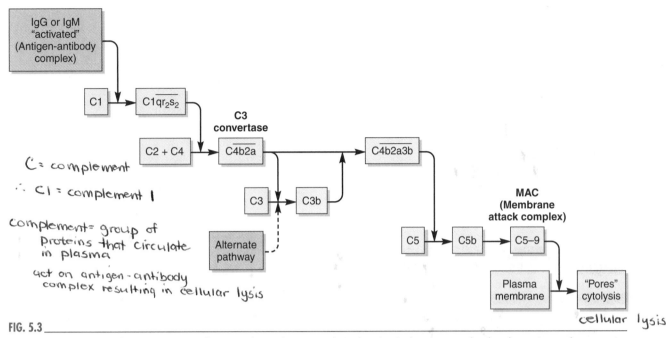

FIG. 5.3 _____

The complement cascade: an overview. The cascade can be initiated via the classical pathway or by the alternative pathway, as indicated. The classical pathway is initiated by binding of antigen to antibody (either IgG or IgM) which produces a conformational change in the Fc portion of the antibody molecule, enabling it to bind to C1. At several steps in the cascade a complex of proteins is formed that has catalytic activity (indicated by the overbar). The final complex of C5–9 (membrane attack complex) forms pores in the plasma membrane causing cellular lysis. (Modified after Abbas AK, Lichtman AH, Pober JS. *Cellular and molecular immunology,* 4th ed. Philadelphia: WB Saunders, 2000.)

tissues where they specialize in the binding of foreign antigens and the initiation of immune induction. Lymphoid tissues, such as the spleen and lymph nodes, have a somewhat complex structure that is beyond the scope of this description. Suffice it to say that macrophages, dendritic cells, and lymphocytes are situated in locations in central lymphoid tissue that optimize their ability to play specific roles in response to immune stimuli.

Each immune cell bears on its surface molecules that equip it to discharge its specialized role in the immune response (Table 5.2). The B cells carry antibody molecules that bind to "epitopes" or individual immune determinants on foreign antigens. Antigens recognized by B cells may be proteins, carbohydrates, or nucleic acids, and their epitopes consist of a small cluster of amino acids, sugars, or nucleic acids, respectively. Thus, an epitope would include about 10 amino acids whereas a protein could be composed of 25–1,000 amino acids.

T lymphocytes carry a TCR that is a heterodimer composed of an α and a β polypeptide. TCRs have a structure that is roughly analogous to

the antibody molecule, but—in contrast to antibodies—recognizes a stretch of 9–20 amino acids bound to a specialized groove on the surface of an MHC class I or II molecule. Individual B and T cells express different antibodies or TCRs, specific for individual epitopes of different antigens. These different receptors are "constructed" by rearrangements of genetic determinants expressed in germline cells, re-

TABLE 5.2		
Comparison of B- and T-cell immune responses		
	B-cell response	**T-cell response**
Type of immunity	Humoral	"Cellular"
Precursor (memory) cell	B lymphocyte	CD8 pCTL
Effector cell	Plasma cell	CD8 CTL
Extent of effect	Widespread, systemic	Local
Mediator molecules	Immunoglobulins	Perforins, granzymes, cytokines
Persistence of effectors	Yes	No
Anamnestic (memory) response	Yes	Yes

pCTL, precursor cytotoxic T lymphocytes.

arrangements that occur during maturation of lymphocytes. Due to this exquisitely specialized developmental arrangement, lymphocytes can encode up to about 10^9 different specificities. In contrast to lymphocytes, macrophages or dendritic cells are equipped with more general receptors for the Fc component of antibody molecules and with other surface molecules that enable them to bind and internalize whole foreign antigens.

T lymphocytes can be divided into two major categories, CD4 and CD8 cells, which serve different immune functions. These two cell types can be identified by the expression on their surface of either CD4 or CD8 molecules, which can be used as markers for the purpose of experimental enumeration or cell sorting. CD4 cells carry TCRs that recognize antigen presented by MHC class II but not class I molecules, and, conversely, CD8 cells recognize antigens presented by class I molecules.

CD4 cells act as "helper" cells by interacting with naïve B lymphocytes or CD8 lymphocytes to provide signals that induce B cells or CD8 cells to proliferate in response to an antigen presented by a professional antigen-presenting cell (APC). These signals involve ancillary surface molecules on the CD4 cells that interact with cognate ligand molecules on the surface of the antigen-responding CD8 or B cells. Thus, there are two interactions involved in optimal immune induction: the recognition by a very small subset of lymphocytes of an epitope that is cognate to its recognition molecule (antibody or TCR) and the nonspecific second signal that initiates clonal proliferation of such antigen-specific lymphocytes. In addition, CD4 cells can be divided into T_H1 and T_H2 cells. T_H1 cells secrete large amounts of IFN-γ and IL-2 and drive the clonal expansion of CD8 cells, whereas T_H2 cells secrete large amounts of IL-4 and drive the clonal expansion of B cells.

CD8 lymphocytes are effector cells that mediate the cellular immune response by their ability to recognize foreign epitopes presented by class I molecules. CD8 cells initiate a local response that includes two components—a lytic attack on target cells carrying a foreign epitope and the production of cytokines that can attract inflammatory cells to the local site—resulting in an indirect attack on the invading parasite.

Immune induction of antibody and of cellular immunity involves different cell types. The B-lymphocyte pathway (Fig. 5.4) begins with the binding of foreign proteins to the immunoglobulin molecules expressed on the surface of naïve B cells.

Naïve B cells are already programmed genetically to express immunoglobulin molecules with a single antigenic specificity, that is, they will recognize only a single epitope of the multitude of determinants on the foreign protein. Bound antigens are then endocytosed, and digested to oligopeptides, which are "loaded" onto the class II MHC molecules on their surface (limited to the very few peptides that bind to the class II molecule of the specific B cell). CD4 (helper) T lymphocytes whose TCRs recognize the specific peptide presented by the class II molecules on the B cell will be stimulated to release cytokines (IL-4, IL-10 in the case of B lymphocytes) that drive the B cell to proliferate and differentiate. Mature B cells (plasma cells) will then secrete the immunoglobulin for which they have been programmed.

The cellular immune response follows a stereotyped sequence of events (Fig. 5.5), although there are many variations. If a foreign antigen—such as a virus protein—is introduced via any portal of entry, some of the molecules are bound by dendritic cells and macrophages. These cells, often known as "professional APCs," are specially equipped to initiate immune induction because they express both MHC class I molecules (expressed at various levels on most cell types) and class II molecules (whose expression is mainly limited to APCs and lymphocytes). The dendritic cell can process protein antigens by two pathways: exogenous and endogenous. The exogenous pathway involves endocytosis of the protein, digestion into small oligopeptides in endolysosomes, and "loading" of these 10–20 mers onto the antigen-binding groove on class II molecules (it is likely that some class I molecules can also be loaded via the exogenous pathway). The endogenous pathway involves the expression of a foreign protein in the cytosol of the professional APC. Viral proteins synthesized in the cytosol of the APC are delivered to the proteosome, a complex cytoplasmic organelle, which digests the protein into peptides and delivers these peptides across internal membranes into endoplasmic reticulum, where they are loaded onto class I molecules. The endogenous pathway may be utilized if the virus can infect the APC.

Once exposed to foreign antigens in the periphery, dendritic cells migrate rapidly via afferent lymphatic channels to draining lymph nodes. Within lymphoid tissue, dendritic cells interact with CD8 cells whose TCRs recognize the epitopes presented by the class I molecules on professional APCs. They undergo clonal expansion by virtue of the second

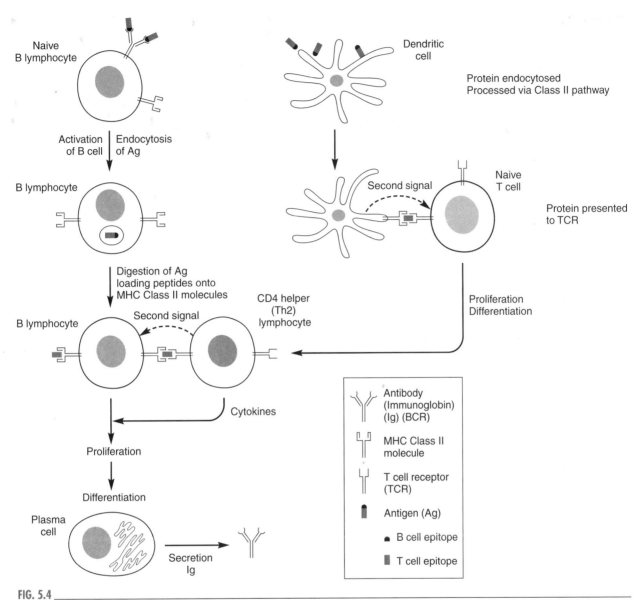

FIG. 5.4

A simplified representation of induction of a B lymphocyte (antibody) response, showing the major cell types involved. (1) A naïve B lymphocyte binds a protein antigen via its surface immunoglobulin (Ig) that recognize a specific epitope on the protein. This protein is then endocytosed and processed through the exogenous major histocompatibility complex (MHC) class II pathway, and the B cell presents (on its surface) a peptide (different from the epitope that binds to the Ig molecules on its surface) bound to its class II molecules. (2) Dendritic cells endocytose the protein, process it through the class II pathway, and present peptides bound to MHC class II molecules. The peptides are recognized by a subset of naïve CD4 T lymphocytes bearing the cognate T-cell receptors, which then proliferate, differentiate, and act as helper cells (CD4 T$_H$2 cells). (3) CD4 T$_H$2 cells recognize the peptides presented by B cells on their class II MHC molecules and stimulate B cells in two ways. First, CD4 cells provide a second signal via the CD40–CD40 ligand pathway; CD40 ligand on CD4 cells binds CD40 on B cells, initiating B-cell intracellular signaling. Second, CD4 cells secrete cytokines (IL-4, IL-10) that stimulate the presenting B lymphocytes to proliferate and differentiate into end-stage plasma cells that secrete their epitope-specific Ig molecules. (Modified after Abbas AK, Lichtman AH, Pober JS. *Cellular and molecular immunology,* 4th ed. Philadelphia: WB Saunders, 2000.)

signal provided by dendritic cells or by CD4 cells, and then mature into effector CD8 cells (Fig. 5.5). CD4 helper cells play a variable role in CD8 induction, and are more important in primary infections or when the antigenic stimulus is minimal. CD4 cells are also important in some anamnestic responses to facilitate the differentiation of CD8 memory (or precursor) cells into effector cells. An example is HIV, a virus that specifically infects CD4 cells but indirectly reduces the ability of CD8 effector cells to contain longstanding persistent opportunistic infections, a phenomenon that is discussed in Chapter 13.

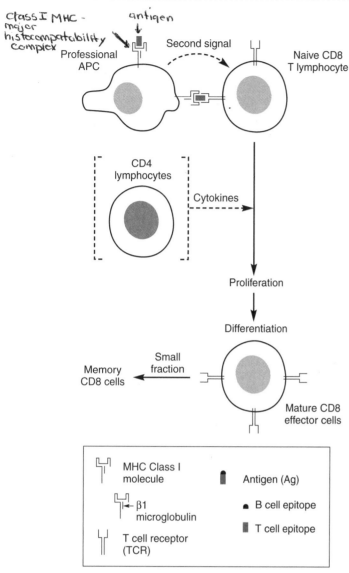

Class I MHC - major histocompatability complex

FIG. 5.5 _____

A simplified representation of induction of a cellular immune response showing the major cell types involved. A professional APC bears major histocompatibility complex (MHC) class I molecules that present a specific peptide bound to the antigen-binding groove. Naïve CD8 T lymphocytes expressing a T-cell receptor (TCR) that recognizes the same peptide will bind to the peptide and be stimulated to proliferate. Proliferation is triggered by two "signals": the recognition of the peptide and cross-linking of interactive accessory proteins on the surfaces of APCs and naïve T cells (such as CD40–CD40 ligand). The ability to provide this second signal distinguishes professional APCs from other most other cells that express class I molecules. Following proliferation, CD8 lymphocytes differentiate into mature CD8 T lymphocytes that are capable of recognizing and "killing" target cells bearing the same epitope on their class I molecules. For many viruses, CD4 helper T lymphocytes (via cytokines such as IL-2 and IL-4) and activated dendritic cells appear to play a limited role in induction of CD8 responses. CD8 memory cells are likely derived from the CD8 effector cell pool. (Based in part on Abbas AK, Lichtman AH, Pober JS. *Cellular and molecular immunology,* 4th ed. Philadelphia: WB Saunders, 2000.)

A virus infection will simultaneously induce antibody and a cellular immune response. The balance between the two responses will be influenced by the relative proportion of T_H1 and T_H2 cells participating in the immune response. As noted above, the effect of T_H1 and T_H2 cells is mediated by the cytokines that they secrete, with IFN-γ and IL-2 driving CD8 cells and IL-4 driving B cells.

ANTIBODY

Measurement of Antibody

There are many methods to measure antibody to viruses, and both the kinetics of the response and its biological significance depend on the assay used. The canonical assay is the neutralization test, in which antibody is tested for its ability to reduce viral infectivity. This test depends on the availability of a convenient method to measure viral infectivity, often a plaque assay, in which each infectious virus particle produces a single lytic plaque in a lawn of susceptible indicator cells. One common technique involves the use of a single viral inoculum, such as 100 plaque-forming units (PFU); serial dilutions of a test antibody are tested to determine the highest dilution that will reduce the plaque count by 50% (Fig. 5.6). An alternative but less frequently used method is the neutralization index, in which the titer of a virus stock is compared in the presence and absence of a test antibody; the index is calculated as the difference in viral titers. Neutralization tests cannot be used for some important viruses, such as hepatitis B and C viruses, that cannot readily be grown in cell culture.

There are many alternative assays that measure the ability of the antibody to bind to viral antigens, including hemagglutination inhibition, immunofluorescence, Western blot, and enzyme-linked immunosorbent assay (ELISA). Of these, the most commonly used is the ELISA assay, which can readily be adapted to quantitation, automation, and rapid throughput. An antigen—either whole virus, a viral protein, or a viral peptide—is bound to a substrate and then incubated with serial dilutions of test antibody; adherence of the test antibody is determined with a conjugated antiserum directed against immunoglobulin of the species under test. The conjugate may carry a fluorescent or other visualizable label or be an enzyme that will convert a substrate from a colorless to a visible form.

In contrast to the ELISA, the Western blot is a qualitative test that provides information about the

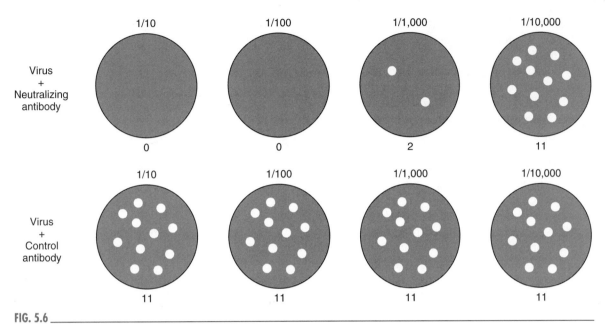

FIG. 5.6

A plaque reduction neutralization test, showing the ability of a test antibody to reduce the infectivity of the virus. (For convenience of illustration, about 10 plaque-forming units are shown although a larger inoculum is usually employed.) The 50% end point is calculated as the dilution of antibody that would reduce the plaque count from the control level of 10 to 5 plaque-forming units. In this example, the end point is about 1/3,750.

specificity of the test antibody (Fig. 5.7). Proteins from a viral lysate are separated, often using polyacrylamide gel electrophoresis, and cross-linked to a cellulose strip, and the antibody is tested for its ability to bind to any of the proteins; the strip is developed with an antiserum against immunoglobulin, as for the ELISA.

Antibody production can also be measured at a cellular level although this is rarely done except for special experimental purposes. In the ELISPOT assay, cells—including plasma cells—are prepared from blood or lymphoid tissues, and overlaid on a surface to which a target antigen has previously been bound. Antigen-specific antibody released by specific plasma cells binds to the cognate antigen, and the "focus" is "developed" using a variation of the method used in ELISA assays. This assay permits the counting of antibody-secreting cells and can be employed for studies of the dynamics of the antibody response.

Effector Functions of Antibody

Antiviral antibodies can subserve host defenses in several different ways. First, they can bind to virus and neutralize it, that is, render the virion noninfectious. There are multiple mechanisms of neutralization, but the most common is to coat the virion and prevent its attachment to receptors on permissive host cells. A minority of neutralizing antibodies bind

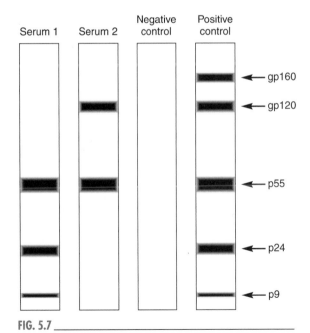

FIG. 5.7

Western blot test for antibody. In the diagram are compared four sera: two containing antibody against viral antigens, and a positive and a negative control. The positive serum reacts with all of the viral proteins on the test strip, whereas each of the test sera shows a different pattern of reactivity.

to the virus but do not prevent attachment to the cell; instead they interfere with later steps in viral entry. Also, it should be noted that many antibodies that bind to viral proteins fail to neutralize, usually because the specific epitope is not expressed on the surface of infectious virions (e.g., if the epitope is part of an internal protein not present on the surface of the virion).

In addition to reducing infectivity of virions, antiviral antibodies can act in several other ways. Many antibodies will bind C1q, the first of the complement proteins, and initiate lysis of the virion. Also, many antibodies will bind to the Fc receptors on Kupffer cells and other sessile macrophages present in the endothelial walls of capillaries. Complexes consisting of virions bound by antibodies will be more efficiently phagocytosed and degraded than free virions, and this process is sometimes called opsonization. Finally, antibodies can "arm" NK cells, which then mediate antibody-dependent cell-mediated cytotoxicity, leading to the destruction of infected cells. NK cells carry receptors for the Fc domain of immunoglobulin molecules. If an antibody with virus specificity binds to the Fc receptor of an NK cell, and the NK cell then contacts a virus-infected cell bearing the cognate antigen, it can initiate a cytolytic attack similar to that mediated by CD8 cytolytic T cells.

In general, the ability of antibodies to protect in vivo correlates with their neutralizing activity. However, this correlation is not absolute, as shown in studies of individual monoclonal antibodies (Table 5.3). The mechanism whereby nonneutralizing antibodies protect is not well understood but, in some instances, may be mediated by the clearance of

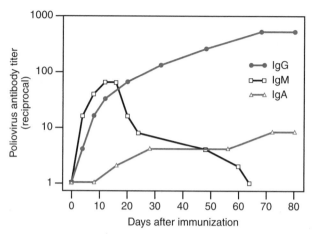

FIG. 5.8

The time course of different classes of antibody response. Subjects were immunized with poliovirus vaccine (IPV, inactivated poliovirus vaccine; OPV, oral poliovirus vaccine) at 0, 32, and 56 days, and followed for their titers of serum antibody (IgG, IgM, IgA). The IgM response is rapid but transient, whereas IgG and IgA are induced more slowly but are long lasting. (After Ogra PL, Karzon DT. Formation and function of poliovirus antibody in different tissues. *Prog Med Virol* 1971;13:156–193.)

virus–antibody complexes by the reticuloendothelial system.

Antibody can also act on the intracellular phase of virus replication. This surprising conclusion is based on studies, both in vivo and in cell culture, demonstrating that antiviral antibody can influence ongoing intracellular replication of several viruses. The mechanism of this phenomenon is unknown, nor is it clear whether it plays a role in many viral infections.

Kinetics of the Antibody Response

Antibodies can be measured in different bodily fluids, such as blood (plasma or serum) and mucosal fluids (nasal, throat, bronchial washes, feces, semen, genital washes). Furthermore, the class of antibody varies. Serum contains IgG (the principal class), IgM, and IgA, most of which is derived from plasma cells in bone marrow, lymph nodes, and spleen. Mucosal fluids contain IgA (locally produced by plasma cells in mucosal-associated lymphoid tissues) and IgG (partly a transudate of serum IgG and partly produced in mucosal-associated lymphoid tissues).

An example of the humoral (serum) antibody response is shown in Fig. 5.8, which illustrates that the dynamics of the responses are quite different, according to antibody class. In general, serum IgM appears rapidly (1–2 weeks) and disappears in about 3 months, whereas IgG appears more slowly (2–4 weeks), peaks at 3–6 months, and then gradually

TABLE 5.3		
Correlation between ability of antibodies to protect *in vivo* and their neutralizing activity		
	Protection in vivo (number of monoclonal antibodies)	
Neutralization in vitro	**Yes**	**No**
Yes	2	1
No	3	3

The ability of monoclonal antibodies to neutralize in vitro does not necessarily correlate with their ability to protect in vivo. In this example, monoclonal antibodies (100 μg per animal) were administered to mice one day prior to injection of Ebola virus (10 PFU, a 100% lethal dose). Protection: 6–10 mice survived challenge (of 10 tested); no protection: 0/10 mice survived. Neutralization required the addition of complement.

After Wilson JA, Hevey M, Bakken R, et al. Epitopes involved in antibody-mediated protection from Ebola virus. *Science* 2000; 287:1664–1667.

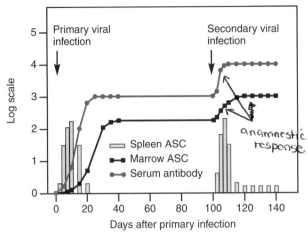

FIG. 5.9 _____

The time course and localization of plasma cells secreting virus-specific antibody. Lymphocytic choriomeningitis virus infection of the mouse induces the expansion of plasma cells (antibody-secreting cells) in the spleen during acute infection. With the waning of the infection, about a month later, the level of antibody production is maintained but is mainly due to plasma cells in the bone marrow. During reinfection there is an increase in the level of serum antibody associated with a transient spurt in the number of plasma cells in the spleen. (After Slifka MK, Matlubian M, Ahmed R. Bone marrow is a major site of long-term antibody production after acute viral infection. *J Virol* 1996;69:1895–1902.)

wanes although it may persist at submaximal levels for a lifetime.

The ELISPOT assay has been used to trace the location and kinetics of plasma cells (mature antibody-producing B cells). As shown in Fig. 5.9, during the primary immune response most antibody-producing plasma cells reside in the spleen but, following acute infection, they are mainly found in the bone marrow. During reinfection, there is another transient increase in plasma cells in the spleen as well as a continuing population in the bone marrow. Importantly, the host continues to produce considerable titers of virus-specific antibody following primary infection, an important distinction from cellular immunity (CD8 effector cells), which is not maintained at high levels after disappearance of the virus.

CELLULAR IMMUNITY

Measurement of Cellular Immune Response

The classical assay for cellular immune responses is the limiting dilution assay or LDA (Fig. 5.10), introduced in the 1960s. Cells are obtained from blood, spleen, or lymphoid tissues of an animal immune to the virus under study (virus "X") and are cultured for 1–2 weeks in the presence of whole inactivated

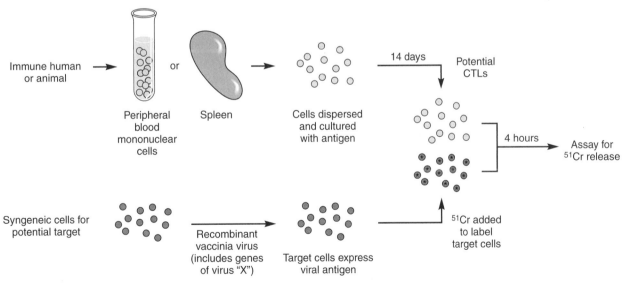

FIG. 5.10 _____

Measurement of CD8 precursor cells, based on their ability to lyse appropriate targets (the limiting dilution assay, LDA). CD8 cells obtained from an immune donor are cultivated in the presence of the viral antigen under study to expand the number of precursors and mature them into effector cytotoxic T lymphocytes. Target cells are prepared from a syngeneic source, infected with a transformed vector so that they express the viral antigens on their surface, and also labeled with ^{51}Cr as a marker of cell death. CD8 cells are incubated with labeled target cells and the lysis of targets is assayed by the degree of ^{51}Cr release. This diagrammatic representation does not show various controls and quantitative features of the assay.

virus X, its proteins, or its peptides. Under these conditions, virus X–specific CD8 cells undergo clonal expansion, presumed proportional to their numbers in the original cultured sample. The expanded virus-specific CD8 lymphocyte population is then tested for its ability to destroy target cells bearing the viral antigens in the context of syngeneic MHC class I molecules.

Production of syngeneic target cells depends on the species. For mice, animals of a single inbred strain are used, whereas for outbred species, such as humans, target cells are usually prepared from each individual to be tested using Epstein–Barr virus to immortalize B cells (B cells are used simply as targets). Target cells are then infected with a vector, such as a transformed vaccinia virus, that expresses the virus X antigens; the targets will now express the antigens of virus X bound to the subject's MHC class I molecules. The target cells are labeled with ^{51}Cr, a marker that binds to intracellular proteins. Cultured CD8 cells are incubated with target cells, and the release of ^{51}Cr is measured as an indicator that the targets have been lysed by the CD8 effector cells. This assay can be quantified by serial dilution of CD8 cells prior to the assay, yielding an estimate of the number of CTL precursors in the original cell suspension. Although the LDA represents a biologically relevant measure of cellular immunity, it is very cumbersome, tedious, capricious, and expensive.

As described above, the CTL assay measures precursor (or memory) CD8 cells, based on their potential for expansion in the presence of their cognate antigen. The same assay can be used on cells freshly obtained from the spleen and is then regarded as a measure of effector cells in contrast to precursor cells.

New assays for cellular immune responses have recently been introduced and are still under active development (Table 5.4). One group of assays are based on a technical innovation, the fluorescent activated cell sorter (FACS). As currently employed, these methods fall into two categories.

One assay, called the tetramer assay, detects antigen-specific CD8 cells by a reagent that consists of a single peptide molecule bound in the antigen-binding groove of an MHC class I molecule that is chemically coupled to a tetramer that also bears a sensitive fluorescent marker. A tetramer reagent will bind to the surface of all CD8 cells bearing TCRs specific for a single cognate peptide, and such labeled cells can be enumerated using the FACS. It is likely that numbers of tetramer-staining cells are proportional to the numbers of cells measured in the classical limiting dilution assay but the relationship is still under study.

The other FACS-based method of titrating CD8 T lymphocytes is based on the response of these cells to the epitope recognized by their TCRs. Epitope exposure (oligopeptides added to the culture) stimulates the production of a number of ILs and cytokines, such as IFN-γ or TNF-α. After epitope exposure, CD8 cells are treated with brefeldin A (to prevent secretion of cytokines) and are subsequently fixed (killed with a chemical that leaves their structure intact, such as glutaraldehyde) and permeabilized so that their intracellular cytokine can be stained by a specific antibody carrying a fluorescent marker. The cells can then be enumerated on a FACS to measure the particular intracellular cy-

TABLE 5.4
Measures of CD8 T-lymphocyte immune responses

Assay	Fresh or cultured cells	Target cell or marker	Readout
Limiting dilution assay	Cultured	Cell presenting epitope bound to syngeneic class I MHC	^{51}Cr release from target cells
Cytotoxic T-lymphocytes Tetramer	Fresh	Cell presenting epitope bound to syngeneic class I MHC	^{51}Cr release from target cells
	Fresh	Fluoresent complex with epitope bound to syngeneic class I MHC	Fluorescent activated cell sorter
Intracellular cytokine	Fresh	Fluoresent complex bound to intracellular cytokine	Fluorescent activated cell sorter
ELISPOT	Fresh	Secreted cytokine	Monocellular focus of released cytokine

MHC, major histocompatibility complex.

tokine, and the staining intensity corresponds to the level of cytokine expression. Using appropriate controls for calibration, it is now possible to determine the number of cells responding to a specific epitope in the context of a specific MHC class I molecule. As with the tetramer assay, the number of cytokine-staining cells is probably proportional to the number of cells measured in the LDA.

Another new method is the ELISPOT assay, which is similar to that described above for plasma cells. In this application, lymphoid cells are plated in medium, stimulated with oligopeptides, and those CD8 T lymphocytes recognizing a specific epitope secrete cytokines. The surface of the culture plate has been prepared by coating with antibody that will recognize a particular cytokine, such as IFN-γ, and the number of foci of bound cytokine are then enumerated.

Effector Functions of CD8 Lymphocytes

CD8 cells exert their effects mainly by two mechanisms: cytolytic attack on target cells or secretion of ILs and cytokines. When CD8 cells—by interaction between the TCR on the CD8 cells and the peptide on the MHC class I molecule of the target cell—are stimulated to "attack" target cells, they release perforin, a molecule that produces channels in the plasma membrane of the target cell leading to lysis. In addition, CD8 cells secrete granzymes (serine esterases), which pass through the channels in the target cell and trigger apoptosis. Effector CD8 cells are not destroyed in this process and survive to kill additional "prey."

CD8 cells also release a number of cytokines, such as IFN-γ, TNF-α, and IL-2. These cytokines can play an essential role in clearing virus-infected tissues, either by initiating apoptosis or by purging cells without cell death—a phenomenon that is not well understood. The relative role of cytolysis versus the cytokines in clearing an infection depends on the virus. Thus, CTLs play an important role in lymphocytic choriomeningitis virus infection, whereas cytokines are more important in hepatitis B virus infection (Fig. 5.11). In addition, cytokines initiate an inflammatory response by exerting a chemotactic effect that draws monocytes and other cells to the site of infection, leading to the elimination of virus and the removal of dead cells.

CD4 cells can also act as effector cells under certain circumstances. Mice in which β_2-microglobulin gene has been inactivated (such mice fail to produce MHC class I molecules or CD8 cells) will nevertheless produce virus-specific CTLs with a CD4

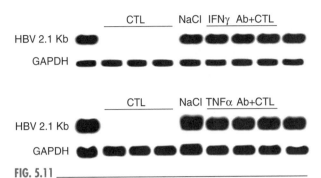

FIG. 5.11

The clearance of some virus infections depends more on cytokines than on the cytolytic activity of CD8 cells. The figure shows the amount of RNA specific for hepatitis B virus (HBV) in the liver of transgenic mice that express the whole genome of HBV. When these mice are adoptively immunized with CD8 cells from a nontransgenic mouse immunized with HBV, the transferred cells clear the virus. This effect is abrogated by antibodies directed against either interferon γ or tumor necrosis factor, implying that those cytokines are responsible for viral clearance. (After Guidotti LG, Ando K, Hobbs MV, et al. Cytotoxic T lymphocytes inhibit hepatitis B virus gene expression by a noncytolytic mechanism in transgenic mice. *Proc Natl Acad Sci USA* 1994;91:3764–3768.)

phenotype. In a few viral infections of humans in which CTLs are induced, the CD4 subset of T lymphocytes can be shown to mediate cytolysis and cloned CD4 cell lines can also be shown to attack virus-specific target cells. However, CD4 cells usually represent only a small fraction of total CTLs and, under physiologic conditions, CD8 lymphocytes usually constitute the major population of effector cells.

Finally, it should be noted that CTLs can initiate a process of viral clearance without necessarily destroying infected cells. Thus, in the hepatitis B virus model mentioned above (Fig. 5.11), cytokines cause a down-regulation of viral mRNA transcripts and their cognate proteins. In this transgenic model 100% of hepatocytes carry the viral genome, and clearance cannot involve cell death because the mice undergo a mild disease from which they recover. A similar phenomenon has been described for clearance of lymphocytic choriomeningitis virus from neurons by transferred virus-specific CD8 T lymphocytes. This ability of an extracellular stimulus to purge intracellular viral RNA and proteins is similar to the effects of antibody and is a remarkable phenomenon that remains to be explained.

Kinetics of the Cellular Immune Response: Effector and Memory T Cells

A number of viral infections have been used as experimental models to follow the kinetics of CD8 responses to viral infection, and similar results have been obtained in humans undergoing natural infec-

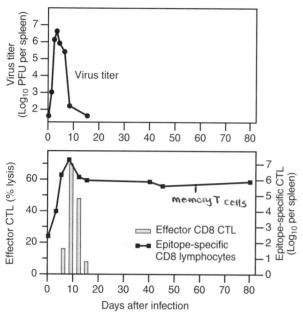

FIG. 5.12

The kinetics of the cellular immune response. In this example, mice were infected with lymphocytic choriomeningitis virus and were followed for their cellular immune responses, assayed in spleen cells. Freshly obtained cells were assayed directly for their cytolytic activity (^{51}Cr release assay) as a measure of effector cytotoxic T lymphocytes. In addition, the fresh cells were assayed for the frequency of CD8 lymphocytes that recognized a single immunodominant epitope (amino acids 118–126 in the nucleoprotein). (After Lau LL, Jamieson BD, Somasundarma T, Ahmed R. Cytotoxic T cell memory without antigen. *Nature* 1994;369:648–652, and Murali-Krishna K, Altman JD, Suresh M, Ahmed R. Counting antigen-specific CD8 T cells: a reevaluation of bystander activation during viral infection. *Immunity* 1998;8:177–187.)

tions. The following description is based on acute viral infections, in which the virus is totally cleared from the host after a period of weeks to months (Fig. 5.12). Virus-specific CD8 effector CTLs appear about 1 week after infection, rapidly increase in number to a high peak at 2–3 weeks after infection, and then quickly wane by 3–6 weeks to levels that are difficult to detect. The CD8 peak often corresponds to the period when virus is being cleared by the host. At maximum, the number of virus-specific CD8 cells can be very high, representing up to 20% of total circulating CD8 cells. These effector T lymphocytes are constantly undergoing apoptosis and have a short half-life, probably no more than a few days, which accounts for their rapid disappearance after the peak of viral infection.

During the acute period of infection, precursor CTLs also appear with roughly the same kinetics as effector CTLs. It is estimated that in different experimental viral infections precursor CTLs increase by 100- to 10,000-fold during infection. Following in-

fection, these virus-specific CD8 cells drop by 10- to 100-fold below their peak but do not completely disappear (Fig. 5.12). This residual population of antigen-committed precursor CD8 cells are usually considered to be "memory" T cells because of their ability to proliferate rapidly if the host is reinfected. In contrast to effector CTLs, which can be difficult to detect after the clearance of infection, memory T cells probably persist for the lifetime of the host. It appears that their maintenance is not dependent on either the persistence of antigen or the presence of virus-specific CD4 T_H2 cells. However, these virus-specific CD8 cells do have a finite lifespan (half-life of several weeks) and proliferate slowly to maintain homeostasis. Memory and effector T lymphocytes can be distinguished from each other by a variety of surface markers, as well as by their content of intracellular cytokines, both of which permit them to be separated in the FACS.

IMMUNITY AS A HOST DEFENSE

Recovery from Initial Infection

A single host may be repeatedly exposed to the same virus. Recovery from the initial infection should be distinguished from recovery from reinfection because the relative role of various host defenses is probably quite different. The cellular immune response can produce a large number of effector cells in a relatively short time (Fig. 5.12), whereas the antibody response develops more slowly (Fig. 5.9). Also, effector CTLs have several mechanisms for destroying or purging virus-infected cells, whereas antibody acts most effectively on free infectious virus that has not yet initiated cellular infection. For these reasons, the cellular immune response is probably the most important component of host defense against a primary infection.

Protection Against Reinfection

There are two salient differences between the naïve host and the host that has been previously infected (or vaccinated) and has developed an immune response to a specific virus. First, the immune host usually maintains a significant titer of antiviral antibodies capable of immediately interacting with virus. Second, the immune host maintains virus-specific memory cells, both CD4 and CD8 lymphocytes, and on reexposure to antigen these cells proliferate and mature to effector cells more rapidly than during primary immune induction (Figs. 5.8, 5.9, and 5.12).

Antibodies circulate in the plasma and are also present in mucosal fluids, with the antibody class and concentration depending on the prior immunizing experience. These preexisting antibodies can play a significant role in protecting the immune host against the virus. Classical experiments that demonstrate the efficacy of antibodies have been conducted for many viruses using a passive protection protocol. Experimental animals are pretreated with virus-specific immune serum and then are challenged with virus, in comparison with appropriate controls (Fig. 5.13). Even a modest level of neutralizing antibody protects recipients against the disease consequences of infection. It should be noted that protected animals do undergo reinfection but the virus fails to disseminate to target organs or to replicate to sufficient titer to exceed a disease threshold. The efficacy of antibody depends on the pathogenesis of individual viral infections, and plasma antibody is particularly effective if viremia plays an essential role in the dissemination of the infection to target tissues.

Not only does prior infection provide antibody that exists at the time of reinfection but the reexposure induces an anamnestic or recall response, so that the antibody titer rises more rapidly than after first exposure to the virus and to higher peak titers.

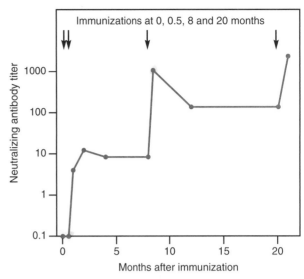

FIG. 5.14

Reexposure to an antigen elicits an anamnestic antibody response. Seronegative subjects were given multiple doses of an inactivated nonreplicating poliovirus (inactivated poliovirus vaccine, IPV): two primary doses, 2 weeks apart, followed by "boosts" 8 and 20 months later, which elicited an anamnestic response that was greater than the primary response. (After Salk J. Considerations in the preparation and use of poliomyelitis virus vaccine. *JAMA* 1955;158:1239–1248, with permission.)

Furthermore, restimulation leads to "maturation" of the antibody response, due to a selection of B cells that synthesize antibodies with increased affinity for the stimulating antibody. Such antibodies have greater biological activity because of their enhanced ability to bind viral antigens. The anamnestic response is readily demonstrated in studies of vaccines in which the timing and amount of antigen exposure can be precisely controlled (Fig. 5.14).

The cellular response to virus is also more rapid and greater upon reinfection than during primary immune induction and may play a role in resistance to reinfection. In most infections where the immune host has preexisting neutralizing antibody and where such antibody can offer significant protection by itself, the cellular response is probably less vital than antibody.

Mucosal Immune Responses

Most viruses infect via a mucosal portal of entry, so that mucosal immunity is potentially important, particularly in protection against reinfection. Both antibody and CD8 cells participate in the mucosal immune response, which differs in certain important ways from the systemic response.

Plasma cells are found in all submucosal tissues, that is, both in the connective tissue and in lymphoid

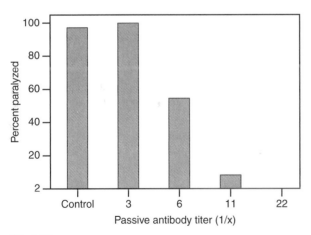

FIG. 5.13

Passive antibody provides protection against a pathogenic virus challenge. In this instance, the challenge is the intramuscular injection of the virulent Mahoney strain of type 1 poliovirus, and almost all of the control animals developed paralytic disease. A titration was conducted of different doses of anti-poliovirus antiserum, given 1 day prior to virus. The results indicate that protection is dose dependent and that a titer of about 1:6 in the recipient provided 50% protection. (After Nathanson N, Bodian D. Experimental poliomyelitis following intramuscular virus injection. III. The effect of passive antibody on paralysis and viremia. *Bull Johns Hopkins Hosp* 1962;111:198–220.)

TABLE 5.5				
Mucosal vs. systemic immunization				
		IMMUNIZATION		
		Subcutaneous	Rectal	Not immunized
CTL response (% lysis)	Spleen	68%	45%	ND
	Peyer's patches	3%	35%	ND
	Lamina propria	5%	30%	ND
Rectal challenge	Virus titer, log 10 per ovary	8.3	4.3	8.3

Mucosal immunization induces an immune response with a different distribution of effector cells than systemic immunization. Mice were immunized with a gp160 peptide of HIV either by subcutaneous or intrarectal route. Assays showed that antigen-specific CTLs were found in the spleen regardless of route of immunization but in intestinal mucosal sites only after rectal immunization. Correspondingly, resistance to rectal challenge with a vaccinia virus incorporating the gp160 antigen was greater in animals immunized by the intrarectal route. Readout was based on the titer of vaccinia virus in the ovary 6 days after virus challenge.

CTL, cytotoxic T lymphocyte; ND, not done.

After Belyakov IM, Ahlers JD, Brandwein BY, et al. The importance of local mucosal HIV-specific CD8 cytotoxic T lymphocytes for resistance to mucosal viral transmission in mice and enhancement of resistance by local administration of IL-12. *J Clin Invest* 1998;102:2072–2081.

tissues such as the tonsil and Peyer's patches in the intestinal wall. A high proportion of mucosal plasma cells secrete IgA, which is assembled as a dimer bound to secretory piece (another peptide). In this form, IgA is preferentially targeted for secretion into the lumen lined by mucosal tissue. In addition, serum IgG escapes across the capillary endothelium and is secreted by mucosal tissues as a passive transudate. Both types of antibody can provide a barrier at the mucosal surface.

Mucosal antibody tends to be relatively transient, lasting only months to years, in contrast to long-lasting serum antibody. This may be due to the persistence of antibody producing plasma cells in the bone marrow, which account for persistent serum antibody (Fig. 5.9), whereas actively secreting plasma cells do not persist in peripheral sites, such as the mucosa. It is speculated that the bone marrow may provide a high concentration of cytokines that stimulate the survival of mature plasma cells and the continual proliferation of plasma cell precursors.

Mucosal tissues contain both CD8 and CD4 cells capable of mounting a cellular immune response. Furthermore, mucosal T cells are distinguished from those in central lymphoid organs by their tendency to home to mucosal sites. As a result, presentation of an antigen to a mucosal surface results in a different distribution of CD8 effector CTLs than does systemic presentation (intravenous, intraperitoneal, intramuscular, subcutaneous) of the same antigen. Furthermore, mucosal immune induction can produce greater protection than systemic immunization against a mucosal challenge with a virus (Table 5.5).

Escape from Immune Surveillance

The contest between virus and host is a never-ending one, but the virus has the advantage that it can undergo more rapid evolution. Not surprisingly, viruses have developed a number of mechanisms to evade immune surveillance and persist in the infected host. These include tolerance in which viruses are treated as self antigens, latency in which the viral genome persists in the absence of gene products and therefore cannot be recognized by the immune system, and other still more byzantine strategies. These will be discussed in Chapter 9 dealing with viral persistence.

FURTHER READING

Reviews and Chapters

Abbas AK, Lichtman AH, Pober JS. *Cellular and molecular immunology,* 4th ed. Philadelphia: WB Saunders, 2000.

Ahmed R, Biron CA. Immunity to viruses. In: Paul WE, ed. *Fundamental immunology.* Philadelphia: Lippincott–Raven Publishers, 1999:1295–1334.

Parren PW, Burton DR. The anti-viral activity of antibodies in vitro and in vivo. *Adv Virus Res* 2001;77:195–262.

Original Contributions

Allen TM, Vogel TU, Fuller DH, et al. Induction of AIDS virus-specific CTL activity in fresh unstimulated peripheral blood lymphocytes from rhesus macaques vaccinated with a DNA prime/modified vacccinia virus Ankara boost regimen. *J Immunol* 2000;164:4968–4978.

Belyakov IM, Ahlers JD, Brandwein BY, et al. The importance of local mucosal HIV-specific CD8 cytotoxic T lymphocytes for resistance to mucosal viral transmission in mice and enhancement of resistance by local administration of IL-12. *J Clin Invest* 1998;102:2072–2081.

Belz GT, Stevenson PG, Castrucci MR, et al. Postexposure vaccination massively increases the prevalence of γ-herpesvirus-specific CD8$^+$ T cells but confers minimal survival advantage on CD4-deficient mice. *Proc Natl Acad Sci USA* 2000;97: 2725–2730.

Binder GK, Griffin DE. Interferon g mediated site specific clearance of alphavirus from CNS neurons. *Science* 2001;293: 303–306.

Fujinami RS, Oldstone MBA. Antiviral antibody reacting on the plasma membrane alters measles virus expression inside the cell. *Nature* 1979;279:529–530.

Guidotti LG, Ando K, Hobbs MV, et al. Cytotoxic T lymphocytes inhibit hepatitis B virus gene expression by a noncytolytic mechanism in transgenic mice. *Proc Natl Acad Sci USA* 1994; 91:3764–3768.

Guidotti LG, Rochford R, Chung J, et al. Viral clearance without destruction of infected cells during acute HBV infection. *Science* 1994;284:825–829.

Hou S, Mo XY, Hyland L, Doherty PC. Host response to Sendai virus in mice lacking class II MHC glycoproteins. *J Virol* 1995; 69:1429–1434.

Huang S, Hendriks W, Althage A, et al. Immune response in mice that lack the IFN-γ receptor. *Science* 1993;259:1742–1745.

Jennings SR, Bonneau RH, Smith PM, et al. CD4-positive T lymphocytes are required for the generation of the primary but not the secondary CD8-positive cytolytic T lymphocyte response to herpes simplex virus in C57BL/6 mice. *Cell Immunol* 1991;133:234–252.

Kagi D, Ledermann B, Burkl K, et al. Cytotoxicity mediated by T cells and natural killer cells is greatly impaired in perforin-deficient mice. *Nature* 1994;369:31–37.

Kaech SM, Ahmed R. Memory CD8+ T cell differentiation: intial antigen encounter triggers a development aprogram in naïve cells. *Nature Immunology* 2001;2:415–422.

Lau LL, Jamieson BD, Somasundarma T, Ahmed R. Cytotoxic T cell memory without antigen. *Nature* 1994;369:648–652.

Muller D, Koller BH, Whitton JL, et al. LCMV-specific class II-restricted cytotoxic T cells in B2-microglobulin-deficient mice. *Science* 1992;255:1576–1578.

Murali-Krishna K, Altman JD, Suresh M, Ahmed R. Counting antigen-specific CD8 T cells: a reevaluation of bystander activation during viral infection. *Immunity* 1998;8:177–187.

Nathanson N, Bodian D. Experimental poliomyelitis following intramuscular virus injection. III. The effect of passive antibody on paralysis and viremia. *Bull Johns Hopkins Hosp* 1962;111: 198–220.

Ochsenbein AF, Fehr T, Lutz C, et al. Control of early viral and bacterial distribution and disease by natural antibodies. *Science* 1999;286:2156–2159.

Ogra PL, Karzon DT. Formation and function of poliovirus antibody in different tissues. *Prog Med Virol* 1971;13:156–193.

Rahemtulla A, Fung-Leung WP, Schilham MW, et al. Normal development and function of CD8+ cells but markedly decreased helper cell activity in mice lacking CD4. *Nature* 1991; 353:180–184.

Salk J. Considerations in the preparation and use of poliomyelitis virus vaccine. *JAMA* 1955;158:1239–1248.

Slifka MK, Matlubian M, Ahmed R. Bone marrow is a major site of longterm antibody production after acute viral infection. *J Virol* 1996;69:1895–1902.

Takeuchi Y, Porter CD, Strahan KM, et al. Sensitization of cells and retroviruses to human serum by α(1–3) galactosyltransferase. *Nature* 1996;379:85–88.

Tishon A, Eddleston M, de la Torre JC, Oldstone MBA. Cytotoxic T lymphocytes cleanse viral gene products from individually infected neurons and lymphocytes in mice persistently infected with lymphocytic choriomeningitis virus. *Virology* 1993; 197:463–467.

Wilson JA, Hevey M, Bakken R, et al. Epitopes involved in antibody-mediated protection from Ebola virus. *Science* 2000; 287:1664–1667.

Chapter 6
Virus-Induced Immunopathology

IMMUNE MECHANISMS OF VIRUS-INDUCED DISEASE

Most viral diseases are caused by virus–cell interactions that either lead to cytolysis or cause cellular dysfunction, as described in Chapter 4. However, sometimes viral disease is mediated by the immune response to infection rather than by the infection itself. This is particularly true of viruses that are relatively noncytopathic, so that infected cells are not immediately destroyed. In such instances, the very same cellular or humoral immune responses to viral antigens that serve as host defenses, by containing and clearing the virus invader, can also mediate a pathologic response. Although at first glance this appears paradoxical, it reflects the dynamic balance between virus and host (Table 6.1). Thus, if the immune response clears the infection by destroying a small number of virus-infected cells, the host survives with minimal symptoms and no permanent damage. If a large number of cells are infected before immune induction, the same immune-mediated destruction can cause severe or fatal pathologic consequences.

Every component of the immune response is capable of causing disease, although the mechanisms and manifestations will differ. Antibody–antigen complexes, cell-mediated cytolysis, and inflammation are individually responsible for some instances of virus-induced immunopathology. Table 6.2 lists selected examples of immune-mediated viral diseases of animals or humans.

What is the evidence that a virus disease is immune mediated? In some examples, such as hepatitis B virus (HBV), the first clue is that the specific virus can cause persistent infections without apparent illness, suggesting that disease is caused by an indirect mechanism. In other examples, pathologic examination reveals the hallmarks of immunopathology, such as lesions containing antigen–antibody complexes, complement, and inflammatory cells. If the disease can be reproduced in an experimental animal, then it is possible to dissect the mechanism by use of such manipulations as immunosuppression, genetic knockout of components of the immune system, adoptive immunization, expression of viral genes by transgenic animals, and the like. All of these methods are illustrated in the remainder of this chapter.

CELL-MEDIATED IMMUNOPATHOLOGY

Mechanisms of Cell-Mediated Immunopathology
$CD8^+$ cytolytic T lymphocytes (CTLs) are the most important cellular mediator of virus-induced immunopathology. As described in Chapter 5, they attack virus-infected cells in a series of steps, including (a) contact of the CTL and its target cell; (b) recognition by the T-cell receptor on the CTL of its

70

TABLE 6.1

Protective vs. pathogenic effects of the immune response to a noncytopathic virus

Type of infection	Rate of virus replication	Rate of immune response	Tissue destruction	Outcome of infection	Role of immune response
Acute	Moderate	Robust	Restricted	Recovery	Protects
Acute	Slow	Moderate	Restricted	Recovery	Protects
Acute	Rapid	Moderate	Extensive	Severe illness	Causes disease
Acute	Moderate	Slow	Extensive	Severe illness	Causes disease
Persistent	Slow or rapid	None	None	Silent infection	Fails to eliminate virus

In acute infections, the relative rates of virus replication and immune induction determine the balance between the protective and the destructive effects of the immune response. If there is little or no immune response ("tolerance") to the viral antigens, a persistent infection occurs with no tissue

cognate peptide epitope presented by class I major histocompatibility complex (MHC) on the target cell; (c) activation of the CTL; (d) delivery of a lethal attack by release of membrane-bound granules containing perforin and granzyme B; (e) fusion of granules with the plasma membrane of the target cell and polymerization of perforin monomers to form a channel in the plasma membrane of the target cell; and (f) osmotic swelling leading to cytolysis of the target cell. Simultaneously, release of granzyme B (a serine protease) into the cytosol of the target cell triggers an intracellular biochemical cascade involving the interleukin-1 (IL-1)–converting enzyme protease that initiates apoptosis. Although CTLs express Fas ligand that could bind to Fas on the surface of the target cell, this is probably not an important pathway for CTL attack.

In addition to the direct cytopathic attack on target cells, activated CD8$^+$ T lymphocytes release a variety of proinflammatory cytokines that probably play a role in some manifestations of virus-induced immunopathology, such as interferon γ (IFN-γ), tumor necrosis factor α (TNF-α), and several ILs. Also, natural killer (NK) cells, as described in Chapter 5, can attack virus-infected targets in two ways: either as a result of virus infection down-regulating MHC class I expression, thus triggering NK cell attack (elimination of "nonself" cells), or via antibody-dependent cell-mediated cytotoxity. However, it is not clear that NK cells play a significant role in the well-recognized examples of virus-induced immunopathology.

CD4 T lymphocytes may play an ancillary role in enhancing some CD8-mediated immunopatho-

TABLE 6.2

Immune-mediated viral diseases of animals or humans: a selected list[a]

Host species	Virus family	Virus *Disease*	Pathogenic immune modality	Protection from reinfection
Human	Hepadnaviridae	Hepatitis B *Hepatitis*	CD8 Antibody?	Antibody CD8?
	Flaviviridae	Dengue *Hemorrhagic fever*	Immune complexes, T cells?	Neutralizing antibody
	Paramyxoviridae	RSV *Bronchiolitis*	CD4 cells Antibody?	Neutralizing antibody
Mouse	Arenaviridae	LCMV *Choriomeningitis Hepatitis*	CD8	Antibody CD8
Mink	Parvoviridae	ADV *Aleutian disease (glomerulonephritis)*	Immune complexes	?
Sheep	Lentiviridae	Visna Maedi virus *Visna (Encephalitis) Maedi (Interstitial pneumonitis)*	CD8?	Neutralizing antibody?

[a]The pathogenic immune response is also responsible for recovery from acute infection (Table 6.1).

ADV, Aleutian disease virus; LDV, lactic dehydrogenase virus; LCMV, lymphocytic choriomeningitis virus; RSV, Rous sarcoma virus.

logic syndromes. Thus, LCMV-specific CD4 T cells can produce some degree of disease in β_2-microglobulin knockout mice that cannot express MHC class I molecules and therefore lack CD8 effector T cells. Also, depletion of T lymphocyte subsets in immunologically intact mice reveals that CD4 and CD8 virus-specific T cells play different and synergistic effector roles in production of LCMV-induced immunopathologic hepatitis.

Experimental Examples of Cell-Mediated Immunopathology

Lymphocytic choriomeningitis virus (LCMV) is an arenavirus that naturally infects wild mice. Experiments with this somewhat obscure virus first led to the idea that viruses could cause immune-mediated disease, and it still serves as a classical model of immunopathology (Sidebar 6.1). If LCMV is injected by the intracerebral route in adult mice, the animals sicken in about a week and rapidly die while displaying neurologic symptoms, the most striking of which is convulsions. The brain shows severe inflammation of its meninges (the external linings of the brain) but no involvement of the parenchyma (brain tissue), and examination with tagged antibodies shows that the virus replicates in cells in the meninges but not in the parenchyma. Although the pathophysiology of the convulsive state is not completely understood, it is probably mediated both by direct cytopathic attack on infected choroidal cells and by release of molecules by inflammatory cells. For instance, macrophages in the inflammatory exudate express high levels of inducible nitric oxide synthase, and the elevated levels of nitric oxide could play a role in inducing convulsions.

Mice infected with LCMV that are also treated with the immunosuppressive drug cyclophosphamide fail to develop acute convulsive disease (Fig. 6.1). Although the animals appear healthy, the virus replicates at a high level and produces a lifelong persistent infection of the brain and lymphoid tissues. The protective effect of immunosuppression and the persistence of high virus titers in healthy mice taken together strongly suggest that acute convulsive disease is due to an antiviral immune response rather than to virus replication per se.

To test this hypothesis, syngeneic normal mice were "immunized" by intraperitoneal injection with LCMV, which produces a transient infection (sometimes accompanied by hepatitis or other disease) leading to clearance of the virus and development of a cellular immune response. These "immune" mice were then used as donors for antiserum or lympho-

cytes. When immune lymphocytes were transferred (known as adoptive immunization) to immunosuppressed mice with asymptomatic brain infection, the acute convulsive disease was reproduced (Fig. 6.1). Subsequent studies indicated that CD8 lymphocytes mediated choriomeningitis and that these same cells also mediated clearance of the virus in intraperitoneally infected animals. Evidence for this view is provided by experiments in perforin knockout mice whose T lymphocytes cannot mediate a CTL "attack." Both choriomeningitis and viral clearance are abrogated in such mice (Fig. 6.2).

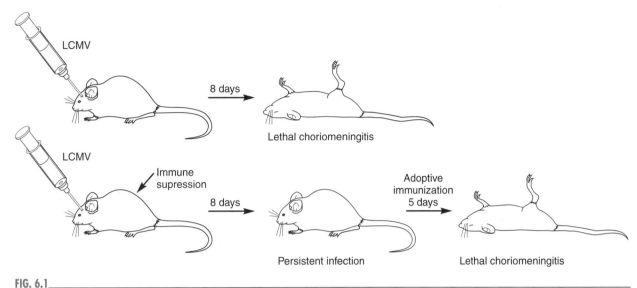

FIG. 6.1

Experimental demonstration that a viral disease is mediated by the immune response, using immunosuppression and immune reconstitution. This illustration describes the protocol and results of intracerebral lymphochoriomeningitis virus injection of adult mice that die as a result of being in an acute convulsive state. Immunosuppression converts the infection to an asymptomatic persistent infection of the meninges of the brain, and adoptive immunization of persistently infected mice reproduces the acute disease. Disease is produced by transfer of immune T lymphocytes (CD8 CTLs) but not immune antiserum or nonimmune lymphocytes. (After Gilden DH, Cole GA, Monjan AA, Nathanson N. Immunopathogenesis of acute central nervous system disease produced by lymphocytic choriomeningitis virus. *J Exp Med* 1972;135:860–889.)

In some instances, immune-mediated disease causes tissue destruction in addition to inflammation. One case is the hepatitis caused by hepatotropic strains of LCMV (Fig. 6.2). Another striking example is destruction of the cerebellum in newborn rats infected by LCMV. When newborn rats are injected intracerebrally with LCMV, they develop a nonlethal infection of the brain concentrated in the cerebellum, a region of the brain associated with integration of locomotion. These animals survive but become severely ataxic (incoordinated). Pathologic examination indicates that the cerebellum has undergone acute necrosis and degeneration as a result of infection localized to this brain region (Fig. 6.3). Immunosuppression protects against brain pathology and ataxia, even though virus replication is undiminished, again indicating an immune-mediated disease process. It is likely that this process is the direct consequence of the cytopathic action of CD8 lymphocytes.

Some virus-induced immunopathologic syndromes occur in the context of immunization. One example is Rous sarcoma virus (RSV), which produces severe, sometimes fatal bronchopneumonia in infants. Early attempts to immunize babies employed a formalin-inactivated RSV vaccine that not only failed to protect but appeared to enhance pulmonary disease following natural exposure to wild-type RSV. The failure of the vaccine was associated with the fact that it induced nonneutralizing antibodies that were directed against only one of the two envelope proteins of the virus (the fusion protein) and not against the other envelope protein (the hemagglutinin).

The lack of neutralizing antibodies explained the failure to protect but not the enhancement of disease. Subsequent depletion studies in experimental animals showed that disease enhancement was mainly mediated by CD4 cells, apparently acting as effectors rather than via their helper function for induction of antibodies (Table 6.3). Thus, disease could be induced by transfer of T lymphocytes but not by antisera from mice that were undergoing immunopathologic responses.

Cell-Mediated Immunopathology: Human Disease

Hepatitis B is probably the most important viral disease of humans that is immune mediated. Following exposure to HBV, the infection can take several different courses, and the first clues about the pathogenesis of HBV hepatitis come from a comparison of viral antigen and antibody markers in patients with diverse outcomes. Figure 6.4 compares a patient undergoing acute hepatitis and another infected subject who remains asymptomatic. The asymptomatic individual has a high titer of virus in the blood, as reflected by the amount of hepatitis B surface antigen (HBsAg) but fails to clear the infection

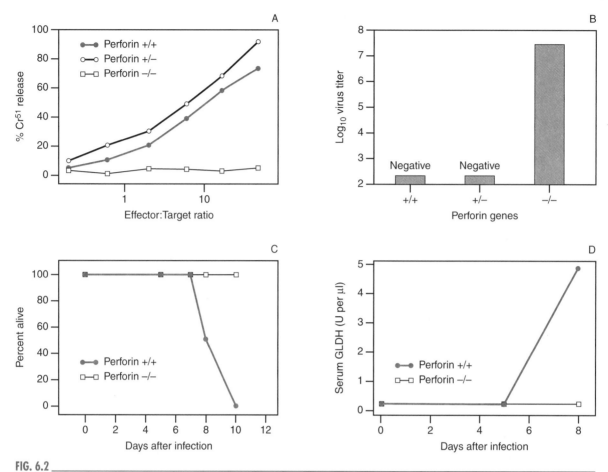

FIG. 6.2

The use of an immunologic "knockout" to demonstrate that viral disease is mediated by a specific component of the immune response, such as the cytopathic action of CD8$^+$ T lymphocytes. In this experimental example, mice with a normal perforin gene (+/+) are compared to mice with the gene knocked out (−/−) and with heterozygous mice (+/−), following infection with lymphocytic choriomeningitis virus (LCMV). The perforin gene is required for CD8 effector function. **A:** Demonstrates the absence of CD8 cytolytic activity in perforin −/− mice compared to perforin +/+ animals. **B:** Following intravenous infection, perforin +/+ mice clear the infection by day 12 after infection while perforin −/− mice fail to clear the virus and are persistently infected. **C:** Following intracerebral infection perforin +/+ mice develop fatal choriomeningitis while perforin −/− mice are infected but do not die. **D:** After intravenous injection of a hepatotropic strain of LCMV, perforin +/+ mice develop hepatitis as reflected by a high serum level of a liver enzyme (glutamic dehydrogenase), whereas perforin −/− mice are infected but do not develop liver disease. (After Kagi D, Ledermann B, Burki K, et al. Cytotoxicity mediated by T cells and natural killer cells is greatly impaired in perforin-deficient mice. *Nature* 1994;369:31–37.)

that persists, often for many years. Such persistently infected persons exhibit little evidence of antibody (anti-HB) against the surface antigen in their serum. In contrast, the patient with acute hepatitis clears the infection during the period when liver disease is at its peak. Shortly after clearing the virus, serum anti-HB appears and rises to high titer. From the association of viral clearance and disease, it may be surmised that the same process that clears the virus also produces disease, another example of the dual role of the immune response (Table 6.1).

Evidence for the basis of hepatic disease and viral clearance can be derived from studies in mice that carry a transgene for the hepatitis B genome and express the major structural proteins of the virus in

hepatocytes. This model has confirmed the dual roles of the immune response and has permitted a detailed dissection of the mechanisms involved (Sidebar 6.2).

If transgenic mice expressing HbsAg in their hepatocytes are injected with syngeneic CTL CD8$^+$ lymphocyte clones directed against a specific epitope of the envelope protein, acute hepatitis is produced. There are two steps in the development of pathology. First, the injected CTLs attach to infected hepatocytes and induce apoptosis, but this is limited to a small number of potential target hepatocytes, probably because CTLs cannot move readily through solid tissue. Second, if the CTL clones produce substantial amounts of IFNγ, large numbers of

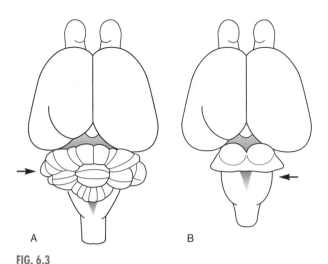

FIG. 6.3

Immunopathologic mechanisms can produce focal tissue destruction. Gross pathology of newborn rats infected with lymphocytic choriomeningitis virus. Although both animals were infected, immunosuppression of animal in panel A protected against the severe cerebellar involution shown in panel B. (After Monjan AA, Gilden DH, Cole GA, Nathanson N. Cerebellar hypoplasia produced by lymphocytic choriomeningitis virus. *Science* 1971;171:194–196.)

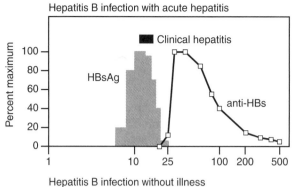

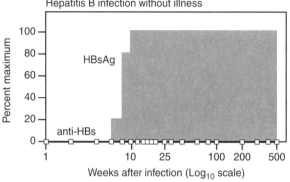

FIG. 6.4

Hepatitis B disease likely is mediated by the immune response to infection rather than a direct effect of viral replication. Among adults, hepatitis B infection takes several different courses. (*Top*) About 20%–35% of infections are accompanied by acute clinical hepatitis. Most of these patients recover completely, clear the virus, and acquire lifelong immunity. (*Bottom*) About 65%–80% of infections are inapparent, and most of these individuals clear the virus and acquire immunity. However, as illustrated in this panel, about 2%–10% of individuals undergoing primary inapparent infection fail to clear their virus that persists for many years. The presence of high titers of virus in the absence of hepatitis—contrasted with the occurrence of acute hepatitis in patients who clear their virus—suggests that disease is not caused by infection per se but by the immune response to the virus. (Redrawn from Robinson WR. Hepatitis B virus and hepatitis D virus. In: Mandell GL, Bennett JE, Dolin R. *Principles and practice of infectious diseases.* New York: Churchill Livingstone, 1999;1652–1683.)

white blood cells (mononuclear and polymorphonuclear cells) are attracted and produce necroinflammatory foci in which much of the cell killing is produced by nonspecific "bystander" cells distant from the CTLs. This process usually induces a limited acute hepatitis from which the mice recover, similar to most cases of human hepatitis B. If the transgenic mice express the large envelope protein, much of which is retained in the endoplasmic reticulum, then the inflammatory response progresses to a third stage, characterized by activation of liver macrophages (Kupffer cells) and progressive inflammation and necrosis that kills many of the mice;

TABLE 6.3			
Virus-induced immunopathology that is more dependent on CD4 than on CD8 T lymphocytes			
Immunization (day 0)	**Depletion (days 18, 20, 23)**	**Intranasal RSV infection (day 21)**	**Alveolitis (% lung area) (day 25)**
Formalin-inactivated RSV	None	Yes	9.6
	CD8	Yes	3.9
	CD4	Yes	0.5
	CD8 + CD4	Yes	0.3

In this protocol, mice were immunized with formalin-inactivated RSV and were subsequently infected with RSV by intranasal spray. Under these circumstances, infection produces a severe bronchiolitis. Mice immunized with the formalin-inactivated RSV were divided into groups and depleted of CD4, CD8, or both CD4 and CD8 cells, prior to and after challenge with intranasal live RSV. CD4 depletion (more effectively than CD8 depletion) protected against virus-induced bronchiolitis. Since alveolitis could not be produced by passive transfer of antibodies alone, it was concluded that this effect is mediated, at least in part, by CD4 effector cells.

RSV, Rous sarcoma virus.

After Connors M, Kulkarni AB, Firestone CY, et al. Pulmonary histopathology induced by Rous sarcoma virus (RSV) challenge of formalin-inactivated RSV-immunized BALB/c mice is abrogated by depletion of CD4+ T cells. *J Virol* 1992;66:7444–7451.

this process resembles fulminant hepatitis B, which is seen in about 1% of human cases of hepatitis B.

The same process that produces hepatitis in transgenic mice also mediates viral clearance but the mechanisms are different. Whereas hepatocellular inflammation and necrosis are produced by the direct action of CTLs, clearance seems to depend on the cytokines IFN-γ and TNF-α, which are produced in abundance by activated CTLs. Treatment with antibodies against either of these two cytokines abrogates viral clearance without altering the course of necroinflammation (Fig. 5.11). The most remarkable finding is that the cytokines down-regulate all aspects of the transcription and translation of the HBV transgenes in 100% of hepatocytes without killing them, perhaps even "curing" them of infection.

The role of anti-HBs should also be mentioned. Based on studies with passive transfer of antibody from immune subjects and on studies with HBV vaccine, it appears that anti-HBs is highly effective as a prophylactic to protect uninfected subjects from future exposure to HBV (discussed in Chapter 14).

Antibody probably also plays a role in acute hepatitis, both in disease causation and in viral clearance. When administered to transgenic mice expressing HbsAg, passive antibody causes a rapid clearance of viral antigen from the blood and a mild, very transient hepatitis. This suggests that in acute infection of humans with HBV, antibody clears viral proteins from the circulation, synergizing the CTL-mediated clearance of virus from hepatocytes. Antibody probably plays a minor role in the pathogenesis of acute hepatitis, since acute hepatic disease can be produced by individual CD8 CTL clones—in the absence of antibody—when transferred to HbsAg-expressing transgenic mice.

ANTIBODY-MEDIATED IMMUNOPATHOLOGY

Experimental Examples of Antibody-Mediated Immunopathology

Some persistent viral infections are characterized by high titers of virus or viral antigens in the blood over a long period. If any antiviral antibodies are produced

---

■ ■ ■
SIDEBAR 6.2

Mouse Model for the Study of Hepatitis B Virus Pathogenesis

A mouse model has been developed for the dissection of the immunologic mechanisms responsible for disease production and viral clearance in hepatitis B virus (HBV) infection. The model focuses on a single viral protein, the envelope protein known as HbsAg. The envelope protein occurs in three nested forms—the large, middle, and major envelope proteins—that are translated from three different RNA transcripts of the viral genome. The major and middle proteins assemble into 22-nm spherical particles, and the large protein assembles into long filamentous particles about 22 nm in diameter; both particles are found at very high concentrations in the serum of HBV-infected persons.

The mouse model has two major variables. First, transgenic mice can be created that express the large, middle, or major envelope proteins to elucidate the different roles of these three proteins. Second, mice can be vaccinated with immunogens that express the envelope proteins, and from the spleens of immune mice CTL clones can be grown that are directed against individual epitopes within each of the three envelope proteins. This exquisite specificity permits a more detailed analysis of mechanisms than is available for other examples of virus-induced immunopathology. ■

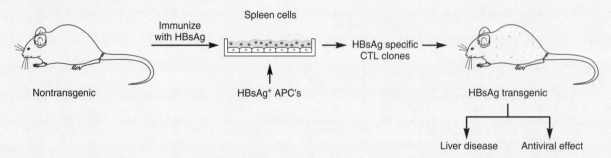

Diagram based on figure 31.5 in Chisari FV, Ferrari C. Viral hepatitis. In: Nathanson N, et a., eds. *Viral pathogenesis*. Philadelphia: Lippincott–Raven Publishers, 1997:745–778.

during such viremic infections, antigen–antibody (immune) complexes may be formed in the circulation. The glomerular filtration system in the kidney readily passes water, electrolytes, and smaller molecules that enter the renal tubules for readsorption or excretion into the urine. Immune complexes, due to their high molecular weight, may fail to transit the glomerular filter and accumulate under the basement membrane that is situated external to the glomerular capillary wall. Immune complexes can fix complement that, in turn, will generate proinflammatory cytokines that draw a variety of leukocytes into the periglomerular space. The resulting chronic inflammatory process causes scarring of the glomerulus, chronic glomerulonephritis, gradual reduction of renal function, and eventual kidney failure.

One example of this process is LCMV-infected mice that develop antibody-mediated glomerulonephritis and kidney failure. Mice infected with LCMV at birth have a lifelong plasma viremia. However, they also produce antiviral antibodies against all of the major structural proteins of LCMV, and the virus in their serum circulates as infectious immune complexes, which can be demonstrated by the fact that the infectious titer is reduced by antibodies directed against mouse immunoglobulin (Table 6.4). Both LCMV antigen and antibodies can be demonstrated in the eluates of kidneys of mice with longstanding infection that have massive deposits of immune complexes in their kidneys (Fig. 6.5).

Long-term high-titer viremias are seen in a number of persistent viral infections (see Chapter 9), and it is likely that most of them are characterized by circulating immune complexes. Depending on the lev-

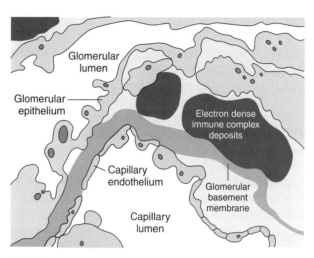

FIG. 6.5

Immune complex deposits in the glomerulus of a mouse with chronic disease associated with persistent lymphocytic choriomeningitis virus infection. The amorphous deposits are located between the epithelial cells of the glomerulus (that line the glomerular filtration space) and the basement membrane that underlies the capillary endothelium. Normally, a filtrate of blood passes across the endothelium and the basement membrane and then crosses the epithelial lining to enter the glomulerular filtration space. Immune complexes are so large that they may become lodged in the interstitial space under the basement membrane, where they gradually accumulate, leading to inflammation, scarring, and loss of functional capacity of the glomerulus, i.e., chronic glomerulonephritis that is eventually fatal. (After Buchmeier MJ, Welsh RM, Dutko FJ, Oldstone MBA. The virology and immunobiology of lymphocytic choriomeningitis virus infection. *Adv Immunol* 1980;30:275–331.)

els of antigen and antibody, such infections may lead to immune complex deposition and kidney damage (Table 6.2). In addition to kidney disease, circulating immune complexes may produce other types of lesions, such as vasculitis.

Antibody-Mediated Immunopathology: Human Disease

Dengue hemorrhagic fever/dengue shock syndrome (DHF/DSS) is an important example of antibody-mediated viral disease in humans. Dengue viruses are members of the flaviviridae and can be grouped into four antigenically distinct types (1–4) based on serologic typing. Dengue is transmitted by *Aedes aegypti* mosquitoes and is confined to specific tropical areas where the vector is indigenous. From time to time, depending on prevalence of the vector, massive outbreaks of dengue occur in southeast Asia or in the Caribbean.

During epidemics, most infections cause an acute but self-limited febrile illness with complete recovery, but a small proportion of patients develop shock and/or a hemorrhagic diathesis, with a mortality up to 10%. Epidemiological studies indicate that

TABLE 6.4	
Viremia presenting as an infectious immune complex	
Treatment of sera from mice persistently infected with LCMV	LCMV titer ($\log_{10}$ LD$_{50}$ per 0.02 mL)
Anti-mouse immunoglobulin	<1.0
Controls	
Normal rabbit serum	3.7
Anti-mouse albumin	3.5

In the example shown, sera from 3-month-old mice infected with LCMV at birth were treated with antibodies against mouse immunoglobulin or control antibodies and were then assayed for infectivity. The inactivation by anti-immunoglobulin antibodies is evidence that the infectious virus is bound to mouse antiviral antibodies.

LCMV, lymphocytic choriomeningitis virus.

After Oldstone MBA, Dixon FJ. Pathogenesis of chronic disease associated with persistent lymphocytic choriomeningitis viral infection. I. Relationship of antibody production to disease in neonatally infected mice. *J Exp Med* 1969;129:483–499.

DHF/DSS occurs almost exclusively in children who are undergoing a second infection with a dengue virus of a serotype different from their primary dengue infection. Also, DHF/DSS is seen in infants undergoing primary infection but who carry maternal antidengue antibody acquired from their mothers. These observations suggested that DHF/DSS might be mediated by an immunologic mechanism. Furthermore, clinical studies have indicated that patients with DHF/DSS have both increased capillary permeability and fragility, and a reduction in thrombocytes (a subcellular element required for normal blood clotting).

Cell culture studies showed that dengue viruses replicated in blood monocytes and that infection of these cells could be enhanced in the presence of small amounts of antibodies against dengue viruses. Enhancement was mediated by binding of the Fc portion of the antibody molecule to Fc receptors on the surface of monocytes, which facilitated entry of virus into their host cells. Enhancement could also be shown in monkeys infected with dengue virus and pretreated with antibodies against dengue viruses (Fig. 6.6).

Although antibody-dependent enhancement may explain enhanced viral replication in secondary dengue infections, the pathophysiology of DSS is still a matter for tentative reconstruction. Monocytes or macrophages infected with dengue virus, particularly in the presence of enhancing antibody, secrete a number of vasoactive cytokines, including TNF-α, IFN-γ,

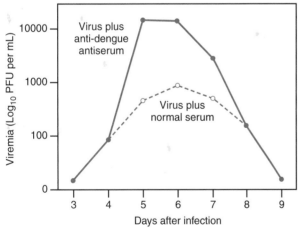

FIG. 6.6

The replication of dengue virus is enhanced by antibody. A pair of monkeys were treated with either human antiserum against dengue virus or a control serum and 15 minutes later were infected by an subcutaneous dose of dengue virus type 2. The subsequent course of viremia is plotted to show that the antibody enhanced the infection. Similar results were obtained in four additional experiments. (After Halstead SB. In vivo enhancement of dengue virus infection in rhesus monkeys by passively transferred antibody. *J Infect Dis* 1979;140:527–533.)

and IL-2. Also, in secondary dengue infections, CD4 and CD8 T lymphocytes specific for dengue virus antigens are activated and secrete a similar array of vasoactive cytokines. This cytokine "storm" could account for increased capillary fragility (with associated hemorrhages) and permeability (with associated shock). In addition, circulating immune complexes (dengue viral antigens and antibodies) initiate the complement cascade evidenced by subnormal levels of several serum complement proteins. Activated complement proteins, particularly C3a and C5a, also enhance vascular permeability and are associated with sequestration of thrombocytes.

In this and the previous chapter, it has been noted that antibodies may play either a protective or a disease-enhancing role in viral infection. This point can be illustrated by experimental circumstances in which the same antibody preparation can protect, enhance, or have no effect on the outcome of infection (Fig. 6.7).

VIRUS-INITIATED AUTOIMMUNITY

It has long been suspected that occasionally a virus infection might induce an autoimmune response, that is, an immune response to antigenic determinants in a host protein. The following sequence of events could mediate such a response. (a) A viral infection elicits an immune response to a number of B- and T-cell epitopes on the virus-encoded proteins. (b) A few of these epitopes are also shared with one or more host proteins. (c) The virus-induced epitope-specific antibodies or T lymphocytes are capable of reacting with the cognate epitopes on the host proteins. (d) These antihost immune responses may or may not elicit an autoimmune disease process.

Is there evidence of epitopes that are shared between virus proteins and host proteins? Sharing of epitopes, sometimes called "molecular mimicry," would be a rare phenomenon ($< 1:10,000,000$) based on chance alone. Nevertheless, there are examples of similar sequences of amino acids in viral and host proteins, as shown in Table 6.5. When antiviral monoclonal antibodies are prepared from mice shortly after infection with a variety of RNA and DNA viruses, a small number of these antibodies can be shown to react with one or more tissues in uninfected mice. In most of these instances, the shared epitope remains to be defined.

These observations suggest that viral infections may, on occasion, induce immune responses that are directed against an epitope present on both a viral

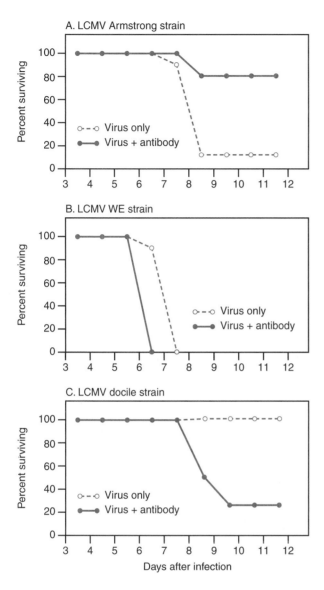

A. LCMV Armstrong strain

Percent surviving

o–-o Virus only
•—• Virus + antibody

B. LCMV WE strain

Percent surviving

o–-o Virus only
•—• Virus + antibody

C. LCMV docile strain

Percent surviving

o–-o Virus only
•—• Virus + antibody

Days after infection

FIG. 6.7

Antibody can have different effects on the outcome of a viral infection depending on the conditions. In this experiment, mice were pretreated with a single preparation of anti-lymphocytic choriomeningitis (anti-LCMV)–neutralizing antibody 4 hours before intracerebral challenge with three different strains of LCMV. This produced acutely fatal choriomeningitis in untreated animals. In this model, the virus replicates in the meninges, producing a "target" for potential immunopathologic attack, and also spreads to the spleen and other lymphoid tissues where it induces a CD8 response that mediates the choriomeningitis. If the strain of LCMV replicates sufficiently rapidly in the lymphoid tissues, it can induce so-called high-dose immune paralysis probably by driving induced CD8 antigen-responsive lymphocytes into apoptosis. **A:** Antibody reduced the hematogenous spread of slowly replicating Armstrong strain of LCMV, preventing the induction of a robust cytotoxic T lymphocyte (CTL) response, and thereby protected most of the animals against the immunopathologic disease. **B:** Antibody did not sufficiently slow the hematogenous spread of intermediate-replicating WE strain of LCMV and had little effect on disease. **C:** Antibody reduced viral replication in the lymphoid tissues sufficiently to convert "high-dose paralysis" to a functional CTL response, which produced lethal choriomeningitis. (After Battegay M, Kyburz D, Hengartner H, Zinkernagel RM. Enhancement of disease by neutralizing antiviral antibodies in the absence of primed antiviral cytotoxic T cells. *Eur J Immunol* 1993;23:3236–3241.)

and a host protein. However, it is important to recognize that an epitope-specific immune response against a host protein will not necessarily induce disease. Autoimmune disease induction probably requires that the immune response directly interferes with an essential activity of a host protein or that the response triggers a cascade of immune responses that leads to inflammation, cell destruction, and loss of function of a tissue or organ.

Myelin basic protein (MBP) can be used to illustrate this point. MBP is a major component of the myelin sheath of neuronal processes in the central and peripheral nervous system, and its loss drastically curtails the conductive ability of neuronal axons and dendrites, producing dysfunction that can be progressive and fatal. If rodents are immunized with the complete protein, a severe disease (experimental

allergic encephalitis, or EAE) is induced. Immunization with different segments of MBP induces immune responses against the corresponding epitopes. However, the ability to induce EAE is confined mainly to one 10-amino-acid sequence (the "encephalitogenic site") of MBP, and immunization with other domains of the protein, although they elicit an immune response, do not induce EAE.

Certain viral proteins, such as the polymerase of HBV, contain sequences similar to the encephalitogenic epitope of MBP. When a peptide based on the viral sequence is used to immunize rabbits, it induces an antibody and cellular response to rabbit MBP, as well as perivascular inflammation confined to the central nervous system. This is an experimental example of an antiviral immune response that is capable of inducing an autoimmune disease process.

TABLE 6.5		
Potential shared epitopes between viral and host proteins		
Protein	Sequence	Immunologic cross-reactivity demonstrated
Poliovirus VP2	STTKESRGTT	Yes
Acetylcholine receptor	TVIKESRGTK	
Papillomavirus E2	SLHLESLKDS	Yes
Insulin receptor	VYGLESLKDL	
Rabies virus glycoprotein	TKESLVIIS	Yes
Insulin receptor	NKESLVISE	
HIV p24	GVETTTPS	Yes
IgG constant region	GVETTTPS	
Measles virus P3	LECIRALK	No
Corticotropin	LECIRACK	

Table lists examples of sequences where there are at least six sequential identical amino acids. In most of these instances, immunologic cross-reactivity has also been demonstrated.

After Oldstone MBA. Molecular mimicry and autoimmune disease. *Cell* 1987;50:819–820.

Although it seems plausible that viral infections trigger some cases of autoimmune disease in humans, it has been difficult to develop definitive evidence for specific diseases.

REPRISE

Although the immune response usually plays a protective role in viral infection, a few viral diseases are immune mediated. The same immune responses that can clear viral infections can also cause immunopathologic processes, and this apparent paradox is explained by a kinetic view of infection as a race between the invading virus and host immune defenses. If the immune response is sufficiently brisk, it will clear the infection with minimal cellular damage, but if the infection is widespread when immune effector cells appear, then the attack on infected cells can produce severe disease.

Immunopathology is most often seen with viruses that are relatively noncytopathic, so that infected cells are not immediately destroyed. Each major effector arm of the antiviral immune response may, on occasion, mediate disease. CD8 cytolytic T lymphocytes cause several distinct pathologic syndromes in mice infected with arenaviruses, such as LCMV, including acutely fatal choriomeningitis, hepatitis, and cerebellar destruction. Likewise, effector T cells are the major cause of hepatitis produced by HBV and perhaps by hepatitis C virus in humans.

In some persistent viral infections of animals, antibody binds to viral proteins to produce immune complexes that can cause progressive glomerulonephritis and vasculitis. Similar immune complex disease probably occurs in some persistent viral infections of humans, such as those caused by human immunodeficiency virus and HBV. Antibodies probably play an important role in at least one acute human viral disease, dengue hemorrhagic fever, although the pathogenesis is not completely understood.

In summary, immune-mediated viral disease is a reflection of the delicate balance between the defensive and destructive effects of the immune response, and of the race between viral invaders and the defenses mounted by the infected host.

FURTHER READING

Reviews and Chapters

Bloom ME, Kanno H, Mori S, Wolfinbarger JB. Aleutian mink disease: puzzles and paradigms. *Infect Agents Dis* 1994;3: 279–301.

Buchmeier MJ, Zajac AJ. Lymphocytic choriomeningitis virus. In: Ahmed R, Chen I, eds. *Persistent viral infections*. New York: John Wiley and Sons, 1999:575–605.

Buchmeier MJ, Welsh RM, Dutko FJ, Oldstone MBA. The virology and immunobiology of lymphocytic choriomeningitis virus infection. *Adv Immunol* 1980;30:275–331.

Chisari FV, Ferrari C. Viral hepatitis. In: Nathanson N, et al., eds. *Viral pathogenesis*. Philadelphia: Lippincott–Raven Publishers, 1997:745–778.

Rothman AL, Ennis FA. Immunopathogenesis of dengue hemorrhagic fever. *Virology* 1999;267:1–6.

Classic Papers

Gilden DH, Cole GA, Monjan AA, Nathanson N. Immunopathogenesis of acute central nervous system disease produced by lymphocytic choriomeningitis virus. *J Exp Med* 1972; 135:860–889.

Guidotti LG, Ando K, Hobbs MV, et al. Cytotoxic T lymphocytes inhibit hepatitis B virus gene expression by a noncytolytic mechanism in transgenic mice. *Proc Natl Acad Sci USA* 1994; 91:3764–3768.

Halstead SB. In vivo enhancement of dengue virus infection in rhesus monkeys by passively transferred antibody. *J Infect Dis* 1979;140:527–533.

Kagi D, Ledermann B, Burki K, et al. Cytotoxicity mediated by T cells and natural killer cells is greatly impaired in perforin-deficient mice. *Nature* 1994;369:31–37.

Oldstone MBA, Dixon FJ. Pathogenesis of chronic disease associated with persistent lymphocytic choriomeningitis viral infection. I. Relationship of antibody production to disease in neonatally infected mice. *J Exp Med* 1969;129:483–499.

Zinkernagel RM, Doherty PC. Restriction of in vitro T cell-mediated cytotoxicity in lymphocytic choriomeningitis with a syngeneic or semiallogeneic system. *Nature* 1974;248:701–702.

Original Contributions

Battegay M, Kyburz D, Hengartner H, Zinkernagel RM. Enhancement of disease by neutralizing antiviral antibodies in the absence of primed antiviral cytotoxic T cells. *Eur J Immunol* 1993;23:3236–3241.

Connors M, Kulkarni AB, Firestone CY, et al. Pulmonary histopathology induced by respiratory syncytial virus (RSV) challenge of formalin-inactivated RSV-immunized BALB/c mice is abrogated by depletion of CD4+ T cells. *J Virol* 1992;66:7444–7451.

Fujinami RS, Oldstone MBA. Amino acid homology between the encephalitogenic site of myelin basic protein and virus: mechanism of autoimmunity. *Science* 1985;230:1043–1045.

Gossman J, Lohler J, Utermohlen O, Lehmann-Grube F. Murine hepatitis caused by lymphocytic choriomeningitis virus. II. Cells involved in pathogenesis. *Lab Invest* 1995;72:559–570.

Guidotti LG, Rochford R, Chung J, et al. Viral clearance without destruction of infected cells during acute HBV infection. *Science* 1999;284:825–830.

Monjan AA, Gilden DH, Cole GA, Nathanson N. Cerebellar hypoplasia produced by lymphocytic choriomeningitis virus. *Science* 1971;171:194–196.

Moriyama T, Guilhot S, Klopchin K, et al. Immunobiology and pathogenesis of hepatocellular injury in hepatitis B virus transgenic mice. *Science* 1990;248:361–364.

Oldstone MBA. Molecular mimicry and autoimmune disease. *Cell* 1987;50:819–820.

Porter DD, Larsen AE, Porter HG. The pathogenesis of Aleutian disease of mink. I. In vivo viral replication and the host antibody response to viral antigen. *J Exp Med* 1969;130:575–593.

Srinivasappa J, Saegusa J, Prabakhar BS, et al. Molecular mimicry: frequency of reactivity of monoclonal antiviral antibodies with normal tissues. *J Virol* 1986;57:397–400.

Zinkernagel RM, Haenseler E, Leist T, et al. T cell-mediated hepatitis in mice infected with lymphocytic choriomeningitis virus. *J Exp Med* 1986;164:1075–1092.

Chapter 7
Virus-Induced Immunosuppression

As a general rule, infections stimulate an immune response specific for the antigens of the virus. However, in a few instances, virus infections suppress the immune response. Immunosuppression is often "global," that is, it affects responses to many antigens. However, in some instances suppression is quite specific for the infecting virus, often associated with antenatal or perinatal infections. The mechanisms of immunosuppression are diverse (Table 7.1) and include (a) replication of the virus in one of the cell types involved in immune induction, such as antigen-presenting cells (particularly macrophages or dendritic cells) or CD4 (helper) T lymphocytes, leading to the induction of cellular apoptosis or aberrant production of cytokines; (b) tolerance, often associated with fetal or newborn infection, produced by clonal deletion of T lymphocytes that respond to viral antigens; (c) perturbation of the immune response due to specific effects of individual viral proteins that act on infected or uninfected cells in a variety of ways.

Immunosuppression associated with a virus infection was first described about 100 years ago by von Pirquet, who noted that patients lost their tuberculin sensitivity (skin test reaction to the antigens of *Mycobacterium tuberculosis*) during and after measles (Fig. 7.1). With the occurrence of AIDS, virus-induced immunosuppression has attracted renewed attention and detailed investigation. Animal models of immunosuppression have provided much useful information about its dynamics and mechanisms.

MECHANISMS OF IMMUNOSUPPRESSION

Viral Infections of Monocytes and Lymphocytes

A number of viruses can replicate in cells of the lymphoreticular system. If the virus destroys a cell type that is critical to immune induction, such as antigen-presenting cells or a specific class of T lymphocytes, immune induction may be impaired resulting in immunosuppression. A few salient examples will illustrate this mechanism of suppression.

			Degree of	Virus specific
TABLE 7.1				
Some mechanisms of virus-induced immunosuppression				
Mechanism	Virus	Host	immunosuppression	or broad
Attack on one or more cells of the lymphoreticular system	HIV	Human	Marked	Broad
	LCMV	Mouse (adult)	Marked	Broad
	CDV	Dog	Marked	Broad
Fetal infection leading to tolerance	Rubella	Human	Moderate	Specific
	LCMV	Mouse	Moderate	Specific
Perturbation of cytokine homeostasis and intracellular signaling	Measles	Human	Moderate	Broad
Viral proteins acting as viroceptors or virokines	HSV	Human	Mild	Specific
	Vaccinia	Human	Mild	Specific
Suppressor T lymphocytes	FLV	Mouse	Moderate	Broad

CDV, canine distemper virus; FLV, Friend leukemia virus; HIV, human immunodeficiency virus; HSV, herpes simplex virus; LCMV, lymphocytic choriomeningitis virus.

AIDS Viruses and CD4 T Lymphocytes Human immunodeficiency virus (HIV) and related viruses (such as simian immunodeficiency virus, SIV) utilize the CD4 molecule as the primary receptor. As a result, the virus replicates in $CD4^+$ T lymphocytes and monocytes/macrophages/dendritic cells, the major cell types that express CD4 on their plasma membranes. Monocytes undergo a relatively noncytopathic infection with HIV but T lymphocytes, if they are proliferating, undergo a cytopathic infection. In addition, indirect mechanisms are involved in destroying $CD4^+$ lymphocytes, as described in Chapter 13. The depletion of CD4 lymphocytes, which act both as T-helper cells and, under some circumstances, as T-effector cells, produces functional immunodeficiency that leads to opportunistic infections (OI) and the other manifestations of AIDS (see the section on retroviruses later in this chapter for a more extensive description).

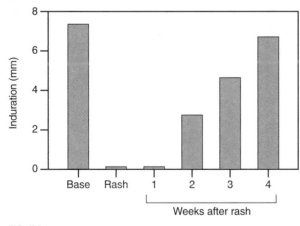

FIG. 7.1

Immunosuppression during a viral infection. The effect of acute measles infection on delayed-type hypersensitivity (tuberculin skin test). Immunosuppression occurs during the incubation period, and the immune response exhibits maximal reduction by the time the rash has developed and gradually returns to normal over then next month. Base: Tests at 1 and 2 weeks prior to measles. (After Tamashiro VG, Perez HH, Griffin DE. Prospective study of the magnitude and duration of changes in tuberculin reactivity during uncomplicated and complicated measles. *Pediatr Infect Dis J* 1987;6:451–454.)

Lymphocytic Choriomeningitis Virus and Antigen-Presenting Cells Certain strains of lymphocytic choriomeningitis virus (LCMV), experimentally selected, are quite immunosuppressive in adult immunocompetent mice. One immunosuppressive strain, clone 13, has been compared with a nonimmunosuppressive strain (Armstrong strain), from which is was derived, to elucidate the mechanism of immunosuppression. It appears that clone 13 preferentially infects macrophages and dendritic cells that are consequently destroyed, resulting in a generalized suppression of immune responses. The preferential replication of clone 13—relative to Arm-

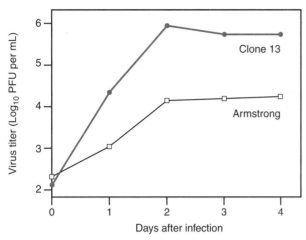

FIG. 7.2

Comparison of the in vitro replication of an immunosuppressive variant (clone 13) and an immunogenic variant (Armstrong) of lymphocytic choriomeningitis virus, showing that the immunosuppressive strain replicates to a higher titer in alveolar macrophages. (After Matloubian M, Kolkekar SR, Somasundaram T, Ahmed R. Molecular determinants of macrophage tropism and viral persistence: importance of single amino acid changes in the polymerase and glycoprotein of lymphocytic choriomeningitis virus. *J Virol* 1993;67:7340–7349.)

strong—in cultured alveolar (lung) macrophages is shown in Fig. 7.2 and in the lung in Fig. 7.3. Immunosuppressive infections often lead to virus persistence, as illustrated in Fig. 7.3, which shows that clone 13 produces a high-titer persistent viremia whereas Armstrong causes an acute viremia that is cleared within 10 days.

The suppressive effect of clone 13 infection in adult mice is demonstrated in Table 7.2, which shows that the immunogenic Armstrong strain induces a cytotoxic T lymphocyte (CTL) response whereas clone 13 does not. The suppressive effect of clone 13 is "global," that is, it compromises the immune response to other viruses as well as to LCMV. Table 7.3 shows that adult mice infected with LCMV clone 13 have markedly reduced responses to vaccinia, influenza, and herpes simplex viruses (HSVs) relative to uninfected control mice.

Tolerance: Virus as Self

Healthy animals are immunologically "tolerant" of their own antigens; that is, their own proteins do not induce antibodies or cellular responses. There are several mechanisms for the induction of tolerance, including the thymic deletion of "forbidden" clones of T lymphocytes, or anergy caused by the peripheral "exhaustion" of such forbidden clones. The potential for responding to a viral antigen as "self" is readily demonstrated in transgenic mice that express a viral antigen but do not produce virus antigen-specific antibodies or T lymphocytes. Tolerogenic virus infections are often characterized by several features: (a) they are most easily induced in fetal or newborn animals; (b) they are more readily induced if virus replication produces high levels of antigen; (c) tolerance is virus specific and responses to other viruses are not impaired; and (d) tolerance usually leads to virus persistence.

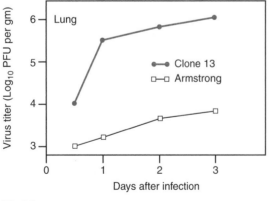

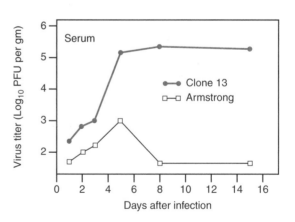

FIG. 7.3

Comparison of the in vivo replication of an immunosuppressive variant (clone 13) and an immunogenic variant (Armstrong strain) of lymphocytic choriomeningitis virus. The immunosuppressive strain replicates to a higher titer in tissues reflecting its greater ability to grow in macrophages and produces a persistent infection, whereas the immunogenic variant produces a short-lived viremia that is cleared by the immune response. (After Matloubian M, Kolkekar SR, Somasundaram T, Ahmed R. Molecular determinants of macrophage tropism and viral persistence: importance of single amino acid changes in the polymerase and glycoprotein of lymphocytic choriomeningitis virus. *J Virol* 1993;67:7340–7349.)

TABLE 7.2				
Comparison of the immune response to an immunosuppressive (clone 13) and an immunogenic clone (Armstrong) of lymphocytic choriomeningitis virus				
	PERCENT ^{51}CR RELEASE FROM TARGET CELLS INFECTED WITH:			
Infecting virus strain	No virus	Armstrong	Clone 13	LCMV antibody ($\log_2$)
Armstrong (immunogenic)	0	51	45	12.7
Clone 13 (immunosuppressive)	0	7	1	12.0

Adult mice were injected on day 0 with $10^{5.2}$ plaque-forming units of either virus and spleen cells were tested on day 8 for their ability to lyse target cells, as measured by the release of radioactive label (^{51}Cr) at an effector target ratio of 50:1. Antibody titers in serum were measured in an enzyme-linked immunosorbent assay at the same time.

LCMV, lymphocytic choriomeningitis virus.

After Ahmed R, Salmi A, Butler LD, et al. Selection of genetic variants of lymphocytic choriomeningitis virus in spleens of persistently infected mice. *J Exp Med* 1984;60:521–540.

Lymphocytic Choriomeningitis Virus and Tolerogenic Infections LCMV provides a well-studied model of tolerogenic infections in mice. Injected into newborn mice, LCMV usually produces a tolerogenic infection, as illustrated in Table 7.3. Such mice exhibit suppression of certain virus-specific immune responses with no evidence of LCMV-specific CTLs even though they produce LCMV antibodies that eventually cause immune complex glomerulonephritis (see Chapter 6). LCMV infects naïve T cells (negative for CD4 and CD8 markers) in the thymus of neonatal mice, and those T cells that carry receptors for LCMV antigens are deleted in the thymus, so that they fail to appear in the blood or peripheral lymphoid system.

Tolerogenic infections can also be induced in adult mice infected with a high dose of a tolerogenic LCMV variant. Table 7.4 and Fig. 7.4 show that a low dose (10^2 plaque-forming units, PFU) of the Docile strain of LCMV induced CTL and the virus was cleared from the spleen within a few weeks. By contrast, a high dose (10^7 PFU) of the Docile strain was tolerogenic, leading to a persistent infection. The high virus dose induced a CTL response that was even more brisk than that induced by the low virus dose, but CTL then waned rapidly and disappeared within 3 weeks of infection. This phenomenon is often called "exhaustion" and is associated with the kinetics of effector CTLs. Effector CTLs are subject to two possible fates; most of them rapidly undergo apoptosis, whereas a minority become "memory" cells that recycle slowly and can be

TABLE 7.3					
Lymphocytic choriomeningitis virus infection in newborn and adult mice					
	CTL RESPONSE (LYTIC UNITS) ON TARGET CELLS INFECTED WITH:				
Infection with clone 13 at age	Herpes simplex	Influenza	Vaccinia	LCMV	Immunosuppression
Not infected	389	586	4000	16246	None
<1 day	378	628	5769	0	LCMV specific
6 weeks	0	56	159	0	Broad

LCMV infection of mice as adults suppresses immune responses to both LCMV and other viruses, whereas infection of newborn mice suppresses responses to LCMV but not to other viruses. Mice were infected with clone 13, an immunosuppressive variant, as newborns or at 6 weeks of age and were subsequently stimulated with vaccinia, herpes simplex, influenza, or the nonimmunosuppressive Armstrong strain of LCMV. Their CTL responses were determined 7 days later against target cells infected with the stimulating virus. A lytic unit represents the number of T lymphocytes needed to lyse 100 target cells and the table shows the number of lytic units per spleen or lymph node.

LCMV, lymphocytic choriomeningitis virus; CTL, cytotoxic T-lymphocyte.

After Tishon A, Borrow P, Evans C, Oldstone MBA. Virus-induced immunosuppression. *Virology* 1993;195: 397–405.

TABLE 7.4				
Virus-specific immunosuppression in mice				
		CTL ACTIVITY, LYTIC UNITS (X 1000) PER SPLEEN		
		ON TARGET CELLS INFECTED WITH:		
LCMV (Docile strain)	Vaccinia virus	LCMV	Vaccinia virus	Uninfected
10^2 PFU	Not infected	228	ND	ND
Not infected	Infected	ND	31	0.3
10^7 PFU	Not infected	<0.2	ND	<0.2
10^7 PFU	Infected	<0.2	26	0.1

Virus-induced high-dose tolerance produces virus-specific immunosuppression. Adult mice were infected intravenously with a tolerogenic (Docile) strain of LCMV at two doses: a low dose (10^2 PFU) that was immunogenic and a high dose (10^7 PFU) that was tolerogenic. Some animals were also infected with vaccinia virus to test their immune responsiveness. Spleen cells were harvested 20 days after infection (immunization) and were tested for pCTLs in a limiting dilution assay (see Chapter 5 for technical details). The LCMV-tolerized mice raised a vaccinia response similar to mice that were not infected with LCMV, indicating that tolerance was virus-specific.

ND, not done; LCMV, lymphocytic choriomeningitis virus; PFU, plaque-forming unit; pCTL, cytotoxic T-lymphocyte.

After Moskophidis D, Lechner F, Pircher H, Zinkernagel RM. Virus persistence in acute infected immunocompetent mice by exhaustion of antiviral cytotoxic effector T cells. *Nature* 1993;362:758–761.

stimulated to proliferate and differentiate into effector cells (see Chapter 5). Apparently, in the presence of excess antigen, all effector cells are driven into apoptosis and none become memory cells, leading to exhaustion of the virus-specific CTL population.

Viral Proteins That Perturb the Immune Response
A number of viruses encode proteins that can perturb the immune response. This is particularly true of larger viruses, such as herpesviruses and poxviruses,

which appear to have "captured" cellular genes that have evolved to act as "decoys" that interfere with antiviral host defenses. These viral proteins can interfere with the immune response through a wide variety of mechanisms, and some of the best defined examples are mentioned below.

Effect on Lymphocyte Proliferation In some instances, a viral protein appears to alter the ability of lymphocytes to respond to an immune stimulus. For exam-

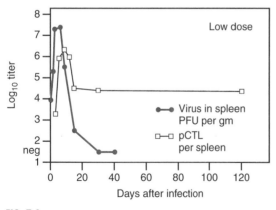 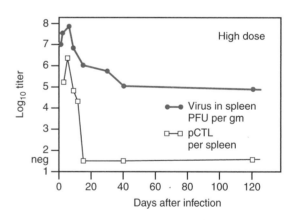

FIG. 7.4

Virus-induced immunosuppression that resembles tolerance to self antigens. Adult mice were infected intravenously with a tolerogenic (Docile) strain of lymphocytic choriomeningitis virus at two doses. The low dose (10^2 plaque-forming units, PFU) was immunogenic, and induced long-lasting CTLs that cleared the infection. The high dose (10^7 PFU) induced rapid development of precursor cytolytic T cells, but these had totally disappeared by day 15 after infection due to the "exhaustion" of these "forbidden" clones, resulting in a tolerogenic persistent infection. (After Moskophidis D, Lechner F, Pircher H, Zinkernagel RM. Virus persistence in acute infected immunocompetent mice by exhaustion of antiviral cytotoxic effector T cells. *Nature* 1993;362:758–761.)

ple, many retroviruses encode the envelope protein p15E, which reduces the proliferative response of T lymphocytes and the activation of monocytes. The effect has been mapped to a small domain within p15E and can be simulated by peptides that have the same immunodepressing sequence.

Effect on Complement As described in Chapter 5, after antibody has bound to its cognate antigen, the Fc portion of the antibody molecule undergoes a conformational change that results in the binding of the C1q complement protein to the Fc domain, initiating the complement cascade. The complement cascade, in turn, can lead to virolysis or cytolysis of virus-infected cells, thereby acting as a host defense. HSV encodes two envelope glycoproteins, gE and gI, that can bind to the Fc moiety of the immunoglobulin (Ig) molecule. This bipolar bridging prevents the secondary effects of antibody binding, such as the activation of the complement cascade (Fig. 7.5). The biological effect of the gE protein can be demonstrated by a comparison of cells infected with wild-type HSV (gE positive) and cells infected with gI- or gE-deficient mutants, showing that the latter are relatively resistant to complement-mediated cytolysis (Table 7.5). In addition, HSV can block the complement cascade in a second way, through another one of its envelope proteins, gC, which binds C3b, one of the intermediates in the cascade (see Chapter 5).

Vaccinia virus, a poxvirus, interferes with the complement system in a different way, by blocking another element in the complement cascade. Vaccinia virus encodes VCP, a vaccinia complement

TABLE 7.5

Comparison of cells infected with wild-type herpes simplex virus and cells infected with gI- or gE-deficient mutants

	SEVERITY OF VIRUS-INDUCED LESIONS		
Strain of HSV	Epithelium of cornea	Corneal stroma	Brain (encephalitis)
Wild type	Severe	Moderate	Fatal
gI negative	Moderate	None	None
gE negative	Minimal	None	None

Viral proteins can perturb the host immune response and increase the pathogenicity of the virus. Wild-type herpes simplex virus is compared with two mutants—gE-negative and gI-negative—that lack the ability to bind the Fc domain of immunoglobulin molecules. Both mutants replicate well in cell culture but have reduced pathogenicity relative to the parent virus. After topical infection of the cornea (surface of the eye) of mice, the epithelium of the cornea was examined 2 days after infection, and the stroma (substance of the cornea) 14 days after infection. HSV, herpes simplex virus; Fc, immunoglobulin fragment c.

After Dingwell KS, Brunetti CR, Hendricks RL, et al. Herpes simplex virus glycoproteins E and I facilitate cell-to-cell spread *in vivo* and across junctions of cultured cells. *J Virol* 1994;68:834–845.

control protein that resembles the human plasma protein C4-BP. The function of C4-BP is to monitor the complement cascade and prevent its unwanted activation, by binding to C4b, one of the intermediaries in the cascade (see Chapter 5). VCP vitiates complement-mediated host defenses by binding C4b and blocking the complement cascade. Its impact can be shown by comparison of wild-type vaccinia virus with VCP-negative mutants that are less pathogenic in vivo (Fig. 7.6).

Down-regulation of MHC Class I Molecules As described in Chapter 5, CD8 cytolytic effector cells, through their T-cell receptors, recognize viral antigens presented as peptides on class I molecules, leading to lysis of virus-infected cells. Poxviruses, adenoviruses, and AIDS viruses encode proteins that down-regulate the expression of class I molecules, which renders virus-infected cells less susceptible to cytolytic T lymphocytes.

The E1A protein of oncogenic adenoviruses reduces transcription of class I molecules. Some adenoviruses are oncogenic in animals and others are not, and the oncogenic adenoviruses down-regulate major histocompatibility complex (MHC) class I expression. The nononcogenic viruses lack this down-regulating activity; they transform cells in culture, but will not produce tumors in normal mice although they are oncogenic in immunocompromised animals (see Chapter 4). Adenovirus also encodes another

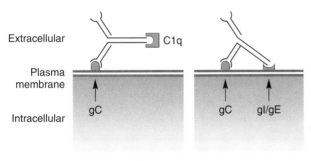

FIG. 7.5

Diagram to illustrate the binding of immunoglobulin (Ig) to the surface of herpes simplex virus-infected cells via glycoprotein C (gC), a viral envelope protein. Two other viral glycoproteins, gE and gI, bind the Fc domain of Ig and prevent the binding of C1q. If cells are infected with mutants defective for gE and gI, the Fc domain is available to bind C1q and initiate the complement cascade.

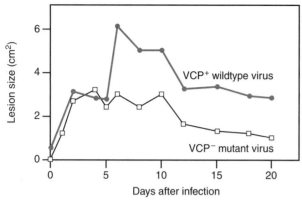

FIG. 7.6

Cell-derived viral genes can perturb the immune response by blocking the complement cascade. In this example, a comparison is made of wild-type vaccinia virus with a mutant that lacks the vaccinia complement control protein (VCP). Rabbits received intradermal injections of 10^6 plaque-forming units (PFU) of the two viruses, and the size of the lesions were compared (medians for three animals in each group). The lesions produced by the VCP-negative mutant were consistently smaller—beginning at 5 days after infection—suggesting that VCP interfered with the ability of the host to contain the vaccinia-induced skin lesions. (After Isaacs SN, Kotwal GJ, Moss B. Vaccinia virus complement control protein prevents antibody-dependent complement-enhanced neutralization of infectivity and contributes to virulence. *Proc Natl Acad Sci USA* 1992;89:628–632.)

protein, E3/19K, which traps MHC class I molecules in the endoplasmic reticulum so that they never reach the cell surface. E3 is a transmembrane protein and its cytoplasmic tail encodes a small domain that retains it in the endoplasmic reticulum while its exodomain encodes a sequence that binds to class I molecules.

Interference with Cytokine Functions Poxviruses produce a number of proteins that act as "viroceptors." These proteins resemble and compete with cellular receptors of the host, binding the cytokine and reducing its physiologic activity. Interferon γ (IFN-γ) or "immune" interferon is produced by antigen-specific T lymphocytes as part of their cytokine response to antigen (see Chapter 5). Furthermore, IFN-γ can play a vital role as an immune defense, as demonstrated by treatment with monoclonal antibodies against IFN-γ that can convert an innocuous into a lethal viral infection in some experimental models (Table 7.6). Since all IFN responses are initiated by binding of IFNs to their cognate cellular receptors, proteins that mimic the IFN-γ receptor can sequester IFN and prevent its binding to the cellular receptor. Many poxviruses have been shown to en-

code a protein (B8R) that is homologous with the cellular IFNγ receptor, and this protein binds IFN and can abrogate the protective effect of IFN in a cell culture system (Fig. 7.7).

ANIMAL VIRUSES ASSOCIATED WITH IMMUNOSUPPRESSION

A number of animal viruses—from many different virus families—can cause immunosuppression of varying degrees of severity and specificity (Table 7.7). Several of these that have served as important models to elucidate the mechanisms of suppression are described below.

Lymphocytic Choriomeningitis Virus

LCMV is an arenavirus that occurs in nature as an indigenous virus of wild mice and has served as an important experimental model for the study of interactions between the immune system and viral infection (see Chapter 6). As described earlier in this chapter, some strains of LCMV are immunosuppressive under certain experimental conditions, and a study of the model elucidates some of the variables that are determinants of immunosuppression.

When adult mice are infected intravenously with LCMV, the outcome depends on the strain of virus used. Table 7.2 illustrates that the Armstrong strain of LCMV induces a virus-specific cellular immune response (CTLs) but clone 13, originally derived from Armstrong virus, is immunosuppressive and fails to induce CTLs, although both

TABLE 7.6			
IFN-γ may mediate CTL control of virus infection			
Spleen cells (day 1)	Treatment with Anti-IFN-γ antibody (days 1–3)	VIRUS TITER (LOG$_{10}$) DAY 4	
		Lungs	Ovaries
No cells	None	4.10	7.49
Naïve cells	None	4.25	7.52
Immune cells	Control antibody	<2.00	4.62
Immune cells	Anti-IFN-γ	4.08	7.02

IFN-γ is essential for the protective effect of virus-specific CD8 T lymphocytes. In this protocol, mice were infected with vaccinia virus on day 0, were adoptively immunized with spleen cells on day 1, were treated with anti-IFN-γ antibody (or control fluid) on days 1–3, and were assayed for virus in lungs and ovaries on day 4. Treatment with monoclonal antibody against IFN-γ abrogates the protective effect of immune spleen cells (spleen cells obtained from mice previously infected with vaccinia virus). IFN, interferon.

After Ruby J, Ramshaw I. The antiviral activity of immune CD8$^+$ T cells is dependent upon interferon-γ. *Lymphokine Cytokine Res* 1991;10:353–358.

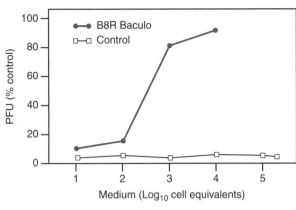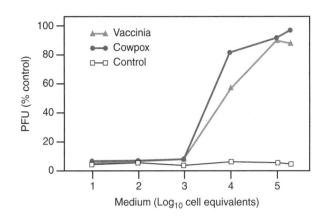

FIG. 7.7 _____

A viral protein that simulates the interferon γ receptor (IFN-γR) can block the antiviral action of IFN-γ. The B8R protein of poxviruses is homologous to the IFN-γR, and this study indicates that it can block the action of IFN-γ. The figure shows an assay in which 10 units of human interferon was used to reduce the titer of Cocal, a vesiculovirus; Cocal virus and human IFN was present in each assay. *Left panel*: The supernates from two baculovirus cultures are compared: a control baculovirus and one that was engineered to express B8R. This experiment indicates that B8R alone can account for the blocking of the interferon effect. *Right panel:* Supernates from cultures infected with two poxviruses, vaccinia and cowpox, are compared with the supernate from a mock-infected culture to show that the media from poxvirus-infected cultures blocks the effect of human interferon, presumably due to the presence in the media of the B8R protein. (After Alcami A, Smith GL. Vaccinia, cowpox, and camelpox viruses encode soluble gamma interferon receptors with novel broad species specificity. *J Virol* 1995;69:4633–4699.)

viruses induce similar antibody responses. Associated with immunosuppression, clone 13 induces a persistent infection in contrast to the immunogenic Armstrong strain that is rapidly cleared (Fig. 7.3). In this model, immunosuppression appears to be due to the differential ability of the immunosuppressive clone 13 variant to replicate in macrophages, both in vivo (Fig. 7.3) and in vitro (Fig. 7.2). Apparently, infected macrophages and dendritic cells are destroyed by virus-specific CTLs, reducing the efficiency of antigen presentation.

TABLE 7.7			
Human and animal viruses that cause immunosuppression: selected examples			
Virus family	Virus *Disease*	Cells infected (lymphoreticular)	Immune manifestation
Human viruses			
Paramyxoviridae	Measles	Monocytes	Reduced DTH
	Measles	Thymic epithelial cells	Enhanced infections
Togaviridae	Rubella	Lymphoid cells	Persistent rubella infection
	Rubella		
Retroviridae	HIV	CD4 T lymphocytes	Opportunistic infections
	AIDS	Monocytes	Enhanced neoplasia
Animal viruses			
Arenaviridae	LCMV	Monocytes	Persistent LCMV infection
	Choriomeningitis	T lymphocytes	
Paramyxoviridae	CDV	Monocytes	Encephalitis
	Canine distemper	Lymphocytes	Bacterial superinfections
	Rinderpest	Monocytes	Lethal gastroenteritis
	RV	Lymphocytes	
Retroviridae	SIV	CD4 T lymphocytes	Opportunistic infections
	AIDS	Monocytes	
	MuLV	B lymphocytes	B-, T-cell dysfunctions
	(defective variant)		Opportunistic infections
	MAIDS		

AIDS, acquired immunodeficiency virus; CDV, canine distemper virus; DTH, delayed-type hypersensitivity; FPV, feline panleukopenia virus; HIV, human immunodeficiency virus; LCMV, lymphocytic choriomeningitis virus; MAIDS, murine AIDS; MuLV, murine leukemia virus; RV, rinderpest virus; SI, stimulation index.

Morbilliviruses: Canine Distemper Virus

Canine distemper virus (CDV) produces an acute febrile and highly fatal disease in dogs and a number of other species of carnivores. Death is due to severe encephalitis, pneumonitis, gastroenteritis, and bacterial superinfections, with hemorrhage and dehydration. A similar virus, rinderpest virus, infects cattle and can cause devastating epidemics with very high mortality. CDV and rinderpest virus are both morbilliviruses, closely related to measles virus.

CDV infects lymphocytes and monocytes and produces a severe leukopenia, that is, a reduction in the number of lymphocytes (both B and T cells) in the circulation and lymphoid tissues, as well as depletion of lymphoid tissues and the thymus. Functionally, there is a loss of the lymphocyte proliferative response to mitogens, and preexisting responses to specific antigens also decrease or disappear. The outcome of infection is variable, and dogs that recover develop prompt antibody and CTL responses to CDV, whereas those that die exhibit reduced or absent immune responses to the virus (Fig. 7.8). The acute suppression of immune responses plays an important role in the high mortality, reducing the ability of the host to contain CDV and also potentiating superinfections with bacteria and other microbial

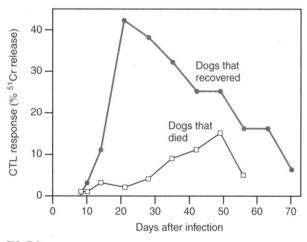

FIG. 7.8

Cellular immune responses in dogs infected with canine distemper virus (CDV). The outcome of this infection is variable, and those animals that undergo a fulminant and fatal course usually show marked depression of their cytotoxic T lymphocyte (CTL) response to CDV compared to dogs that recover. This figure compares six dogs (mean values) that developed a fatal or persistent severe infection with CDV to six dogs that recovered from infection with the same virus (Cornell A75/17 strain), using a CTL assay with autologous target cells. Because this is an assay for effector CTLs, the titers drop as the virus is cleared. (After Appel MJG, Shek WR, Summers BA. Lymphocyte-mediated immune cytotoxicity in dogs infected with virulent canine distemper virus. *Infect Immun* 1982;37:592–600.)

agents. In animals that survive CDV infection, the immune responses generally recovers in 1–3 months.

Retroviruses: Murine Leukemia Virus and Murine AIDS

Murine AIDS (MAIDS) is the name given to a disease produced in mice by a variant of murine leukemia virus (MuLV). Although the disease is called murine "AIDS," it is produced by a replication-incompetent variant of an oncogenic retrovirus and its pathogenesis is completely different from that of human and simian AIDS, which are caused by replication-competent lentiviruses. The most prominent manifestations of MAIDS are splenomegaly and lymphadenopathy with increased susceptibility to experimental infection with many microbial agents, such as ectromelia virus, *Mycobacterium avium*, and *Trypanosoma cruzi*.

The MAIDS virus is a defective variant of MuLV that was isolated in the course of experimental manipulations of laboratory-passaged retroviruses. The MAIDS virus genome lacks most of the retrovirus structural proteins and can only replicate in the presence of a nononcogenic replication-competent murine retrovirus that acts as a "helper" virus by supplying the missing proteins in trans. In other words, cells that are coinfected with the MAIDS virus and a helper virus will produce infectious virions (virus particles) that are formed of proteins encoded by the helper virus encapsulating the genome of the MAIDS virus. Such virus particles can infect target cells but—absent the helper virus—cannot produce infectious progeny virus. MAIDS virus alone can produce disease even though it cannot replicate. The virus mainly infects B lymphocytes and expresses a single protein, a variant form of Pr60^{gag} (a 60-kd protein encoded by the **gag** gene that produces the internal structural proteins of the virus particle). Pr60^{gag} is both necessary and sufficient for the production of MAIDS, binding to the plasma membrane and producing biological effects via biochemical pathways that are not thoroughly understood (see below).

B lymphocytes infected with MAIDS virus proliferate and disseminate widely into both lymphoid and nonlymphoid tissues. Although not the primary target of the MAIDS virus, CD4 T lymphocytes also proliferate excessively (perhaps due to cytokine dysregulation) and play an essential role in disease production. The B and T lymphocytes that undergo excessive proliferation are not capable of responding to neoantigens and may displace normal lymphoid cells populations, resulting in the loss of normal immune induction (sometimes called "anergy").

The detailed pathogenesis of MAIDS is not yet completely understood, and there are gaps in the sequence of events. (a) Pr60^{gag} appears to be the only product of the defective genome of the MAIDS virus, but it is not clear exactly how this leads to the hyperproliferation of B lymphocytes. (b) B-Lymphocyte proliferation initiates immune activation, including proliferation of CD4 T lymphocytes, but the cytokine pathways responsible for activation have not been thoroughly delineated. (c) Finally, the mechanisms by which immune activation interferes with normal immune induction—producing functional immune suppression—remain to be dissected.

VIRAL INFECTIONS OF HUMANS ASSOCIATED WITH IMMUNOSUPPRESSION

There are a number of human viruses that cause immunosuppression to some degree (Table 7.7). Several of the most prominent examples are discussed below.

Measles

Measles is the prototype of an acute virus infection that produces global immunosuppression. Figure 7.1 shows the tuberculin skin test response before, during, and after measles virus infection. Suppression develops during the 10-day incubation period and is most pronounced during the clinical phase of disease, which lasts about a week. Recovery occurs within 3–4 weeks after the rash, and the tuberculin response is essentially normal by 1 month after the acute illness. During the period of measles-induced immune suppression there is an increased susceptibility to other infections, a transient potential for exacerbation of chronic infections such as tuberculosis, and remission of autoimmune diseases such as juvenile rheumatoid arthritis and the nephrotic syndrome.

The mechanisms of measles-induced suppression are complex and probably involve several different cell types and pathways. Monocytes are infected in measles and show aberrant cytokine responses (interleukin-1 production is increased and tumor necrosis factor α production is decreased) compared with uninfected cells. Although blood lymphocytes do not appear to be infected during measles, the numbers of circulating T lymphocytes (CD4 and CD8) are decreased by about 50%. When examined ex vivo, T lymphocytes from acutely in-

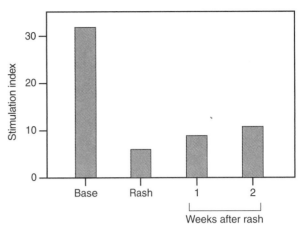

FIG. 7.9

Measles virus infection abrogates the proliferative response of T lymphocytes to mitogens. Peripheral blood mononuclear cells were cultured with mitogen (phytohemagglutinin, a lectin that stimulates most T lymphocytes to divide) and their incorporation of ^{3}H-thymidine determined in comparison with replicate cultures absent mitogen. (After Hirsch RL, Griffin DE, Johnson RT, et al. Cellular immune responses during complicated and uncomplicated measles virus infections of man. *Clin Immunol Immunopathol* 1984;31:1–12.)

fected patients show a decreased proliferative response to mitogens (phytohemagglutinin and other lectins that cause most T cells to divide) (Fig. 7.9). The fact that relatively few monocytes and lymphocytes are infected suggests that indirect mechanisms play a role in immunosuppression.

Cell cultures of monocytes and lymphocytes can be infected with measles virus, and the infection alters their physiologic responses although it does not kill the cell population. When CD4$^+$ T lymphocytes are exposed to cells (or measles virus) expressing the F (fusion) and H (hemagglutinin) proteins, the T cells lose their ability to respond to mitogens even at a ratio of infected to uninfected cells of 1:100. Inhibition is not mediated by soluble factors and requires contact between infected and uninfected cells. The effect entails both F and H proteins and cleavage of the F_0 to F_1 and F_2 peptides. Binding of the F/H complex to a measles virus receptor on the lymphocyte is hypothesized to trigger an intracellular signally pathway that leads to inhibition.

In summary, it appears that the function of antigen-presenting cells (monocytes and macrophages), helper cells (CD4 T lymphocytes), and effector cells (CD8 T lymphocytes) is compromised during measles. A critical aspect of suppression appears to be triggering of an inhibitory pathway in T lymphocytes by the binding of the cleaved F/H complex of measles virus to a lymphocyte receptor.

Rubella

Rubella (German measles) is an example of a virus that only causes immunosuppression in utero or during infancy. In children or adults, rubella is an acute infection, producing a brisk immune response that clears the virus within a few weeks. If primary infection occurs during pregnancy, the virus can cross the placenta and infect the fetus. Fetal infections—which often produce the congenital rubella syndrome, that is, developmental malformations—usually persist throughout pregnancy and are not cleared until about age 6–12 months. Infants or children with the congenital rubella syndrome exhibit markedly diminished cellular immune responses to rubella antigens even when tested many years after in utero infection (Fig. 7.10). However, such infants raise brisk antibody responses to rubella. IgM antirubella antibodies are present at birth and wane from 6–12 months, concomitant with the clearing of the virus, whereas maternal IgG antibodies are predominant from birth to 6 months and are then replaced by the infant's antirubella IgG that is long lasting. It appears that gestational infection with rubella virus leads to a suppression of the cellular

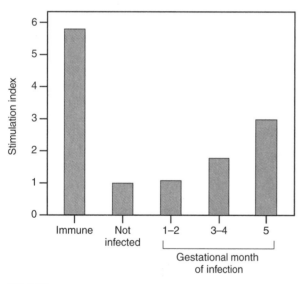

FIG. 7.10

Defective cell-mediated immune responses associated with fetal infection. Children with the congenital rubella syndrome who had been infected at different gestational ages are compared with children infected postnatally and with uninfected children in a lymphocyte proliferation assay using rubella virus as antigen. The children with congenital rubella syndrome exhibit reduced proliferation responses even though they were tested up to 10 years of age. (After Buimovici-Klein E, Lang PB, Ziring PR, Cooper LZ. Impaired cell-mediated immune response in patients with congenital rubella: correlation with gestational age at time of infection. *Pediatrics* 1979;64: 620–626.)

but not the humoral immune response and that the failure to clear virus is associated with suppression of cellular immunity.

This example of immunosuppression associated with fetal infection is reminiscent of LCMV infection of fetal or newborn mice (Table 7.3) and is a human example of tolerance induced by a foreign antigen present during fetal life. However, the exact mechanism of virus-specific suppression (clonal deletion in the thymus, peripheral exhaustion of committed T lymphocytes, or other) has not been elucidated.

HIV and AIDS

HIV is the best-studied example of a human virus infection associated with immunosuppression. The pathogenesis of AIDS is the subject of Chapter 13, and the following sketch focuses on the mechanisms of immunosuppression. The clinical features of AIDS are associated with the loss of host immune defenses and the occurrence of OI or neoplasms. Most OIs, such as *Pneumocystis carinii* infection (a parasitic lung infection), cytomegalovirus retinitis, or tuberculosis, preexist AIDS and represent activation of persistent infections that are latent or contained. Some AIDS-associated neoplasms, such as lymphoma, are seen in immunologically intact individuals but at a lower incidence than in AIDS, whereas others, such as Kaposi's sarcoma, are rare except in patients with AIDS. Cellular immune responses to opportunistic infectious agents are reduced in patients with AIDS in comparison with normal subjects, whereas antibody titers are generally maintained until late in the illness. When patients with OIs receive highly active antiretroviral treatment, a remarkably rapid clearing of the OIs is often seen, implying a return of the effector component of cellular immunity.

The CD4 molecule is the primary receptor for HIV, and the virus mainly infects T lymphocytes and monocytes (and macrophages), the two main cell types that express CD4. HIV is cytolytic for T lymphocytes but not for macrophages, and immunodepression in AIDS is due to the loss of CD4 T lymphocytes whereas monocyte function is not severely compromised (Fig. 7.11). In HIV-infected persons, the reduction of the CD4 lymphocyte population is a complex process that usually occurs gradually over many years. At any time, only a small proportion of CD4 cells are infected, and cytolysis is limited to those infected T cells that are actively proliferating, whereas the virus is relatively latent in quiescent infected T cells. In the first months or

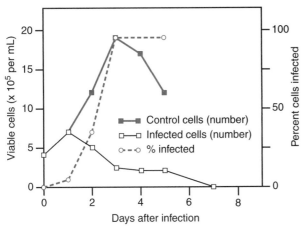

FIG. 7.11

Human immunodeficiency virus (HIV) destroys CD4 T lymphocytes in cell culture. In this example, a T-cell line (MT-4) derived from cord blood was infected with a high multiplicity of HIV and the culture followed to determine the number of viable cells and the proportion infected. The infection spread to 100% of cells in about 3 days, with rapid destruction of the culture, compared with an uninfected sister culture. (After Harada S, Koyanagi Y, Yamamoto N. Infection of HTLV-III/LAV in HTLV-I-carrying cells M-2 and MT-4 and application in a plaque assay. *Science* 1985;563–566.)

years after infection, the destruction of CD4 lymphocytes increases but there is a compensatory increase in the production of naïve T cells from the bone marrow, so that total CD4 lymphocyte concentration in peripheral blood does not decrease. Eventually, destruction exceeds production, and the concentration of circulating CD4 cells decreases. When CD4 counts drop below a critical threshold, patients begin to develop OI. In uninfected normal humans, the concentration of CD4 lymphocytes in peripheral blood is in the range 1,000–1,500/μL and OI are associated with CD4 levels less than 100–400/μL.

CD4 T lymphocytes act as "helper" cells for the induction of both cellular (CD8 cytolytic T cells) and humoral (antibody) immune responses (see Chapter 5). Thus, HIV-infected patients are impaired in their ability to develop de novo immune responses and are particularly susceptible to new infections against which they have no immunity. In addition, persistent latent infections that are contained by cellular immunity, such as tuberculosis, may reactivate. Mature humoral responses that predate the onset of AIDS tend to remain relatively intact because long-lived plasma cells continue to produce antibody without the need for helper T cells.

These observations appear to represent a paradox. Clinically, AIDS usually presents as the activation of preexisting infections, representing a failure of the effector limb of the cellular immune response—mediated by CD8 cells—to contain long-standing latent OIs. But CD4 cells—not CD8 cells—are the direct target of the virus. Likely, the failure of the effector limb of the immune response is partly due to the loss of the helper functions provided by CD4 cells, and partly to dysregulation of CD8 cells, a matter that is discussed in more detail in Chapter 13 on HIV and AIDS.

DETERMINANTS OF IMMUNOSUPPRESSION

The ability of a virus to initiate immunosuppression depends on some rather subtle interactions with the lymphoreticular system. Therefore, it is not surprising that a number of variables such as virus strain, virus dose, route of injection, age of the host, and extraneous immunosuppression will determine whether a virus induces an immunogenic or an immunosuppressive response.

Virus Strain: Lymphocytic Choriomeningitis Virus

The difference between strains of LCMV has already been discussed. In adult mice, the Armstrong strain of LCMV causes an acute, short-term, rapidly cleared infection and is not immunosuppressive, whereas the clone 13 variant causes a persistent infection associated with immunosuppression. This difference is associated with the ability of clone 13 to replicate in and destroy macrophages and dendritic cells that act as "professional" antigen-presenting cells. The immunosuppressive property of clone 13 has been mapped to two amino acids, in the viral envelope protein and the viral polymerase, respectively.

Virus Strain: Simian Immunodeficiency Virus

SIV, like HIV, produces immunodepression by depleting CD4 T lymphocytes. As described above, there is an unstable equilibrium between destruction and replacement of CD4 lymphocytes. SIV strains that are engineered to delete their *nef* gene, a regulatory protein, maintain their replicative capacity in cultured CD4 lymphocytes but are much less immunosuppressive in rhesus macaques than wild-type SIV. The differences are due to a function of the nef protein that down-modulates the expression of MHC class I molecules on the surface of infected lymphocytes (see Chapter 4), so that infected cells are less susceptible targets for CTL attack. Although this has no effect in cell culture, it does have a dra-

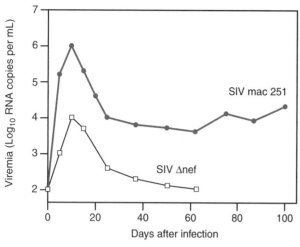

FIG. 7.12 _____

Viral genes that influence immune clearance can alter in vivo infection. Simian immunodeficiency virus (SIV) mac251 wild-type is compared with a mutant engineered to delete the *nef* gene, a regulatory gene that reduces expression of major histocompatibility complex class I molecules on the plasma membrane, permitting escape of virus-infected cells from immune attack. The figure shows the typical course of viremia in individual animals infected with wild-type and Δ nef viruses. (After Connor RI, Montefiori DC, Binley JM, et al. Temporal analyses of virus replication, immune responses, and efficacy in rhesus macaques immunized with a live attenuated simian immunodeficiency virus vaccine. *J Virol* 1998;72:7501–7509.)

matic effect on virus titers in vivo that, in turn, influence pathogenicity (Fig. 7.12).

Age and Immunosuppression

The LCMV model also illustrates that the mechanism of immunosuppression is different in newborn and adult mice. LCMV infection of adult mice, described above, can result in a "global" immunodepression that affects responses both to LCMV and to other viruses. On the other hand, LCMV infection of newborn mice results in suppression of responses to LCMV but not to other viruses. This striking difference is illustrated in Table 7.3, and it probably reflects different mechanisms of suppression. It has already been suggested that, in adult mice, clone 13 infects and destroys macrophages (and dendritic cells), thereby interfering with processing and presentation of many antigens. In newborn mice, clone 13 produces a very widespread infection of many organs and tissues. The virus is treated as a "self" antigen with the induction of "tolerance" by clonal deletion of LCMV-specific T lymphocytes, by the same mechanism that produces tolerance to self antigens.

Other Determinants of Immunosuppression

A number of other variables, such as dose of virus, route of injection, and chemical or physical im-

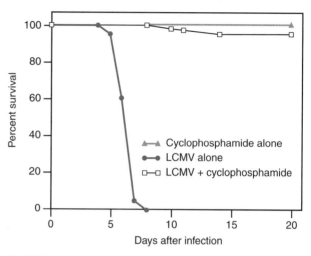

FIG. 7.13 _____

Chemical or physical immunosuppressive treatment can initiate prolonged immunosuppression with certain viruses. Mice injected intracerebrally with lymphocytic choriomeningitis virus (Armstrong strain) raise a brisk cytotoxic T lymphocyte (CTL) response that causes a lethal choriomeningitis. If the animals are treated with a single dose of cyclophosphamide (150 mg/kg), on day 3 of infection, most of them survive. The survivors fail to raise antiviral CTLs and develop a persistent infection. (After Gilden DH, Cole GA, Nathanson N. Immunopathogenesis of acute central nervous system disease produced by lymphocytic choriomeningitis virus. I. Cyclophosphamide-mediated induction of the virus-carrier state in adult mice. *J Exp Med* 1972;135:860–872.)

munosuppressive treatments, can influence whether a virus causes immunosuppression. In some circumstances, a large virus dose produces "high-dose tolerance" whereas a small inoculum initiates an immunogenic response (Table 7.4). Intravenous virus injection may initiate immunosuppression, whereas intracerebral infection is immunogenic. Treatment with X irradiation or the immunosuppressive drug cyclophosphamide can convert an immunogenic infection to an immunosuppressive one (Fig. 7.13).

REPRISE

Most viral infections induce immune responses to the viral antigens, but under some circumstances a virus can be immunosuppressive. Virus-induced immunosuppression is often "global," affecting the response to many antigens, but in some instances suppression is limited to the infecting agent. Immunosuppression can be induced through several mechanisms, including (a) destruction of monocytes (or other antigen-presenting cells) or subsets of T lymphocytes; (b) infection of fetal or newborn animals leading to tolerance; (c) perturbation of cytokine homoeostasis; or (d) viral proteins that may act as virokines (simulating cytokines) or viroceptors (simulating cellular receptors). Virus-induced immunosuppression reflects the delicate balance between host and parasite, and is influenced by a number of variables such as strain of virus, dose and route of infection, and immunocompetent status of the host at the time of infection.

FURTHER READING

Reviews and Chapters

Buchmeier MJ, Zajac AJ. Lymphocytic choriomeningitis virus. In: Ahmed R, Chen I, eds. *Persistent virus infections*. New York: John Wiley and Sons, 1999.

Ricciardi RP. Adenovirus transformation and tumorigenicity. In: Seth P, ed. *Adenoviruses: basic biology to gene therapy*. Austin, TX: RG Landes, 1999:219–228.

Original Contributions

Ahmed R, Salmi A, Butler LD, et al. Selection of genetic variants of lymphocytic choriomeningitis virus in spleens of persistently infected mice. *J Exp Med* 1984;60:521–540.

Alcami A, Smith GL. Vaccinia, cowpox, and camelpox viruses encode soluble gamma interferon receptors with novel broad species specificity. *J Virol* 1995;69:4633–4699.

Appel MJG, Shek WR, Summers BA. Lymphocyte-mediated immune cytotoxicity in dogs infected with virulent canine distemper virus. *Infect Immun* 1982;37:592–600.

Avota E, Avots A, Niewiesk S, et al. Disruption of Akt kinase activation is important for immunosuppression induced by measles virus. *Nat Med* 2001;7:725–731.

Borrow P, Evans CF, Oldstone MBA. Virus-induced immunosuppression: immune system–mediated distribution of virus-infected dendritic cells results in generalized immune suppression. *J Virol* 1995;69:1059-1070.

Buimovici-Klein E, Lang PB, Ziring PR, Cooper LZ. Impaired cell-mediated immune response in patients with congenital rubella: correlation with gestational age at time of infection. *Pediatrics* 1979;64:620–626.

Connor RI, Montefiori DC, Binley JM, et al. Temporal analyses of virus replication, immune responses, and efficacy in rhesus macaques immunized with a live attenuated simian immunodeficiency virus vaccine. *J Virol* 1998;72:7501–7509.

Dingwell KS, Brunetti CR, Hendricks RL, et al. Herpes simplex virus glycoproteins E and I facilitate cell-to-cell spread in vivo and across junctions of cultured cells. *J Virol* 1994;68:834–845.

Dubin G, Socolof E, Frank I, Friedman HM. Herpes simplex virus type 1 Fc receptor protects infected cells from antibody-dependent cellular cytotoxicity. *J Virol* 1991;65:7046–7050.

Gilden DH, Cole GA, Nathanson N. Immunopathogenesis of acute central nervous system disease produced by lymphocytic choriomeningitis virus. I. Cyclophosphamide-mediated induction of the virus-carrier state in adult mice. *J Exp Med* 1972;135:860–872.

Harada S, KoyanagI Y, Yamamoto N. Infection of HTLV-III/LAV in HTLV-I-carrying cells M-2 and MT-4 and application in a plaque assay. *Science* 1985;229:563–566.

Hirsch RL, Griffin DE, Johnson RT, et al. Cellular immune responses during complicated and uncomplicated measles virus infections of man. *Clin Immunol Immunopathol* 1984;31:1–12.

Isaacs SN, Kotwal GJ, Moss B. Vaccinia virus complement control protein prevents antibody-dependent complement-enhanced neutralization of infectivity and contributes to virulence. *Proc Natl Acad Sci USA* 1992;89:628–632.

Iwashiro M, Messer R, Peterson KE, et al. Immunosuppression by CD4$^+$ regulatory T cells induced by chronic retroviral infection. *Proc Natl Acad Sci USA* 2001;98:9226–9230.

Matloubian M, Kolkekar SR, Somasundaram T, Ahmed R. Molecular determinants of macrophage tropism and viral persistence: importance of single amino acid changes in the polymerase and glycoprotein of lymphocytic choriomeningitis virus. *J Virol* 1993;67:7340–7349.

Moskophidis D, Lechner F, Pircher H, Zinkernagel RM. Virus persistence in acute infected immunocompetent mice by exhaustion of antiviral cytotoxic effector T cells. *Nature* 1993;362:758–761.

Paabo S, Bhat BM, Wold WSM, Peterson PA. A short sequence in the COOH-terminus makes an adenovirus membrane glycoprotein a resident of the endoplasmic reticulum. *Cell* 1987;50:311–317.

Ruby J, Ramshaw I. The antiviral activity of immune CD8+ T cells is dependent upon interferon-γ. *Lymphokine Cytokine Res* 1991;10:353–358.

Tamashiro VG, Perez HH, Griffin DE. Prospective study of the magnitude and duration of changes in tuberculin reactivity during uncomplicated and complicated measles. *Pediatr Infect Dis J* 1987;6:451–454.

Tishon A, Borrow P, Evans C, Oldstone MBA. Virus-induced immunosuppression. *Virology* 1993;195:397–405.

Weidman A, Maisner A, Garten W, et al. Proteloytic cleavage of the fusion protein but not membrane fusion is required for measles virus-induced immunosuppresion in vitro. *J Virol* 2000;74:1985–1993.

PART III
Virus–Host Interactions

12. Host Susceptibility to Viral Diseases

Genetic differences, particularly in mice, in susceptibility to specific viruses; nonimmmune and immune mechanisms of resistance and susceptibility; determinants of susceptibility in humans

13. HIV and AIDS

HIV receptors and coreceptors, CD4 T lymphocytes and macrophages as target cells, direct and indirect destruction of T lymphocytes; transmission of HIV; sequential steps in infection and disease evolution; static and dynamic aspects of virus replication and cellular replacment; HIV variation; mechanisms of CD4 depletion and of immunosuppression

Chapter 8
Viral Virulence

Virulence (or pathogenicity) refers to the ability of a virus to cause disease in an infected host. Virulence is a variable biological property of the virus, and different clones or strains of a single virus can differ widely in their pathogenicity. Virulence is central to the study of pathogenesis because variants of a single virus exhibit different patterns of pathogenesis, a consideration that if overlooked can confound studies of pathogenesis. Furthermore, virulence has important practical implications, as avirulent (also called attenuated) variants of a virus can be used as live virus vaccines. Examples are vaccines for smallpox, poliomyelitis, measles, and yellow fever.

There are two distinct yet complementary approaches to the study of virulence. Virulence can be considered a property of the virus, in which case it is possible to utilize genetic methods to define the role of specific viral genes and proteins in determining pathogenic phenotypes. This approach can be extended to the molecular and structural level because in some instances single amino acids can exert a marked influence on pathogenicity. For instance, variation in a single amino acid in the envelope of a

viral protein can determine how readily it is cleaved by a cellular protease, and this property, in turn, influences the range of tissues that can be infected and the corresponding disease pattern.

The other approach to virulence is to compare pathogeneses of virulent and avirulent virus strains to elucidate the biological mechanisms underlying their phenotypes. Variation in pathogenicity must be regarded as a multidimensional phenomenon that is both quantitative and qualitative. Qualitative variation may be manifested in various ways. For instance, viral clones may exhibit differences in their tropism, so that one clone replicates well in the brain whereas another clone replicates well in the liver or intestinal tract. Alternatively, different clones may spread by different routes, with one clone producing viremia and another clone spreading by the neural route. In some instances, viral clones vary in the immune responses that they induce, or in their susceptibility to antibody or to cellular immune defenses. For any given expression of pathogenicity, it is possible to quantify virulence by comparing, for different viral strains, the number of infectious units required to produce a specified outcome, such as mortality.

In summary, virulence is intimately intertwined with pathogenesis. On the one hand, it is important to understand the characteristics of the viral strain that will be used to characterize the pathogenesis of a given virus and to recognize that the pathogenesis might vary for another strain of the same virus. On the other hand, it is critical to specify which aspect of pathogenesis will be evaluated in order to compare the virulence of different viral strains because relative pathogenicity depends upon the parameter that is assessed. These points are illustrated below.

MEASUREMENT OF VIRULENCE

Relative Nature of Virulence

Virulence is not an absolute property of a virus, but depends on many variables, such as viral strain, route of infection, and dose of virus, as well as species, age, and genetic susceptibility of the host. The relative nature of virulence is illustrated in Table 8.1, which compares a virulent wild-type strain of La Crosse virus with an attenuated mutant clone (B.5) derived from the wild-type strain. Suckling mice are highly susceptible to intracerebral injection of both viruses (1 PFU [plaque-forming unit] will cause a fatal infection), and there is no difference between the two viruses. However, after subcutaneous injection of suckling mice, the wild-type virus retains its pathogenicity whereas clone B.5 is much less virulent due to its minimal ability to produce a viremia and reach the brain. La Crosse virus is virulent for adult mice after intracerebral injec-

tion, but clone B.5 is attenuated because of its reduced ability to replicate in neurons. After subcutaneous injection of adult mice, neither virus will cause illness at any dose, so the two viruses appear equally innocuous. To profile the differences between the virulent and attenuated viruses, it is necessary to choose a combination of host and route that is between the most susceptible end of the scale (intracerebral injection of suckling mice) and the most resistant end of the scale (subcutaneous injection of adult mice).

Measures of Virulence

There are many measures that can be used to quantify virulence or attenuation. Most commonly, death or a constellation of symptoms and signs is used, such as paralysis (poliovirus), jaundice (hepatitis viruses), rash (measles), and the like. Many viruses cause a large number of inapparent (asymptomatic) infections as well a few cases of overt illness. For such viruses, the case/infection ratio can be used as a measure of virulence. Table 8.2 compares the ratio for wild-type poliovirus to that for attenuated poliovirus vaccine (oral poliovirus vaccine, OPV), showing that the vaccine virus is about 10,000-fold more attenuated than wild-type virus.

The severity of illness or the incubation period can also be used as a measure of virulence. Domestic rabbits were introduced into Australia in the early 20th century and escaped captivity to become feral. Due to a lack of natural predators, the rabbit population expanded astronomically and threatened farming in some regions of the country. To control these pests, myxoma virus, a naturally occurring poxvirus of wild rabbits (but not of domestic rabbits), was introduced and caused a rapid die-off of the popula-

---

TABLE 8.1
Comparison of two variants of La Crosse virus

La Crosse virus strain	INTRACEREBRAL INFECTION PFU PER LD_{50}		SUBCUTANEOUS INFECTION PFU PER LD_{50}	
	Suckling mice	Adult mice	Suckling mice	Adult mice
Virulent wild type	~1	~1	~1	>10^7
Attenuated clone B.5	~1	>10^6	>10^5	>10^7

Virulence can be expressed as the number of infectious units (PFU) required to kill 50% of animals (ratio $\log_{10}$ PFU per LD_{50}). This table illustrates that virulence is a relative phenomenon and depends on the method of assessment.

PFU, plaque-forming unit.

After Endres MJ, Valsamakis A, Gonzalez-Scarano F, Nathanson N. Neuroattenuated bunyavirus variant: derivation, characterization, and revertant clones. *J Virol* 1990;64:1927–1933.

TABLE 8.2
Comparison of paralytic rate for wild-type poliovirus and attenuated poliovirus vaccine

Virus	Study period	Paralytic rate per 100 primary infections	Relative rates
Wild type	1931–1954	0.7	~10,000
OPV	1961–1978	0.000062	1

For viruses that have a low case infection ratio, virulence can be expressed as the ratio of cases of disease to total infections. Poliovirus is an example since most human infections are inapparent, with less than 1 paralytic case per 100 infections. In this example, two strains of poliovirus are compared, wild type and attenuated vaccine virus (OPV), which is highly attenuated but still causes a small number of cases of paralysis.

OPV, oral poliovirus vaccine.

After Nathanson N, McFadden G. Viral virulence. In: Nathanson N, ed. *Viral pathogenesis.* Philadelphia: Lippincott Raven Publishers, 1997.

TABLE 8.3

Virulence based on case fatality rate and incubation period

Virulence grade	Case fatality rate (%)	Mean survival time (days)	Percentage of isolates
I	>99	<13	4
II	95–99	14–16	18
IIIA	90–95	17–22	39
IIIB	70–90	23–28	25
IV	50–70	29–50	14
V	<50	NC	1

Virulence can be measured as the proportion of infections that are fatal (case fatality rate) or by the survival time until death. In this example, various strains of myxoma virus, a poxvirus of rabbits, were compared after subcutaneous inoculation.

NC, not calculable.

After Marshall ID, Fenner F. Studies in the epidemiology of infectious myxomatosis of rabbits. VII. The virulence of strains of myxoma virus recovered from Australian wild rabbits between 1951 and 1959. *J Hygiene* 1960;58:485–488.

tion. Over the following years, the virus underwent natural attenuation and a large group of isolates were compared for their degree of virulence, based on the case/fatality rate and incubation period as shown in Table 8.3.

Virulence can also be assessed by the quantitation of pathologic lesions in an organ that is attacked by the virus under study. Figure 8.1 shows a comparison of the central nervous system (CNS) lesions

caused by five neurotropic flaviviruses. Some viruses produced most severe lesions in the brain, others targeted the spinal cord, and some produced similar lesions throughout the neuraxis, reflecting the multidimensional nature of virulence. Laboratory tests can be used as indirect indicators of the disease severity in a particular organ. Thus, the severity of hepatitis can be measured by the serum titer of alanine transaminase, which reflects the release of intracellular proteins from dying hepatocytes, and the severity of AIDS can be assessed by the blood concentration of CD4 lymphocytes.

Quantitation of Virulence

Virulence can be quantitated in experimental models by determining the number of infectious units required to produce a specific end point. For instance, calculation of the PFU per LD_{50} makes it possible to distinguish different degrees of virulence as shown in Figure 8.2, which separates viruses of high, intermediate, and low virulence. Other semi-quantitative methods are based on the severity of pathologic

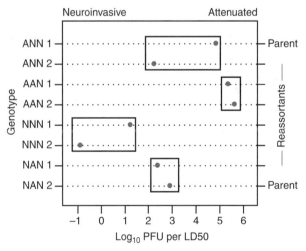

FIG. 8.2

Virulence can be expressed in a quantitative manner by determining the number of infectious units to produce a given disease phenotype. In this example, a set of clones of La Crosse, a bunyavirus, were each titrated by subcutaneous injection in suckling mice and for PFU in cell culture, and the ratio PFU/LD_{50} was computed (the lower the ratio the greater the neuroinvasiveness or virulence of the virus). Genotype: bunyaviruses are trisegmented and the three letters refer to the large, middle, and small gene segments; A, segment derived from attenuated Tahyna virus; N, segment derived from neuroinvasive La Crosse virus. Reassortant viruses with a genotype NNN were the most virulent whereas viruses with a genotype AAN were the most attenuated. (After Griot C, Pekosz A, Davidson R, et al. Replication in cultured C2C12 muscle cells correlates with the neuroinvasiveness of California serogroup bunyaviruses. *J Virol* 1994;201:399–403.)

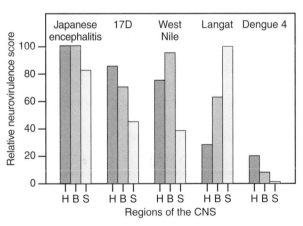

FIG. 8.1

Pathologic lesions can be used to measure virulence. In this example, monkeys were injected intracerebrally with five different neurotropic viruses and the severity of lesions in different CNS regions were graded using a standard scale. H, hemispheres; B, brainstem, S, spinal cord. This figure also shows that different viruses cause varying lesion profiles, so that lesion severity is a qualitative as well as a quantitative phenomenon. (After Nathanson N, Gittelsohn AM, Thind IS, Price WH. Histological studies of the monkey neurovirulence of group B arboviruses. III. Relative virulence of selected viruses. *Am J Epidemiol* 1967;85:503–517.)

lesions (Fig. 8.1), the case/infection ratio (Table 8.2), and the case/fatality rate (Table 8.3).

EXPERIMENTAL MANIPULATION OF VIRULENCE

To study virulence, it is necessary to possess viral variants that differ in their pathogenicity. For some viruses, naturally occurring isolates exhibit marked differences in virulence. For instance, type 1 and type 3 reoviruses differ sharply in tropism and yet are genetically so similar that reassortants (containing a mix of gene segments from types 1 and 3 viruses) can be readily obtained. For most viruses, however, it is necessary to use laboratory manipulation to obtain viral clones that are attenuated relative to the wild-type viruses isolated from nature. A number of procedures have been utilized to obtain attenuated viral variants.

Passage in Animals

In the pioneering days of virology, viruses were often maintained by serial passage in animal hosts. It was found that pathogenicity would change during the course of passage, and this adventitious finding was exploited to obtain viruses of different pathogenicity. In general, during repeated animal-to-animal transmission a virus will adapt to replicate optimally and become virulent under the conditions of passage. Yellow fever, a flavivirus, produces fatal hepatitis in monkeys; if passaged intracerebrally in mice it becomes highly neurovirulent for mice but loses its ability to cause hepatitis in monkeys.

Passage in Cell Culture

With the advent of cell culture, viruses were usually maintained by cell culture passage, and it was soon observed that serial transmission altered the biological phenotype, often reducing virulence for animals or humans. This observation was exploited in the deliberate search for attenuated variants that could be used as prophylactic vaccines. A typical example of the effect of passage on virulence is shown in Table 8.4. An uncloned series of La Crosse virus strains were passaged 25 times in BHK-21 cells and the resulting uncloned viruses tested for neurovirulence in mice. Of the ten passage lines, one (RFC/B) appeared to be the most attenuated, and ten clones from this virus stock were tested in mice. Of these ten clones, one (RFC/B.5) appeared to be the most attenuated and was selected for further study. This experiment illustrates several underlying principles.

TABLE 8.4	
Effect of passage in cell culture on the virulence of La Crosse bunyavirus	
Passage line	**Mortality (dead/tested)**
Each Virus Passed 25 Times, Then Tested In Mice	
La Crosse/original, pass A	5/5
La Crosse/original, pass B	3/5
La Crosse p10, pass A	3/5
La Crosse p10, pass B	4/5
La Crosse/pp31, pass A	5/5
La Crosse/pp31, pass B	4/5
Tahyna/181-57, pass A	5/5
Tahyna/181-57, pass B	5/5
La Crosse/RFC, pass A	1/5
La Crosse/RFC, pass B	0/5
RFC B Clones Tested in Mice most attenuated	
1	5/5
2	4/5
3	5/5
4	3/5
5	0/5
6	4/5
7	2/5
8	1/5
9	5/5
10	1/5

A number of different strains of La Crosse virus were passed 25 times in BHK-21 cells, and were then tested for virulence by intracerebral injection of 100 PFU into adult mice. One of these passaged lines (RFC/B) appeared relatively attenuated, and 10 plaques were further tested by intracerebral injection of 10,000 in adult mice. Plaque B.5 was selected as a highly attenuated clone.

After Endres MJ, Valsamakis A, Gonzalez-Scarano F, Nathanson N. Neuroattenuated bunyavirus variant: derivation, characterization, and revertant clones. *J Virol* 1990;64:1927–1933.

- Apparently identical passage lines can yield virus stocks differing in their virulence.
- An RNA virus stock represents a "swarm" of viral clones of different phenotypes.
- Passage probably selects for viral clones already in the population that grow preferentially and thus alter the phenotype of the virus swarm.

Passage and Viral Phenotype

Historically, the failure to recognize the influence of passage on the biological phenotype has led to some important errors in virology research. Thus, intracerebral passage of poliovirus in monkeys leads to selection of variants that are highly neurotropic but have lost much of their infectivity and pathogenicity when administered by the oral (natural) route. Studies with neuroadapted poliovirus resulted in the mistaken conclusion that poliovirus was not an enterovirus but was naturally transmitted by the intranasal route, and this misapprehension led to trials of nasal astringent sprays as a method to protect children against paralysis.

More recently, passage of human immunodeficiency virus (HIV) in T-cell lines selected for laboratory variants that differed from wild-type virus in the exclusive use of the CXCR4 coreceptor, inability to infect macrophages, and ability to plaque in MT-2 cells (a T-cell line). When used for serologic assays, the adapted viruses are readily neutralized by sera from patients naturally infected with HIV. These findings resulted in the mistaken conclusion that HIV could be readily neutralized and the prediction that it would be relatively easy to develop a prophylactic vaccine. Once it was recognized that viral isolates only maintained their natural phenotype if passaged in primary blood mononuclear cells, it became clear that wild-type HIV isolates were very resistant to neutralization, presenting a daunting challenge for vaccine development, a problem that has yet to be solved.

Selection of Attenuated Variant Viruses

There are a number of methods that have been used to enhance the selection of attenuated virus variants from an uncloned virus stock.

Temperature-Sensitive Mutants Wild-type viruses replicate well at the standard temperature of 37°C and at temperatures up to 40°C (most cell cultures do not thrive above 40°C). On the other hand, temperature-sensitive (ts) variants replicate well at 37°C but poorly (if at all) at 40°C. It is relatively easy to select for ts variants, which often exhibit an attenuated phenotype when tested in animals. The attenuated phenotype is not due to the body temperature of the test animal (which is usually around 37°C, a permissive temperature for ts mutants). Instead, ts mutants usually have a restricted range in host tissues relative to wild-type virus. Figure 8.3 illustrates these points for the attenuated RBC/B.5 clone whose derivation is described in Table 8.4. Clone B.5 replicates well in BHK-21 cells at 37°C but hardly at all at 40°C, whereas wild-type La Crosse virus replicates equally well at both temperatures. Furthermore, clone B.5 replicates very poorly in mouse neuroblastoma cell cultures (NA cells) even at 37°C, which is further evidence of its host restriction. Finally, clone B.5 replicates very poorly after intracerebral injection by comparison with neurovirulent wild-type La Crosse virus.

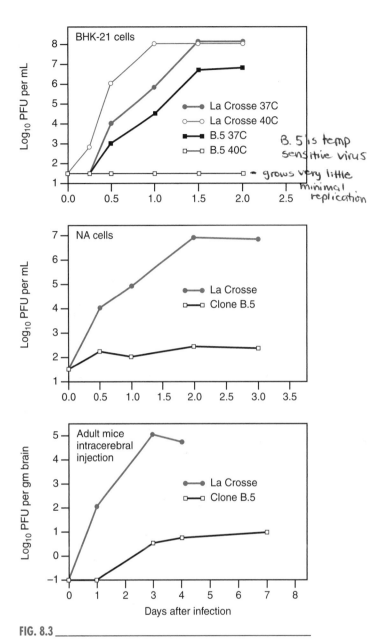

(handwritten annotations on figure: "B.5 is temp sensitive virus", "grows very little minimal replication")

FIG. 8.3

Phenotype of an attenuated variant virus showing its temperature sensitivity, restricted host range, and reduced virulence in animals. La Crosse virus, clone RFC/B.5 is temperature sensitive (*top*), restricted in its ability to replicate in neuroblastoma cells (*middle*), and has reduced neurovirulence after intracerebral injection of 700 PFU in adult mice (*bottom*). (After Endres MJ, Valsamakis A, Gonzalez-Scarano F, Nathanson N. Neuroattenuated bunyavirus variant: derivation, characterization, and revertant clones. *J Virol* 1990;64:1927–1933.)

Cold-Adapted Variant Viruses Another method for the selection of attenuated mutants is passage at a low temperature, such as 25°C (about the lowest temperature at which mammalian cell cultures can be maintained). Table 8.5 shows an example of a cold-adapted influenza virus, compared with its

TABLE 8.5
Selection of attenuated virus by cold adaptation

Character	Not adapted	Cold adapted
Replication (PFU per mL) at		
37°C	$10^{7.3}$	$10^{7.3}$
30°C	$10^{4.6}$	$10^{6.4}$
25°C	$10^{2.0}$	$10^{6.5}$
Temperature sensitive at 40°C	No	Yes
Virulence for ferrets (intranasal infection)	Yes	No

An influenza virus was passed six times in cell culture at 25°C, and its properties compared with the nonadapted parent virus from which it was derived.

After Maassab HF, De Borde DC. Development and characterization of cold-adapted viruses for use as live virus vaccines. *Vaccine* 1985;3:355–369.

wild-type parent virus stock. The cold-adapted variant is temperature sensitive and exhibits restricted pneumotropism in ferrets, an animal that develops severe pneumonia after intranasal infection with wild-type human influenza viruses.

Monoclonal Antibody–Resistant Viruses Monoclonal antibodies with neutralizing activity can readily be obtained for most viruses. When an uncloned RNA virus stock is treated with a neutralizing monoclonal antibody, there is a residual fraction—10^{-4} to 10^{-6}—that resists neutralization. The resistant fraction yields clones that are resistant to the selecting monoclonal antibody and probably represent naturally occurring point mutants (mutants with a single nucleotide change), which tend to occur at the rate of about 10^{-5} in most virus populations. When tested in animals, some monoclonal antibody–resistant (MAR) mutants are attenuated, and these have been used to study the molecular basis of attenuation because their phenotype can be mapped to specific viral protein (the virus attachment protein) and often to a single amino acid in that protein.

Mutagenized Viruses To increase the likelihood of an altered phenotype it is possible to use irradiation or chemical agents to mutagenize a virus stock, following which it can be tested for altered phenotype. Although mutagenesis is an effective way to increase the frequency of viruses with an altered phenotype, it usually introduces multiple mutations, making it difficult to determine the viral gene or protein responsible for the altered phenotype. In contrast, clones with altered phenotypes obtained from nonmutagenized viral stocks are more likely to possess a limited number of genetic changes.

With the introduction of modern molecular genetics into virology, it has become relatively easy to alter the viral genome in a controlled manner. Genetic changes that can be deliberately introduced include point mutations that alter function of individual proteins, or inactivation of nonessential viral genes by introduction of stop codons or deletions of substantial gene segments. Viruses with "designer" mutations can then be compared with the parent clones from which they were derived.

Choice of Attenuated Viruses for Study Before investing effort in characterization of attenuated mutants, it is worthwhile to consider several points.

1. First, it is important to select viruses that are as genetically pure as possible. This is usually accomplished by repeated plaque purification of a viral clone prior to study. In addition to biological cloning, genetic cloning may be justified if the virus is a candidate for use as an attenuated vaccine where maximal genetic stability is essential.

2. Second, if a clone is to be used for determination of the genetic basis of viral phenotypes, then it is preferable to use variant viruses that are likely to have a minimal number of mutations relative to the parent virus with which they will be compared.

3. Finally, the most informative variants are host range mutants, that is, viruses that can replicate well in reference permissive cell cultures but are restricted in animals or in specialized cell cultures (Fig. 8.3). Attenuated viruses that are "wimpy" (unable to replicate well in any cellular substrate) do not provide much information about the properties of the virus that determine its in vivo virulence or attenuation.

COMPARATIVE PATHOGENESIS OF VIRULENT AND ATTENUATED VIRUSES

The sequential steps in viral pathogenesis are described in Chapter 2. In theory, two viral clones could differ at any of these steps, thereby affecting virulence or attenuation. There are examples of viruses that differ in their infectivity at the portal of entry, in their ability to disseminate, or in their replicative capacity in target organs or tissues. Also, pathogenicity may be qualitative rather than quantitative because viral clones may differ in their relative tropism for different tissues or organs. Finally, in some instances, pathogenicity is determined by the ability to induce or evade host defenses, such as the immune response.

Portal of Entry

Many respiratory viruses are naturally somewhat "cold adapted" because they can replicate in the nasal mucosa and upper respiratory tract where the temperature is a few degrees lower than core body temperature. Those respiratory viruses, like influenza, that cause severe disease usually can also replicate well in the lower respiratory tract where the temperature is closer to 37°C. When influenza virus is cold adapted (Table 8.5) it can no longer replicate well in the lower respiratory tract and is thereby attenuated.

Viremia

As described in Chapter 2, most systemic viruses spread via the blood. If viral strains differ in their viremogenicity (duration and titer of viremia) this may alter their ability to reach critical target organs and thereby influence virulence. Poliovirus is an enterovirus that is basically innocuous unless it invades the CNS where it destroys anterior horn cells (neurons in the spinal cord that innervate striated muscle), thereby producing a lower motor neuron flaccid paralysis. Poliovirus strains vary in the degree of viremia that they produce, and this correlates with their paralytogenicity after extraneural infection (Fig. 8.4).

The extent of viremia is usually an indication of the ability of the virus to replicate in cells that shed virus into the circulation. In those instances where the cellular source of viremia is known, it may be possible to correlate viremogenicity with replication in a specific organ or tissue. An example is shown in Figure 8.5, which compares two bunyaviruses. After subcutaneous injection in suckling mice, La Crosse virus can replicate well in striated muscle, produces a considerable viremia, invades the brain, and is lethal. Tahyna virus strain 181-57 replicates poorly in striated muscle, fails to produce viremia, and does not kill suckling mice after subcutaneous injection.

Neural Spread

Some viruses spread along neural pathways rather than by viremia (see Chapter 2), and neurally spreading viruses can also be attenuated. Rabies virus is a good example of an "obligatory" neurotrope for which attenuated strains have been derived for use as live vaccines. Vaccine strains of rabies virus, obtained by passage of wild-type virus in

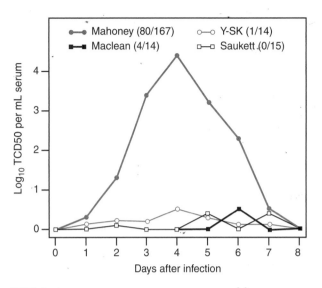

FIG. 8.4 _____

Viremia potential can be a determinant of pathogenicity. In this example, different wild-type strains of human poliovirus were injected intravenously in cynomolgus monkeys. The different isolates varied in the level of viremia produced, and the Mahoney strain, the most viremogenic, also caused the greatest frequency of paralysis. (After Bodian D. Viremia in experimental poliomyelitis. I. General aspects of infection after intravascular inoculation with strains of high and low invasiveness. *Am J Hyg* 1954;60:339–357.)

nonneural cultured cells, show a marked reduction in their virulence when tested by intracerebral injection in mice. The attenuated phenotype is maintained on passage in nonneural BHK-21 cells but reverts to greater virulence on passage in cultured neural cells or in the brains of suckling mice (Table 8.6). This suggests that attenuated strains probably include some virulent viral clones that are rapidly expanded in cultures where they have a growth advantage.

The basis for the attenuation of rabies viruses has been investigated using MAR variants selected by growing wild-type virus in the presence of monoclonal antibodies. Some of the MAR variants have markedly reduced virulence in mice, and these have been studied to elucidate the mechanisms of attenuation. In one set of investigations, MAR variants with a mutation at amino acid 333 of the envelope glycoprotein were found to be attenuated; the virulent phenotype was associated with arginine or lysine (both positively charged amino acids), whereas the attenuated phenotype was associated with glycine, glutamine, or methionine (uncharged).

Comparative pathogenesis studies showed that after intramuscular injection, two of these attenuated variants, RV194-2 and Av01, spread to the CNS at

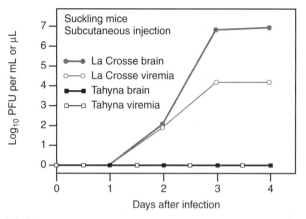
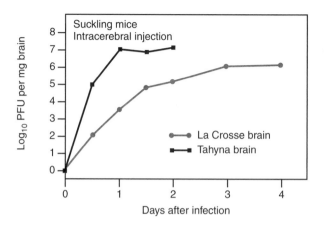

FIG. 8.5 _____

Virulence after parenteral virus infection can be determined by the ability of a virus to replicate in a peripheral tissue that serves as a source for viremia. This example compares two bunyaviruses: La Crosse virus, which is fatal for suckling mice after subcutaneous injection, and Tahyna virus strain 181/57, which produces an inapparent infection after subcutaneous infection. (*Left*) La Crosse virus produces a viremia, reaches the target organ (CNS), and replicates to high titer in the brain. Tahyna virus fails to produce viremia and does not invade the brain. (*Right*) On intracerebral injection of 700 PFU Tahyna virus replicates to a higher titer than La Crosse virus, indicating that its peripheral attenuation was not due to a reduction in its ability to replicate in the target organ but rather to its inability to produce a viremia after subcutaneous injection. (After Janssen R, Gonzalez-Scarano F, Nathanson N. Mechanisms of bunyavirus virulence: comparative pathogenesis of a virulent strain of La Crosse and an attenuated strain of Tahyna virus. *Lab Invest* 1984;50:447–455.)

about the same rate as the virulent parent virus, but once disseminated did not spread to contiguous neurons as rapidly as the virulent virus. Primary dissemination involved transsynaptic transmission to only two or three sequential neurons, following which the attenuated viruses failed to spread further. Furthermore, when the same viruses were compared

after intraocular injection, the virulent virus utilized several different neural pathways (sensory and motor) whereas the avirulent virus was restricted to a single pathway, indicating that attenuation involved qualitative as well as quantitative differences. Tissue culture studies showed that both virulent rabies virus and attenuated RV194-2 spread with equal speed through nonneural BHK-21 cells, but that they differed when used to infect NA neuroblastoma cells. The avirulent virus spread more slowly than the virulent rabies virus, possibly due to quantitative differences in entry into neurons (Fig. 8.6).

Target Organ

Variants of a virus may differ in their virulence for a target organ, such as lung, liver, or CNS, and this property is distinct from the ability of different variants to invade the target organ from the portal of entry. Figure 8.7 compares four bunyaviruses for their respective pathogenicity for the CNS after intracerebral injection.

Attenuated polioviruses offer another illustration of differences in virulence for the target organ that is distinct from peripheral infectivity. In the search for attenuated poliovirus strains for potential use as vaccines, Sabin tested many wild-type and laboratory-passaged viruses and found that viruses that replicated well in cell culture (primary monkey kidney fibroblasts) varied widely in their ability to replicate in the CNS of monkeys and chimpanzees. Furthermore, neurovirulence was distinct from enterogenicity (infectivity after virus feeding) (Table 8.7).

TABLE 8.6		
Virulence phenotype can be altered by passage		
Passage history HEP Flury stock	PFU per LD$_{50}$	Neurovirulence (relative to HEP Flury virus)
None	>500,000	1
BHK-21 cells Nonneural (5 passages)	>3,000,000 *higher titer*	~0.16 *less virulent*
NA cells Neural (3 passages)	500 *smaller titer*	>1,000 *much more virulent*
Suckling mouse Intracerebral (4 passages)	2,000 *relatively sm titer*	>250 ↑ *virulence*

The virulence phenotype of viruses that spread by the neural route can be altered by conditions of passage. In this study the HEP Flury strain, a neuroattenuated strain of rabiesvirus used as a vaccine, was passed in neural cells (NA cells, a line derived from a neuroblastoma), nonneural cells (BHK), or in suckling mouse brain. The resulting stocks were titrated intracerebrally in 4-week-old mice and in BHK cells to determine PFU per LD$_{50}$ (50% lethal dose).

HEP, high egg passage; BHK, baby hamster kidney cell line; PFU, plaque-forming unit.

After Clark HF. Rabies viruses increase in virulence when propagated in neuroblastoma cell culture. *Science* 1978;199:1072–1075.

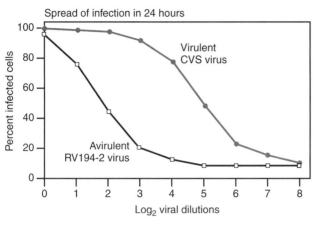

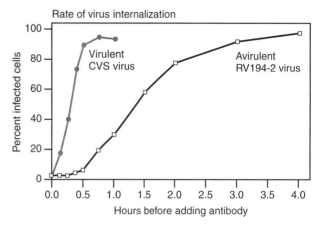

FIG. 8.6

Attenuation of virus that spreads by the neural route correlates with the spread in cultured NA (neuroblastoma) cells. This graph compares virulent rabies virus strain CVS (challenge virus standard) with the attenuated MAR mutant RV194-2. The attenuated virus shows marked reduction in its ability to spread in neuroblastoma cells relative to the virulent virus, but both viruses spread equally fast in BHK-21 cells (data not shown). (After Dietzschold B, Wiktor TJ, Trojanowski JQ, et al. Differences in cell-to-cell spread of pathogenic and apathogenic rabies virus in vivo and in vitro. *J Virol* 1985;56:12–18.)

Relative Pathogenicity for Different Tissues

Variants of a single virus can differ in their relative pathogenicity for different tissues or organs, which confers a multidimensional character upon virulence. A variant virus may show decreased pathogenicity for one tissue or cell type but enhanced pathogenic-ity for another cell type. For instance, after intracere-bral injection in adult mice, reovirus type 1 Lang strain infects the ependymal lining of the brain and causes hydrocephalus, whereas reovirus type 3 Dear-ing strain infects neurons and causes fatal encephali-tis. Nevertheless, these two viruses are genetically compatible, so that reassortants can be made that in-clude any possible combination of gene segments from type 1 and type 3 viruses. Wild-type mouse hepatitis virus (a coronavirus) infects neurons and causes lethal encephalitis upon intracerebral injec-tion of adult mice. Attenuated mouse hepatitis virus

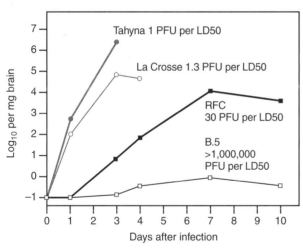

FIG. 8.7

Relative pathogenicity for the CNS of four California serogroup bunyaviruses. Adult mice were injected intracerebrally with 700 PFU of different viruses each of which replicated at a dif-ferent rate, consistent with the quantitative differences in their virulence expressed as PFU per LD_{50}. Neurovirulence is dis-tinct from neuroinvasiveness; for instance, Tahyna virus, the most neurovirulent strain, failed to kill suckling mice after sub-cutaneous injection whereas the less neurovirulent La Crosse virus was also neuroinvasive (Fig. 8.5). (After Endres MJ, Vals-makis A, Gonzalez-Scarano F, Nathanson N. Neuroattenuated bunyavirus variant: derivation, characterization, and revertant clones. *J Virol* 1990;64:1927–1933.)

TABLE 8.7

Relative virulence for a target organ

Type 1 poliovirus strain	TCD_{50} per mL	Enterotropism TCD_{50} per PO ID_{50}	Neurotropism TCD_{50} per IC PD_{50}
Virulent Mahoney (CNS suspension)	10^6	$10^{3.3}$ (monkeys)	$10^{1.9}$
Attenuated LSc (tissue culture fluid)	$10^{7.6}$	$\sim 10^4$ (humans)	$>10^{7.6}$

Relative virulence for a target organ can vary widely among virus vari-ants that replicate well in cell culture. A virulent and attenuated type 1 poliovirus are compared, to show that they both replicate well in cell culture (primary monkey kidney cells) and are enterotropic (infectious after oral administration) but differ markedly in their neurovirulence af-ter intracerebral injection in cynomolgus monkeys.

PO, per os; PD_{50}, 50% paralytic dose; ID_{50}, 50% infectious dose; TCD_{50}, 50% tissue culture dose; CNS, central nervous system.

After Sabin AB, Hennessen WA, Winsser J. Studies on variants of po-liomyelitis virus. *J Exp Med* 1954;99:551–576; Sabin AB. Properties and behavior of orally administered attenuated poliovirus vaccine. *JAMA* 1957;164:1216–1223.

variants are less infectious for neurons and fail to kill mice acutely, but infect oligodendroglia (cells that produce myelin in the CNS) and cause persistent infection and chronic demyelination.

HIV-1 presents another example of differences in tropism. All HIV-1 strains replicate well in primary cultures of peripheral blood mononuclear cells (consisting mainly of T lymphocytes). Some wild-type strains replicate in primary cultures of monocyte-derived macrophages but not in transformed lines of T lymphocytes. Other HIV-1 strains (including laboratory-adapted strains that have been repeatedly passed in T-lymphocyte cell lines) will not replicate in macrophages but grow well in T-lymphocyte cell lines (Table 8.8). These patterns are explained by the expression of two alternate coreceptors for HIV-1 on different cells and by the restricted ability of many HIV-1 strains to use only one of these two coreceptors, in conjunction with their primary receptor, CD4.

Host Immune Response

Altered pathogenicity of variant viruses may be mediated through the host immune response (see Chapter 7). For instance, clone 13 of lymphocytic choriomeningitis virus (LCMV) differed from the Armstrong strain of LCMV from which it was derived by virtue of its ability to replicate more rapidly in macrophages and less rapidly in the CNS. The rapid destruction of macrophages interferes with antigen presentation thereby suppressing the immune response, which in turn permits the virus to escape clearance. As a result, clone 13 initiates a persistent infection without acute illness, in contrast to Armstrong, which produces a benign immunizing infection with rapid virus clearance (see Figs. 7.2, 7.3, Tables 7.2, 7.3).

Simian immunodeficiency virus (SIV) produces a high-titer viremia and rapidly progressive AIDS, whereas the Δnef variant produces a modest viremia and a relatively benign infection with occasional AIDS after many years (Fig. 7.12). The *nef* gene expressed in wild-type SIV down-regulates the expression of major histocompatibility complex class I molecules on the surface of infected CD4 lymphocytes, protecting them against elimination by immune surveillance (antiviral CD8 cytotoxic T lymphocytes, CTLs) and thereby enhancing virus replication. In the absence of the *nef* gene, virus-infected T lymphocytes are more rapidly cleared, enhancing the ability of the host to contain the infection and attenuating the pathogenicity of the virus. In this example, virulence is enhanced by a viral gene that interferes with host immune defenses.

GENETIC DETERMINANTS OF VIRULENCE

Viral virulence, like other viral phenotypes, is encoded in the viral genome and is expressed through the structural and nonstructural viral proteins, as well as the noncoding part of the genome. Over the last few decades, a large body of information has been assembled regarding the genetic determinants of virulence. Some of the generalizations that have

TABLE 8.8

Variation in cellular tropism among isolates of the same virus

Viral biotype (Coreceptor usage)	HIV-1 isolate	GROWTH IN EACH CELL TYPE (CORECEPTOR EXPRESSION)		
		PBMC (CXCR4/CCR5)	Monocyte-derived macrophages (CCR5)	T-cell line Sup-T1 (CXCR4)
T-tropic	IIIB	++++	+	++++
(CXCR4)	DV	++++	++	++++
M-tropic	SF162	++++	++++	−
(CCR5)	89.6	++++	++++	−

Isolates of the same virus can vary in their cellular tropism. All HIV-1 strains replicate well in PBMCs. Laboratory-adapted isolates of HIV-1 are T-cell–tropic and do not replicate well in primary macrophage cultures, whereas wild-type isolates replicate well in macrophages but not in T-cell lines. These patterns can be explained by the requirement of HIV-1 strains for different coreceptors that mediate cell entry in conjunction with CD4. Replication was assessed by peak level of p24 antigen production: ++++ >100 ng/mL, +++ 10–100 ng/mL, ++ 1–10 ng/mL, + <1 ng/ml, − <0.01 ng/mL.

HIV, human immunodeficiency virus; PBMC, primary blood mononuclear cell; CXCR, receptor for cysteine-X-cysteine cytokines; CCR, receptor for cysteine–cysteine cytokines.

After Collman R, Hassan NH, Walker R, et al. Infection of monocyte-derived macrophages with human immunodeficiency virus type 1 (HIV-1): monocyte- and lymphocyte-tropic strains of HIV-1 show distinctive patterns of replication in a panel of cell types. *J Exp Med* 1989;170:1149–1163.

emerged are summarized in Sidebar 8.1. These points are illustrated by the examples that follow.

Reovirus

Reoviruses (family Reoviridae) have a genome consisting of ten double-stranded RNA segments. Reoviruses fall into three serotypes that possess quite different pathogenic characteristics in mice, yet they reassort readily. Reassortants are prepared by dually infecting a permissive cell culture with reoviruses of two different serotypes and preparing biological clones by plaque purification. Although the gene segments from different reovirus types are interchangeable, the segments from different serotypes can be distinguished by their slightly different sizes that lead to variation in migration rate on polyacry-

lamide gel electrophoresis, which is used to genotype individual reassortant clones. An example of the use of reassortant viruses is shown in Table 8.9. A number of different pathogenic phenotypes have been described for reovirus infection of mice, and many of these have been mapped to specific viral gene segments (Table 8.10). This table also indicates that some of the phenotypes are under multigenic control, an important principle to keep in mind when investigating the viral determinants of virulence.

Bunyavirus

Bunyaviruses are negative or ambisense RNA viruses with a trisegmented genome. The large (L) RNA segment encodes the polymerase (L) protein, the middle-sized (M) encodes two glycoproteins (G1 and G2) and a nonstructural protein (NS_m), and the small (S) segment encodes the nucleocapsid (N) protein and a nonstructural (NS_s) protein. The California serogroup of the bunyavirus genus of the family Bunyaviridae includes about 12 members that are genetically compatible, so that reassortants among these member viruses can be selected. These reassortants have been used to identify genetic determinants of virulence and attenuation, although these studies have not yet been extended to the level of individual amino acids.

Middle RNA Segment Comparison of a wild-type clone of La Crosse virus with an attenuated strain of Tahyna virus (clone 181-57) documented that the attenuated virus was highly neurovirulent in mice but was not neuroinvasive after subcutaneous injection of suckling mice (Fig. 8.5). Studies of reassortants between the two viruses showed that neuroinvasiveness cosegregated with the M RNA segment.

Large RNA Segment Comparison of a neuroattenuated clone (B.5) derived from a reassortant of La Crosse and Tahyna virus showed that the attenuated variant had reduced ability to replicate in neuroblastoma cell cultures and reduced ability to replicate after intracerebral injection in adult mice (Fig. 8.3). Reassortants between the attenuated virus and a neurovirulent virus showed that neurovirulence cosegregated with the L RNA segment (Fig. 8.8).

Poliovirus

Polioviruses, of the family Picornaviridae, are single-stranded RNA viruses of positive polarity (their genome can act as mRNA and be translated directly into proteins). Furthermore, cDNA clones con-

		ORIGIN OF GENE SEGMENT										
		OUTER CAPSID				CORE				NS		Pattern of spread
Virus	Clone	M2	S1	S4	L2	L1	L3	M1	S2	M3	S3	
Parent	T1L	L	L	L	L	L	L	L	L	L	L	V
	T3D	D	D	D	D	D	D	D	D	D	D	N
Reassortant	R1	D	L	L	L	L	L	L	L	D	D	V
	R2	L	L	L	D	D	D	D	D	L	D	V
	R3	L	L	D	L	L	L	L	L	L	L	V
	R4	D	L	D	D	D	D	D	D	D	D	V
	R5	D	L	L	D	D	D	L	D	D	D	V
	R6	D	L	D	D	D	D	D	D	L	D	V
Reassortant	R7	D	D	D	D	D	D	D	D	D	L	N
	R8	L	D	L	L	D	L	L	L	L	L	N
	R9	L	D	L	L	L	L	L	L	L	L	N
	R10	L	D	L	D	L	L	L	L	L	L	N

TABLE 8.9

Use of reassortant viruses to determine the gene segment (and corresponding viral protein) that encodes a specific pathogenic phenotype

In this example, reovirus 1 T1L was crossed with reovirus T3D to determine the mode of spread to the central nervous system after footpad (subcutaneous) injection of neonatal mice. T3D spread exclusively by the neural route while T1L spread by the hematogenous route (viremia) and the different phenotypes could be determined by sciatic nerve section which prevented T3D from reaching the spinal cord but did not block the spread of T1L. Comparison of a number of reassortant clones shows that the route of spread cosegregates with a single gene segment, the S1 segment that encodes the σ1 outer capsid protein.

T1L, Type 1. Lang; T3D, type 3 Dearing; V, viremia; N, neural spread; NS, nonstructural proteins; L, T1L; D, T3D.

After Tyler KL, McPhee DA, Fields BN. Distinct pathways of viral spread in the host determined by reovirus S1 gene segment. *Science* 1986;233:770–774.

structed from the viral RNA can be used for manipulation of the genome. The altered cDNA clones are infectious, that is, they can be transfected into permissive cells with the production of infectious virus. These features have facilitated the mapping of poliovirus virulence.

OPV is comprised of attenuated clones of each of the three poliovirus serotypes that, compared to virulent wild-type polioviruses, have markedly re-

TABLE 8.10

An example of viral genes associated with different pathogenic phenotypes

	NEONATAL MICE	
Organ	Phenotype	Gene segment
Whole animal	Virulence (lethality)	M2
	Spread to CNS from periphery	S1
CNS	Neurocellular tropism	S1
Heart	Myocarditis severity	S1, M1, L1, L2
Intestine	Viral titer	S1
	Animal-to-animal spread	L2

The table shows various reovirus virulence phenotypes that have been mapped to specific gene segments. In some instances, several different genes play a role in the phenotype.

CNS, central nervous system.

After Virgin HV IV, Tyler KL, Dermody TS. Reovirus. In: Nathanson N, et al, eds. *Viral pathogenesis.* Philadelphia: Lippincott Raven Publishers, 1997.

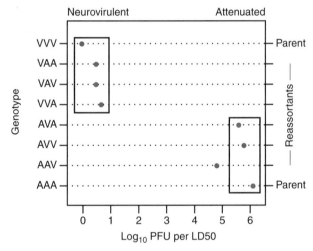

FIG. 8.8

Reassortants between a virulent and an attenuated virus identify the viral gene segment responsible for attenuation. In this example, an attenuated reassortant clone (between La Crosse and Tahyna viruses) was crossed with a neurovirulent reassortant clone, and viruses of all eight possible genotypes were tested in a quantitative assay after intracerebral injection in adult mice. The neurovirulent phenotype segregated with the large RNA segment that encodes the viral polymerase. One reassortant, genotype AAV, was slightly less attenuated, suggesting that other gene segments might modify the effect of the attenuated phenotype. Genotype: bunyaviruses are trisegmented and the three letters refer to the large, middle, and small gene segments; A, attenuated; V, virulent. (After Endres MJ, Griot C, Gonzalez-Scarano F, Nathanson N. Neuroattenuation of an avirulent bunyavirus variant maps to the L RNA segment. *J Virol* 1991;65:5465–5470.)

duced neurovirulence on direct intrathalamic or intraspinal injection in macaques (Table 8.7). There are a number of differences between the OPV strains and their wild-type parents, acquired during many successive passages in cell culture, and only a subset of these are relevant to the neurovirulent phenotype. More specific information can be obtained from OPV viruses that have reverted to virulence after feeding to humans, which is a common consequence of replication in the human gastrointestinal tract. Such virulent revertants have a smaller number of nucleotide differences from the attenuated vaccine strains than do the parent wild-type viruses. Finally, engineered recombinant viruses have been used to deliberately test the influence of individual critical nucleotides (Table 8.11).

For all three of the attenuated OPV strains, there are two sets of neurovirulence determinants. For each strain, there is a critical nucleotide in the 5′ nontranslated region of the genome, at positions 480, 481, or 472, for types 1, 2, and 3, respectively. Based on computer modeling of their secondary structure, the 5′ RNA is predicted to form stem loops at positions 470–540. This region is thought to be involved with initiation of translation at an internal ribosomal entry site, and mutations at the critical site are predicted to perturb the secondary structure and alter the initiation of translation.

The second group of critical sites involves the structural (VP1 to VP4) and nonstructural viral proteins, but is disparate for the three attenuated strains of OPV. For type 1, point mutations associated with virulence are located in several capsid proteins and the nonstructural polymerase; for type 2 OPV, mutations are in VP1; and for type 3 OPV, mutations in VP3 or VP4 are involved. It is not clear how mutations in the structural proteins alter virulence, but it has been hypothesized that these are implicated in viral uncoating or in viral assembly.

OPV strains are temperature sensitive and represent host range mutants because they replicate well in primate fibroblasts but poorly in neuroblastoma cells, in comparison with their virulent counterparts. Thus, it must be presumed that viral determinants of uncoating, translation, or assembly are cell specific and must ultimately be explained in terms of the subtle variations in the activity of poliovirus proteins in various cell types.

VIRULENCE GENES OF CELLULAR ORIGIN

In the last decade, a new class of virus-encoded proteins has been recognized that contribute to virulence of viruses by mimicking cellular proteins. It is hypothesized that viruses have derived the genes

TABLE 8.11

Analysis of genetic determinants of virulence at the nucleotide level

Virus clone	NUCLEOTIDE										Score
	220	472	871	2034	3333	3464	4064	6127	7165	7432	
P3/L											2.71
SV1/L					A						2.68
SP2/L						A					2.51
S3′/L								A	A	A	2.40
SLR2					A	A	A	A	A	A	2.39
L472V3/S	A		A		A	A	A	A	A	A	2.07
SCC/L			A	A	A	A					1.93
SV3/L				A							1.74
L472/S	A		A	A	A	A	A	A	A	A	1.58
LV3/S	A	A	A		A	A	A	A	A	A	1.32
ST/L	A	A	A								1.14
P3/S	A	A	A	A	A	A	A	A	A	A	0.41
SLR1	A	A	A	A							0.28

This example compares wild-type virulent type 3 poliovirus (P3/Leon) with the attenuated type 3 oral poliovirus vaccine (P3/Sabin) derived therefrom. There are 10 nucleotide differences between the two viruses (in the whole genome of >7,000 nucleotides) and a set of recombinant viruses were constructed to determine which divergent nucleotides influenced virulence. In this instance, it appeared that nucleotides 220 and 472 in the 5′ nontranslated region and nucleotides 871 (VP4) and 2034 (amino acid 91 in VP3) were the most significant critical sites. Boxes indicate that the five most virulent recombinant viruses had the "virulent" nucleotide at all four of these sites while the two most attenuated recombinant viruses had the "attenuated" nucleotide at these four sites.

Virus clone: P3/L, virulent parental P3 Leon strain; P3/S, attenuated P3 Sabin strain. Score: neurovirulence score after intracerebral injection of macaques. A, nucleotides found in the attenuated but not the virulent virus.

After Minor PD. The molecular biology of poliovaccines. *J Gen Virolo* 1992;73:3065–3077.

that encode these proteins from the cells in which they replicate, by recombination and subsequent modification. Most of these cell-derived genes have been identified in DNA viruses with large genomes, such as the herpesviruses and poxviruses, which have a greater capacity to accommodate accessory genetic information than do viruses with small genomes. From an evolutionary viewpoint, the cell-derived genes resemble oncogenes but, rather than endowing the virus with transforming properties, they act to subvert the antiviral defenses of the infected host.

Cell-derived viral genes, which include "virokines" and "viroceptors," enhance virulence through many different mechanisms. Virokines secreted from infected cells can mimic the action of cytokines, causing host cells to proliferate, thereby increasing virus production. Viroceptors resemble receptors for cytokines and can act as decoys, binding and sequestering cellular cytokines. Some virus-encoded proteins bind antibodies or complement components, reducing the lysis of virus-infected cells, whereas others perturb antigen presentation or immune induction. A few selected examples are described below.

Poxviruses

Vaccinia Complement Control Protein   The complement cascade, described in Chapter 5, consists of a group of proteins that produce a membrane attack complex that can destroy microbial pathogens or virus-infected cells. Poxvirus infections induce antiviral antibodies that can trigger the complement cascade, and such antibodies destroy virus-infected cells thereby acting as a host defense. Vaccinia, a poxvirus, encodes vaccinia complement control protein (VCP), a protein that abrogates the complement-mediated attack on virus-infected cells. VCP resembles a human plasma protein, C4-BP, that monitors the complement cascade and prevents its unwanted activation, by binding to C4b, one of the intermediaries in the cascade. VCP also binds C4b, thereby blocking the complement cascade and vitiating complement-mediated host defenses. VCP-negative mutants are less pathogenic in vivo than wild-type vaccinia virus, illustrating the in vivo effect of VCP (Fig. 7.6).

Tumor Necrosis Factor Viroceptors   Tumor necrosis factor (TNF) is a family of proinflammatory cytokines that are produced by activated macrophages and T lymphocytes. TNFs bind to cognate receptors expressed on myeloid and lymphoid cells, produc-

ing multiple effects on immune networks and host responses to infection. A number of poxviruses encode a viroceptor that is a soluble homolog of the cellular p75TNF receptor. It is presumed that the soluble receptor binds and sequesters TNF, thus modulating TNF-mediated cellular responses to infection. Poxviruses encode a large number of soluble proteins, some of which have been shown to act as receptors for other cytokines, such as interferon γ, interferon α/β, and interleukin-1β (IL-1β).

Interleukin-4   During immune induction, IL-4 is produced mainly by the T_H2 subset of CD4$^+$ helper cells and down-regulates Th1 activation, deviating the immune response toward B lymphocytes and away from the induction of CD8$_+$ CTLs. When mousepox virus is engineered to express IL-4, the virus becomes much more virulent, associated with a reduced induction of virus-specific CTLs and also a reduced induction of IFN-γ. Although an experimental artifact, this demonstrates the potential consequence of incorporation of a cellular gene into a viral genome.

Herpesviruses

Herpes Simplex Virus Fc Receptors   Two of the glycoproteins of HSV, gE and gI, act together as a receptor for the Fc domain of immunoglobulins

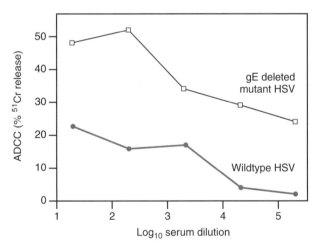

FIG. 8.9

Cell-derived virus-encoded proteins can enhance viral virulence. In this example, wild-type herpes simplex virus (HSV) is compared with a mutant in which gE, a viral glycoprotein that acts as a receptor for immunoglobulin (FcR), has been deleted. Cells infected with each virus are compared for their sensitivity to antibody-dependent cell-mediated cytotoxicity, and those infected with the mutant HSV are much more sensitive to lytic attack. (After Dubin G, Socolof E, Frank I, Friedman H. Herpes simplex virus type 1 Fc receptor protects infected cells from antibody-dependent cellular cytotoxicity. *J Virol* 1991;65: 7046–7050.)

(FcR). HSV infection induces the host to produce antiviral antibodies that can initiate the lysis of HSV-infected cells by antibody-directed cell-mediated cytotoxicity (ADCC). HSV counters this host defense by the action of gE and GI. When antiviral antibodies bind via their Fab domains to virus-infected cells, their Fc domain is bound to gE/gI and cannot trigger ADCC (Fig. 7.5). This protects virus-infected cells from lytic attack, thereby prolonging their production of virus. The effect of these HSV-encoded FcRs is illustrated in Fig. 8.9 which compares cells infected by wild-type HSV with cells infected by a gE-deleted mutant of HSV that lacks the ability to act as an FcR. Cells infected with the mutant HSV are much more sensitive to ADCC than are cells infected with wild-type HSV.

FURTHER READING

Reviews and Chapters

Nathanson N, McFadden G. Viral virulence. In: Nathanson N, ed. *Viral pathogenesis*. Philadelphia: Lippincott–Raven Publishers, 1997.

Virgin HV IV, Tyler KL, Dermody TS. Reovirus. In: Nathanson N, et al., eds. *Viral pathogenesis*. Philadelphia: Lippincott Raven Publishers, 1997.

Original Contributions

Bodian D. Viremia in experimental poliomyelitis. I. General aspects of infection after intravascular inoculation with strains of high and low invasiveness. *Am J Hyg* 1954;60:339–357.

Clark HF. Rabies viruses increase in virulence when propagated in neuroblastoma cell culture. *Science* 1978;199:1072–1075.

Collman R, Hassan NH, Walker R, et al. Infection of monocyte-derived macrophages with human immunodeficiency virus type 1 (HIV-1): monocyte- and lymphocyte-tropic strains of HIV-1 show distinctive patterns of replication in a panel of cell types. *J Exp Med* 1989;170:1149–1163.

Coulon P, Derbin C, Kucera P, et al. Invasion of the peripheral nervous systems of adult mice by the CVS strain of rabies virus and its avirulent derivative Av01. *J Virol* 1989;3:3550–3554.

Dietzschold B, Wiktor TJ, Trojanowski JQ, et al. Differences in cell-to-cell spread of pathogenic and apathogenic rabies virus in vivo and in vitro. *J Virol* 1985;56:12–18.

Dubin G, Socolof E, Frank I, Friedman H. Herpes simplex virus type 1 Fc receptor protects infected cells from antibody-dependent cellular cytotoxicity. *J Virol* 1991;65:7046–7050.

Endres MJ, Valsamakis A, Gonzalez-Scarano F, Nathanson N. Neuroattenuated bunyavirus variant: derivation, characterization, and revertant clones. *J Virol* 1990;64:1927–1933.

Endres MJ, Griot C, Gonzalez-Scarano F, Nathanson N. Neuroattenuation of an avirulent bunyavirus variant maps to the L RNA segment. *J Virol* 1991;65:5465–5470.

Griot C, Pekosz A, Davidson R, et al. Replication in cultured CdC12 muscle cells correlates with the neuroinvasiveness of California serogroup bunyaviruses. *J Virol* 1994;201:399–403.

Jackson RJ, Ramsay AJ, Christensen CD, et al. Expression of mouse interleukin-4 by a recombinant ectromelia virus suppresses cytolytic lymphocyte responses and overcomes genetic resistance of mousepox. *J Virol* 2001;75:1205–1210.

Janssen R, Gonzalez-Scarano F, Nathanson N. Mechanisms of bunyavirus virulence: comparative pathogenesis of a virulent strain of La Crosse and an attenuated strain of Tahyna virus. *Lab Invest* 1984;50:447–455.

Janssen R, Endres MJ, Gonzalez-Scarano F, Nathanson N. Virulence of La Crosse virus is under polygenic control. *J Virol* 1986;59:1–7.

Maassab HF, De Border DC. Development and characterization of cold-adapted viruses for use as live virus vaccines. *Vaccine* 1985;3:355–369.

Marshall ID, Fenner F. Studies in the epidemiology of infectious myxomatosis of rabbits. VII. The virulence of strains of myxoma virus recovered from Australian wild rabbits between 1951 and 1959. *J Hyg* 1960;58:485–488.

Minor PD. The molecular biology of poliovaccines. *J Gen Virol* 1992;73:3065–3077.

Nathanson N, Gittelsohn AM, Thind IS, Price WH. Histological studies of the monkey neurovirulence of group B arboviruses. III. Relative virulence of selected viruses. *Am J Epidemiol* 1967;85:503–517.

Sabin AB, Hennessen WA, Winsser J. Studies on variants of poliomyelitis virus. *J Exp Med* 1954;99:551–576.

Sabin AB. Properties and behavior of orally administered attenuated poliovirus vaccine. *JAMA* 1957;164:1216–1223.

Teng MN, Borrow P, Oldstone MBA, de la Torre J. A single amino acid change in the glycoprotein of lymphocytic choriomeningitis virus is associated with the ability to cause growth hormone deficiency syndrome. *J Virol* 1996;70:8438–8443.

Tyler KL, McPhee DA, Fields BN. Distinct pathways of viral spread in the host determined by reovirus S1 gene segment. *Science* 1986;233:770–774.

Chapter 9
Viral Persistence

The prototypical viral infections are acute and induce host defensive responses that clear the foreign agent and leave the host with long-lasting virus-specific immunity. However, a significant number of viruses are capable of persisting for the lifetime of the host. In order to persist, a delicate balance must be achieved so that, on the one hand, the host is not killed by the destructive effects of the virus while, on the other hand, the virus is able to evade the multitude of immune defenses that act to eliminate it. How this happens is the theme of this chapter.

The mechanisms of persistence range along a spectrum (Sidebar 9.1). At one extreme are viruses that continue to *replicate at high titers* over long periods of time, whereas at the other extreme are viruses that become *latent*, emerging at rare intervals to replicate for short periods of time. Between the ends of the spectrum are examples of *smoldering infections* that share characteristics of both replication and latency. Viruses employ a variety of strategies to escape immune surveillance, and these tend to be specific for different styles of persistence. Thus, immune tolerance often characterizes high-titer persistent infections whereas active immune responses are seen in many latent infections. A few selected examples of each style of persistence are listed in Table 9.1. The viruses selected for discussion represent some of the best studied examples, and the mechanisms of persistence are less

well analyzed or more complex in many uncited instances.

For some viruses, persistence may be partially determined by the conditions of infection. For instance, lymphocytic choriomeningitis virus (LCMV) causes a persistent infection in newborn mice but not in adult mice, as is true of several viruses of humans such as hepatitis B virus (HBV) (see Chapter 7). On the other hand, numerous herpesviruses cause persistent infections regardless of age at infection.

CELL CULTURE MODELS OF VIRAL PERSISTENCE

Many viruses can establish persistent infections in cultured mammalian cells. Although artificial, these models illustrate certain principles of virus–cell interaction in the absence of the complexities associated with infection of animals. Persistently infected cell cultures (sometimes called "carrier cultures") fall into two categories, those involving noncytocidal viruses and those involving lytic viruses (Table 9.2).

Nonlytic Viruses
There are several hallmarks of carrier cultures produced by nonlytic viruses. All cells in the culture are infected, and all daughter cultures established by the cloning of single cells are also infected. Conversely,

> **■ ■ ■**
> **SIDEBAR 9.1**
>
> **Rules of Persistence in Vivo**
>
> - RNA and DNA viruses from many different families can persist in vivo.
> - Strategies for persistence vary widely, from full-blown "high titer" replication to "latency," with intermediate examples of "smoldering" infection.
> - For each viral strategy, there is a corresponding strategy for evasion of host immune surveillance.
> - High-titer replication requires either that virus be noncytocidal or that there be rapid replacement by cellular proliferation of target cells. Immune surveillance cannot eliminate the virus, due to tolerance, immune complex formation, viral mutation, and other mechanisms.
> - Latency requires that the viral genome can persist in a nonreplicating mode, either integrated into the genome of the host cell or as an episome. Immune surveillance may be competent to eliminate the viral infection, which can only recur for brief intervals, or recurrence may be associated with intermittent reduction of immune defenses.
> - Smoldering infections involve continuous productive infection and cell-to-cell transmission at a low level. Usually, potentially effective immune surveillance is circumvented by mechanisms such as antigenic variation, infectious immune complexes, or intercellular bridges. **■**

it is usually impossible to "cure" the cultures by antiviral treatments, such as interferon or virus-specific antibodies. During extended passaging, free infectious virus may wane or even disappear, due to evolution from a replication-competent virus to defective genomes that may or may not produce budding viral particles. The continued presence of viral genomes can be detected by the presence of intracellular viral antigens or viral genetic sequences, or by the resistance of carrier cultures to superinfection with the same virus.

Lytic Viruses

Surprisingly, lytic viruses can often establish carrier cultures, although not as readily as nonlytic viruses. Characteristically, such carrier cultures are composed of a mix of infected and uninfected cells. The persistence of the culture depends on an equilibrium between the loss of infected cells and the increase of uninfected cells by cell division. Often, this equilibrium requires the presence of antiviral agents, such as interferon, which act as governors on virus replication. If this antiviral activity is removed by frequent changes of the culture medium, the culture may be destroyed. Conversely, such cultures can be "cured" by the addition of antiviral agents, such as specific antibody. If single-cell clones are derived from such carrier cultures, some of them will usually give rise to uninfected cultures.

TABLE 9.1

Human and animal viruses that employ different styles of persistence: a selected list

Virus family Example	Host(s)	Site of persistence	Cytocidal in permissive cells	Immune response
High-Titer Replication				
Arenaviridae LCMV	Mouse	Macrophage	No	Restricted
Hepadnaviridae HBV	Human	Hepatocyte	No	Restricted
Latent Infection				
Herpesviridae HSV	Human	Sensory neuron	Yes	Brisk
Polyomaviridae Papilloma	Human	Epidermal cells	Yes	Brisk
Smoldering Infection				
Picornaviridae TMEV	Mouse	CNS Glial cells	Yes	Normal
Paramyxoviridae Measles	Human	Neurons	Yes	Super normal
Lentiviridae HIV	Human	CD4 lymphocyte	Yes	Variable

LCMV, lymphocytic choriomeningitis virus; HBV, hepatitis B virus; HSV, herpes simplex virus; TMEV, Theiler's murine encephalomyelitis; HIV, human immunodeficiency virus.

TABLE 9.2		
Persistent viral infection of cell cultures: an experimental classification		
Characteristics of the carrier culture	Nonlytic virus	Lytic virus
What fraction of the cells are infected?	~100%	<100%
Are single-cell clones always infected?	Yes	No
Must antiviral factors be present in the culture medium to protect the cells?	No	Yes[1]
Can the culture be "cured" by adding antiviral antibody or interferon?	No	Yes[1]
Does the carrier culture resist superinfection with the same virus?	Yes[1] *receptors already saturated*	No[2]

The generalities above apply to many but not all instances. Notes: (1) Usual finding but there are exceptions. (2) Depends on experimental conditions and method of assay.

After Walker DL. The viral carrier state in animal cell cultures. *Prog Med Virol* 1964;6:111–148.

Variation in Virus and Cells

Continuous passage of carrier cultures may result in evolution of either virus or cells. As mentioned above, the infecting virus may be replaced by defective virus or by replication-competent virus with reduced cytopathogenicity. Carrier cultures established by lytic viruses may lead to the selection of cells that are relatively resistant to destruction by the virus. Likely, the artificial conditions of persistent infection select for preexisting viral or cellular variants that have a survival advantage in carrier cultures. Virus variants may exhibit reduced pathogenicity due to reduced efficiency at any of the steps in the replication cycle or due to diminished triggering of apoptosis. Cellular variants may be resistant because of differences in any of the multiple cellular molecules that are involved in steps from virus entry to release.

Using genetic reassortants, it is possible to map the variant viral genes and proteins selected during continuous passage. Reovirus is usually lytic in L929 cells (a continuous murine cell line) but carrier cultures can be established under controlled conditions. Virus isolated from these cultures is less lytic than wild-type virus, and changes in the S4 and S1 genes, which encode the σ3 and σ1 proteins, are responsible for establishing and maintaining the persistent phenotype (Table 9.3). The σ3 protein is the major outer capsid protein and the σ1 protein is the virus attachment protein. It appears that the variant capsid proteins associated with persistent viral variants reduce the efficiency of viral entry and the likelihood of overwhelming lytic infection, thus promoting the establishment of persistence.

Certain aspects of cell culture models are relevant to persistence in animals. Studies with nonlytic viruses demonstrate that long-term infections can readily be established, that they spread widely through the population of susceptible cells, and that variant viruses, often with reduced rates of replication, may be selected during long-term persistence. Observations on lytic viruses indicate that, surprisingly, they often can persist in a population of susceptible cells but that this usually requires an extraneous antiviral modulator, which prevents total destruction of the cell population. Furthermore, during persistence there is a tendency for evolution of a less lytic virus–cell relationship, both through selection of variant viruses that are less destructive and by the selection of cells that are less permissive for virus replication. These observations foretell some of the characteristic features of viral persistence in animal hosts.

TABLE 9.3		
Genetic determinants of viral persistence in cell culture		
	PREDOMINANT GENE SEGMENTS (TYPE 2wt or TY PE 3ts)	
Gene segments	Day 16	Day 230
L1	2	2
L2	2/3	2
L3	2	2
M1	2	2
M2	2	2
M3	2	2
S1	3/2	3
S2	2	2
S3	2	2
S4	3/2	3

Reovirus is a lytic virus but can be induced to cause persistent infections in cell culture by coinfecting cultures with a lytic wild-type virus (type 2wt) and a temperature-sensitive variant (type 3ts). The virus isolated late after persistent infection is a reassortant carrying the genes of the T2wt virus except for the S4 and S1 genes of the T3ts variant virus, and it appears that these two gene segments are responsible for the persistent phenotype.

After Ahmed R, Fields BN. Role of the S4 gene in the establishment of persistent reovirus infections in L cells. *Cell* 1982;28:605–612.

IMMUNE CLEARANCE OF VIRAL INFECTION

As a prelude to consideration of persistence, it is useful to recapitulate briefly the mechanisms by which the immune response controls and eliminates an acute virus infection (see Chapter 5). Effector T lymphocytes can destroy virus-infected cells, produce antiviral cytokines, and recruit mononuclear cells to sites of viral replication and destruction, whereas antibody neutralizes and opsonizes free infectious virions. In some instances, both antibody and virus-specific effector lymphocytes can purge virus-infected cells without destroying them. It is these mechanisms that a virus must evade in order to persist.

An example where viral clearance appears to be mediated by antibody is shown in Figure 2.6, which illustrates the disappearance of poliovirus from the plasma concomitant with the appearance of serum-neutralizing antibody. Evidence for the role of the immune response in the elimination of virus from solid tissues is based on the effect of experimental

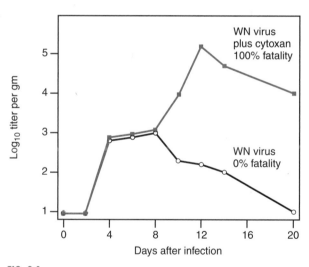

FIG. 9.1

Enhancement of an acute infection by chemical immunosuppression in order to implicate the role of the immune response in termination of an acute viral infection. Adult rats were injected intracerebrally with $10^{6.3}$ suckling mouse LD_{50} of West Nile virus (WNV), a flavivirus, and one group was immunosuppressed by three doses of cyclophosphamide on days 1, 8, and 14 after infection. In normal rats, WNV causes a minimal nonlethal infection of the brain and the virus is completely cleared by 20 days after infection. Immunosuppression potentiates the infection, which is not cleared but spreads slowly to kill all animals by 20 days. The curves indicate that suppression does not alter the replication of the virus for the first week of infection but interferes with the clearance process thereafter. (After Cole GA, Nathanson N. Potentiation of experimental arbovirus encephalitis by immunosuppressive doses of cyclophosphamide. *Nature* 1968;220:399–401.)

immunosuppression. Figure 9.1 summarizes an experiment in which an immunosuppressive drug abrogates the clearance of a neurotropic virus from the brain, indicating the ability of intact host immune defenses to terminate a viral infection and eliminate the causal agent.

MECHANISMS OF PERSISTENCE AND ESCAPE FROM IMMUNE SURVEILLANCE

Three patterns of persistence are diagrammed and contrasted with an acute infection in Figure 9.2. If the length of an acute virus infection is defined as the period from acquisition of infection to total elimination, then the duration varies from about 1 week (for rhinoviruses) to about 6 months (for HBV). Beyond those limits infections may be considered persistent. Some persistent infections undergo spontaneous termination, so that the duration of persistence in individual hosts varies from months to lifelong. This variability reflects the delicate balance between parasite and host. (Rabies virus, a virus that produces acute infections with incubation periods ranging up to months or years, is an exception.)

High-Titer Persistent Infections and Immune Tolerance

For a persistent virus to replicate at high titer, it must avoid catastrophic pathogenic effects by one of two mechanisms: either because it is not acutely cytocidal or because it attacks target cells that can be replenished by a very high rate of proliferation. Many viruses can replicate productively without causing cell death (see Chapter 4), and a number of them can cause persistent infections. In such instances, the initial dynamics resemble those of an acute infection, following which the virus titer decreases somewhat but then reaches a set point that may be maintained indefinitely or gradually decline. Examples of this pattern are HBV (Fig. 6.4), LCMV (Fig. 7.4), and human immunodeficiency virus (HIV) (Chapter 13).

Nonlytic Viruses High-titer persistence is often characterized by "tolerance," an apparent absence of virus-specific immunity. The mechanisms by which the tolerance can be induced include deletion of "forbidden" clones of naïve T lymphocytes in the thymus and exhaustion of peripheral virus-specific T lymphocytes in the presence of excess antigen (discussed in Chapter 7). Tolerance may be limited to specific components of the effector limb of the immune response. For instance, hepatitis B persis-

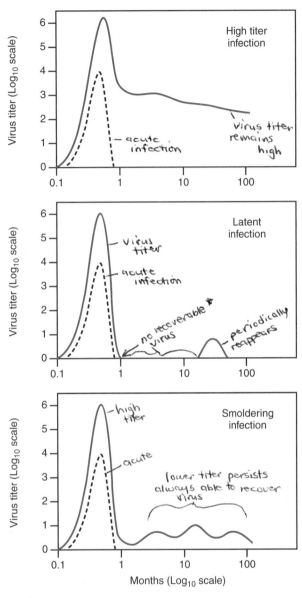

tence is characterized by absence of antiviral antibody against the surface protein of HBV (HBsAg), whereas anti-hepatitis B core antigen (anti-HBcAg) may be induced. LCMV persistence is characterized by absence of cellular immune responses whereas virus-specific antibody is produced. The presence of persistent high titers of virions and viral antigens may overwhelm either antibody or CD8 lymphocytes, so that assays must be interpreted with caution. Thus, in persistently infected mice, anti-LCMV antibodies circulate as immune complexes and not as free antibody, which originally led to the mistaken view that antibody was not induced in persistently infected mice.

Evidence for the role of immune tolerance in maintaining viral persistence is provided by the experimental termination of persistence by intravenous injection of virus-specific CD8 cells. Figure 9.3, summarizing an experiment with LCMV, exemplifies this phenomenon. Similar results have been obtained with HBV, where cytotoxic T lymphocytes (CTL) specific for hepatitis B surface anti-

FIG. 9.2

Patterns of virus persistence. For reference, an acute infection is shown by dashed line in all panels. (*Top*) High-titer infection. An acute phase is often apparent, following which the titer drops but persists at a high level for a long time, during which virus titers may remain stable or gradually wane. (*Middle*) Latent infection. An acute infection is followed by disappearance of replicating virus, which persists only as a latent genome. Periodic recrudescence occurs, with replication of the virus that may be accompanied by signs of disease, followed by another period of latency. Intervals between recrudescences may last from weeks to many years, and some infected individuals may never experience a recrudescence. (*Bottom*) Smoldering infection. An acute infection of varying extent is followed by marked reduction in overt virus replication, but infectious virus may be frequently recovered indicating that true latency has not occurred. (Modified after Johnson R. *Neurotropic virus diseases*. New York: Raven Press, 1985.)

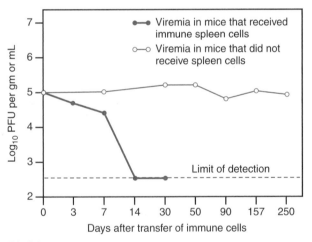

FIG. 9.3

Immune T lymphocytes can clear a persistent virus infection, implying that (in some instances) persistence involves escape from the cellular arm of the immune response. In this example, lymphocytic choriomeningitis virus causes persistent lifelong infection of mice. When such mice are treated by adoptive immunization of cytotoxic T lymphocytes from an immune donor ($10^{7.3}$ spleen cells from adult mice immunized by infection 60 days prior to transfer), the virus is cleared. The specificity of the process is shown by the requirement for syngeneic donor lymphocytes and by the ability of clones of CD8 T lymphocytes specific for an individual viral epitope to clear infection. The rate of clearance differs for different organs and tissues, implying some differences in the action of effector cells and in the mechanisms of immune evasion. Viremia: virus titer in serum of mice that received immune spleen cells in contrast to virus titer in mice that did not receive spleen cells. (After Ahmed R, Jamieson B, Porter DD. Immune therapy of a persistent and disseminated viral infection. *J Virol* 1987;61:3920–3929.)

gen (HBs) epitopes cleared virus from hepatocytes (Fig. 5.11).

It was noted that the expression of the viral genome may evolve in carrier cultures, and the same phenomenon has been observed in vivo. For instance, following the infection of newborn mice with LCMV, the nucleoprotein and envelope glycoproteins are both expressed in infected neurons for the first week, but over the subsequent 10 weeks the expression of glycoproteins gradually wanes while there is no diminution in nucleoprotein levels. This phenomenon may play a role in persistence because the absence of glycoproteins (the only viral protein expressed on the cell surface) would make neurons poor targets for antibody recognition.

Lytic Viruses It is unusual for high-titer persistence to be produced by a cytolytic virus, but the primate lentiviruses represent an important exception. Figure 7.12 shows the persistent viremia produced by a particularly virulent strain of simian immunodeficiency virus (SIV). The main target cells for these lentiviruses are CD4 lymphocytes that undergo lytic infection. It has been calculated that the continuous destruction of CD4 cells results in a reduction of the average half-life of these cells from 75 to 25 days. However, the bone marrow is able to respond to the abnormal rate of destruction by increasing the production of naïve CD4 cells, at a rate sufficient to maintain a reasonable concentration of circulating CD4 cells. This permits this generally lytic virus to persist at a high titer for an extended period of time, although eventually the bone marrow is unable to compensate and CD4 levels drop, leading to functional immunodeficiency. The pathogenesis of HIV and AIDS is discussed in Chapter 13.

In contrast to most high-titer persistent infections, lentiviruses induce immune responses rather than tolerance. The immune response to lentiviruses is quite effective, as judged by its rapid containment of the acute phase of infection, resulting in a reduction from peak viremia at about 6 weeks to a set point at about 3–6 months that may be as much as 1,000-fold lower than the peak viremia. Once this set point is reached, a dynamic equilibrium is established between virus production and clearance. The half-life of individual SIV virions is less than 30 minutes in the absence of immunity and about 10 minutes in infected animals with an established immune response. It has been calculated that to maintain virus titers of 10^2 to 10^4 infectious virions per milliliter of plasma requires the production of 10^{10} to 10^{12} new infectious virions daily. In this instance, high-titer persistence is maintained by an extraordinary rate of virus production that exceeds the rate at which a potent immune response can clear the virus.

Latency

Latent infections are produced by a considerable number of herpesviruses, including herpes simplex viruses (HSV), varicella-zoster virus (VZV), Epstein–Barr virus, and cytomegalovirus (CMV) of humans. There is a characteristic sequence of events following primary infection. Initially the virus replicates in permissive cells at the portal of entry. The virus is lytic and destroys permissive target cells. Once immune induction has occurred, the virus is cleared and appears to be eliminated.

However, the viral genome persists in a latent form. Latency occurs in one or more cell types—such as neurons—that are distinct from the permissive cell types that support productive lytic infection. Neurons appear to be restrictive or permissive, depending on their physiologic state. Under conditions of restriction, the virus undergoes the early steps of entry and uncoating, but further steps in replication are blocked (see discussion of HSV below). In the absence of active replication, the viral genome is maintained in one of several forms, depending on the specific virus and host cell. In some instances, the double-stranded DNA genome integrates into the host genome, whereas in other examples the genome persists as a nonintegrated episome in the nucleus or cytoplasm.

If latency occurs in cell types such as neurons that do not divide, then there is no need to replicate the latent genome. If the latent genome is maintained in dividing cells, then the genome must be replicated or it will be diluted to extinction. If the genome is integrated into the host genome, as with retroviruses, then it will be automatically replicated during the cell cycle. Episomal DNA can also be replicated by the enzymes involved in copying cellular genomes. However, there are no parallel mechanisms for RNA, so RNA viruses cannot assume a latent state unless they undergo reverse transcription to DNA intermediates. Latent viral genomes can be detected by in situ polymerase chain reaction methods that—at their most sensitive—can detect as little as one genome per infected cell (Fig. 9.4).

Latency maintains the viral genome for the lifetime of the infected host. Activation of latent infections occurs at irregular intervals, and may never occur in some infected individuals. Activation of latent genomes can be initiated by a number of stimuli, characteristic for each virus. For instance, HSV can

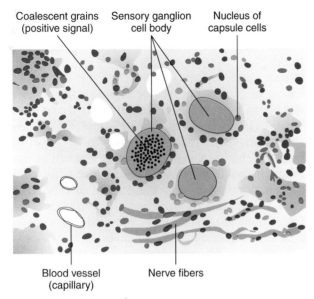

Coalescent grains (positive signal)   Sensory ganglion cell body   Nucleus of capsule cells

Blood vessel (capillary)   Nerve fibers

FIG. 9.4

Demonstration of latent genomes. Herpes simplex virus (HSV) is maintained as a latent genome in first-order sensory neurons, whose cell bodies are located in sensory ganglia. This photomicrograph shows the large cell bodies of three sensory neurons, and the grains mark the nucleus of one of these cells that contains HSV gene sequences as detected by in situ hybridization. The probe detects RNA transcripts from the latency-associated transcript domain of HSV. (After photograph supplied by Nigel Fraser, University of Pennsylvania.)

be activated by fever, sunburn, and trigeminal nerve injury. Most of these stimuli appear to act on the primary sensory neurons in which latent HSV genomes are maintained. However, on occasion, waning of the immune response can serve as a trigger for activation of some herpesviruses.

Following reactivation of HSV, the viral genome may be transported by axoplasmic spread in both centripetal and centrifugal directions. Centrifugal spread, toward the periphery, conducts the virus to the skin where it may replicate and spread, causing herpes labialis ("fever blister" or "cold sore"). After spreading for a few days, host defenses prevent further spread, and the skin lesion heals. Centripetal spread from the trigeminal ganglion conducts the HSV genome to the central nervous system (CNS), where, in some instances, it can spread to cause an encephalitis that may be devastating.

Typically, viruses that cause latent infections induce a brisk and potent immune response that clears the initial infection. However, latently infected cells do not express viral proteins, permitting escape from immune surveillance. Furthermore, when the latent infection is activated, immune surveillance limits its spread. Latent viruses cannot be spread from host to

host, but virus produced during activation may be spread to another host. For instance, activation of latent VZV produces characteristic skin lesions in older adults; seronegative children exposed to virus aerosolized from these lesions can develop chicken pox, the primary form of VZV infection.

Smoldering Infections

"Smoldering" infections fall between the extremes of high-titer persistence and latency. Infectious virus is produced, but at minimal levels that may require special methods for detection and isolation. Virus continues to spread from infected to uninfected cells, but often at an indolent tempo. If the virus is pathogenic, it may produce a gradually progressive chronic disease. There is a detectable immune response to the virus, and in some instances the response may be hypernormal due to the continuous presence of viral antigens. The ability of a virus to spread in the presence of a potentially effective immune response is a paradoxical phenomenon, and involves a variety of strategies several of which may operate in any given example. Some of the more important strategies are described below.

Immunologically Privileged Sites   There are a few organs and tissues that appear to favor virus persistence, particularly the brain and kidney. The brain has classically been considered an immunologically "privileged" site because immunologic effector mechanisms may spare cells bearing foreign antigens if these cells are located in the brain (in contrast to foreign cells in other sites). This concept was originally enunciated by Medawar and collaborators who observed that grafts of allogeneic or xenogeneic tissues were more likely to survive in the brain than on the skin or at other sites. There are at least two factors that account for virus persistence in the brain. First, the blood–brain barrier limits the trafficking of lymphocytes through the brain and, second, neurons express little if any major histocompatibility complex (MHC) class I molecules rendering them relatively poor targets for virus-specific CTL.

The kidney is the other tissue that frequently harbors persistent viruses, such as JC and BK polyoma viruses and CMV. Consistent with this observation, LCMV is cleared more slowly from the kidney than from other tissues, even the brain. However, there is no clear explanation why virus in the kidney should be able to evade immunologic surveillance, although it has been speculated that lymphocytes may not readily cross the subendothelial basement membrane to access infected glomerular epithelial cells (see Fig. 6.5 for anatomic relationships).

Intercellular Bridges　In some instances, the process of entry of viruses into cells can be short circuited, so that a transient intercellular bridge is formed permitting the viral genome to pass from cell to cell without having to survive in the extracellular environment (Fig. 9. 5), thus providing a means of avoiding neutralizing antibody. This phenomenon probably is operative in a progressive fatal disease called subacute sclerosing panencephalitis (SSPE). In SSPE a defective variant of measles or rubella virus spreads gradually from neuron to neuron in spite of extraordinarily high titers of neutralizing antibody in the extracellular fluid of the brain parenchyma.

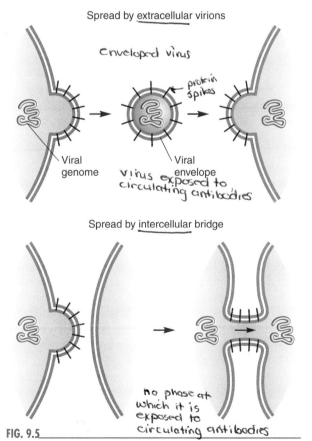

FIG. 9.5

Cell-to-cell transmission of a virus by intercellular bridges, a hypothetical reconstruction. (*Top*) Normal cell-to-cell spread of an enveloped virus as free extracellular virions. (*Bottom*) Cell-to-cell spread of the viral genome through a transient intercellular bridge. This phenomenon has been invoked to explain the spread of measles and rubella viruses during persistent infection of the brain.

Suppression of MHC Class I Expression　A number of viruses, including adenoviruses and lentiviruses, encode specific proteins that are capable of down-regulating the expression of MHC class I molecules (see Chapter 4). Virus-infected target cells are rendered relatively less sensitive to virus-specific CTL attack, permitting them to continue to produce virions for an extended period. Experimental deletion of the down-regulating protein appears to reduce the ability of the virus to persist in vivo, suggesting that MHC class I damping may be a significant factor in persistence (Fig. 7.12).

Infectious Immune Complexes　In some instances where a virus persists in the presence of an active immune response, infectivity in the blood circulates in the form of immune complexes that are composed of infectious virions coated by virus-specific antibodies. Immune complexes can be demonstrated by the addition of anti-IgG antisera that "neutralize" the infectivity (Table 6.4). The molecular mechanism by which an antibody-coated virion can retain its infectivity has never been well elucidated. One possibility is that the complex is bound to Fc receptors on macrophages and internalized in vacuoles in which the complex dissociates, followed by infection of the macrophage. Consistent with this hypothesis, several of the persistent viruses (LCMV, lactic dehydrogenase virus, and Aleutian disease virus) for which infectious complexes have been demonstrated target macrophages as a major host cell.

Impaired CTL Function　Recent advances have made it possible to quantify the function of CTL according to their intracellular content of effector molecules, such as interferon γ (IFN-γ), tumor necrosis factor α (TNF-α), and perforin (described in Chapter 5). If CTL that recognize individual viral epitopes are sorted using tetramer staining, they vary in their content of effector molecules. It appears that HIV-specific CTL may be deficient in perforin content, and this could provide an additional mechanism for escape from immune surveillance.

Antigenic Variation　During the course of persistent infection, there may be a selection for viral variants that are able to escape neutralization. The ability of antibody to select for "escape" mutants has been repeatedly documented in cell cultures. When a virus is plaqued in the presence of a single neutralizing monoclonal antibody, the titer is reduced drastically but, characteristically, some plaques occur at the frequency of about 10^{-5}. When these plaques are grown into virus stocks, they are rendered resistant to neutralization (Fig. 9.6) by the selecting monoclonal antibody (but not necessarily by other neutralizing monoclonal antibodies). Such resistant virus variants usually represent point mutations, often in the viral attachment protein.

Neutralization escape mutants also play a role in some persistent infections of animals. This phenomenon has been observed with several persistent lentiviruses, such as visna maedi virus of sheep and

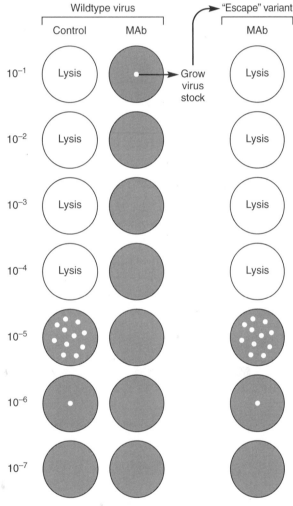

FIG. 9.6

In cell culture, antibody selects for "escape" mutants that are nonneutralizable. In this example, a virus is titrated in the absence and presence of a single monoclonal antibody, which reduces the viral titer by 10^5. When the virus that escapes neutralization is grown into a stock, its titer is not reduced in the presence of the selecting antibody, indicating that it is an "escape" mutant. Shaded area: cellular monolayer; "holes," individual viral plaques; "lysis," viral destruction of monolayer. (After Gonzalez-Scarano F, Shope RE, Calisher CH, Nathanson N. Monoclonal antibodies against the G1 and nucleocapsid proteins of La Crosse and Tahyna viruses. In: Calisher CH, ed. *California serogroup viruses*. New York: Alan R Liss, 1983.)

equine infectious anemia virus. Equine infectious anemia virus is a lentivirus that produces a lifelong persistent infection of horses; the virus may be isolated from the blood even though the animals develop neutralizing antibody. Newly infected horses undergo discrete episodes of acute anemia, associated with bursts of viral replication. In each instance, the virus isolated during the episode of illness resists neutralization by serum obtained at the time of virus isolation (Table 9.4). However, the same serum can neutralize virus isolated at earlier times in infection. In other words, there is sequential replacement of virus with newly emerging variants that can escape neutralization. Ponies that survive repeated episodes of illness finally develop such a broad neutralization response that they can suppress all potential variants, thereby modulating but not clearing the smoldering infection.

The circumstances under which antigenic variants are selected in vivo have been studied using LCMV infection of mice. Ordinarily, when adult mice are infected with LCMV, the virus is cleared by CD8 CTL (Fig. 9.3). However, CD8 "knockout" mice (CD8$^{-/-}$) also clear virus, although less efficiently than intact animals. In these mice, clearance is mediated by neutralizing antibody, illustrating the synergistic role of different arms of the immune response. However, some weeks after clearance the virus may reappear (a phenomenon only seen in CD8$^{-/-}$ mice), and the reemergence is due to the outgrowth of escape variants that resist neutralization (Fig. 9.7). When cloned, such variants can be shown to have a few mutations at sites that alter their neutralizing epitopes.

Escape mutants can also be selected by the cellular immune response. Most evidence for this phenomenon has been derived from somewhat contrived experimental models, but it is possible that similar mechanisms play a role in some naturally occurring persistent infections. Figure 9.8 shows an example where variants of LCMV were selected under pressure from specific antiviral CTL clones. The variants were sequenced and shown to bear mutations in the epitope against which the CTL clone was directed. In this example, there are three immunodominant CTL epitopes, and a variant virus with mutations in two of these epitopes was compared with the wild-type virus for the ability to persist after infection of adult mice. The double escape mutant (but not single mutants) persisted for much longer than wild-type virus, although it was eventually eliminated. In early SIV infection of macaques—a model for

		NEUTRALIZATION INDEX ($\log_{10}$) OF SERUM COLLECTED ON THE INDICATED DAY AFTER INFECTION					
Virus isolate (day of infection)	Fever spike (day of infection)	0 d	20 d	44 d	62 d	83 d	155 d
0 d	—	0	0	0.7	2.5	3.2	3.2
20 d	21	0	0	1.0	1.5	1.5	2.5
44 d	44	0	0	0	3.5	5.4	>5.4
62 d	62	0	0	0	0	2.0	2.0
83 d	83	0	0	0	0	0	3.5
155 d	155	0	0	0	0	0	0

TABLE 9.4

A smoldering infection associated with antigenic variation of persistent virus

Equine infectious anemia virus infection is associated with periodic febrile episodes. The table shows data from one infected horse that experienced several fever spikes each lasting about 5 days. During each fever spike, virus isolated from the blood was tested for neutralizability. Each virus isolate was neutralized by sera collected after the time of the isolate but not by sera collected at or prior to the time of isolation. Likewise, each serum neutralized all the virus isolates made prior to the date of the serum but none of the isolates made thereafter. This isevidence of continual antigenic drift of the virus, which probably explains the burst of replication associated with each fever spike as well as the persistence of the virus in the face of an active neutralization response.

After Kono Y, Kobayashi K, Fukunaga Y. Antigenic drift of equine infectious anemia virus in chronically infected horses. *Arch Virusforschung* 1973;41:1–10.

HIV in humans—a cloned virus undergoes mutations in the Tat protein in apparent response to the induction of Tat-specific CTL, suggesting that viral variation may play an important role in this persistent infection.

Evolution of virus during persistent infection probably plays a role in virus survival and disease production in certain human viral diseases, such as HIV/AIDS. During SIV infection of macaques, there is continual evolution of the viral phenotype due to the selection of viral clones that are increasingly adapted to the host as defined by their increasing capacity for in vivo replication and rapid production of immunodeficiency. In some instances, such viruses also are neutralization escape mutants.

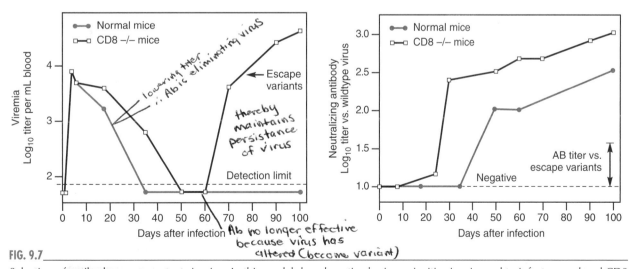

FIG. 9.7

Selection of antibody escape mutants in vivo. In this model, lymphocytic choriomeningitis virus is used to infect normal and CD8 knockout (CD8$^{-/-}$) mice (the latter cannot produce effector cytotoxic T lymphocytes). The normal mice clear virus rapidly whereas the knockout mice clear virus more slowly, mediated by neutralizing antibody. However, in some mice, the virus reemerges weeks later. The reemerging virus is an antibody escape variant that resists neutralization. (After Ciurea A, Klenerman P, Hunziker L, et al. Viral persistence in vivo through selection of neutralizing antibody-escape variants. *Proc Natl Acad Sci USA* 2000;97:2749–2754.)

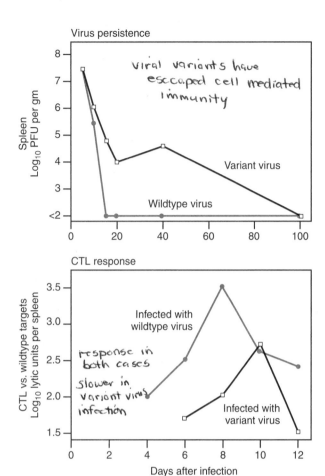

Virus persistence

viral variants have escaped cell mediated immunity

Variant virus

Wildtype virus

CTL response

Infected with wildtype virus

response in both cases

slower in variant virus infection

Infected with variant virus

Days after infection

FIG. 9.8

Persistent infection associated with viral variants that escape the cellular immune response. In this example, variant lymphocytic choriomeningitis viruses were selected in cell culture and compared with the wild-type parent virus for their ability to persist in adult mice after intravenous injection. (*Top*) The variant virus persists in the spleen for much longer than the wild-type virus. (*Bottom*) T lymphocytes from mice infected with the variant virus have much less cytotoxic T-lymphocyte activity than do lymphocytes from mice infected with the wild type virus, when both are tested against target cells infected with wild-type virus.(After Aebischer T, Moskophidis D, Rohrer UH, et al. In vitro selection of lymphocytic choriomeningitis virus escape mutants by cytotoxic T lymphocytes. *Proc Natl Acad Sci USA* 1991;88:11047–11051.)

EXAMPLES OF PERSISTENCE

To provide a flavor of the diverse and subtle nature of persistent viruses, a few well-studied examples are described below. They illustrate different modes of persistence, including latency (HSV), smoldering infection with demyelination (Theiler's virus), and smoldering infection with progressive encephalitis (SSPE caused by measles virus).

Herpes Simplex Virus

HSV represents a classic example of a virus that persists because of its ability to establish latent infections (Fig. 9.9). HSV is an alphaherpesvirus of humans that includes two serotypes: HSV-1 infects skin and mucous membranes of the mouth, whereas HSV-2 infects the genital mucosa. HSV is transmitted by skin-to-skin contact and initiates a productive lytic infection of epithelial cells. In the lytic cycle, three sets of viral genes are expressed in sequential order: IE (immediate early), E (early), and L (late).

During epithelial infection, the virus may enter the peripheral dendrites of first-order sensory neurons, whose peripheral processes interdigitate among the epithelial cells. Fusion of the viral envelope with the plasma membrane of the neuronal process releases the viral nucleocapsid into the neuron, where it can be passively transmitted by retrograde axoplasmic flow to the cell body located in a sensory ganglion, such as the trigeminal ganglion. Sensory neurons are permissive for HSV, and the virus may replicate in the ganglion with destruction of some cells. After 1–2 weeks, host immune defenses terminate productive infection in both skin and ganglion, and the virus appears to have been cleared.

However, in some sensory neurons—but not epithelial cells—the virus establishes latency. The latent state is defined experimentally by the failure to recover infectious virus from a homogenized ganglion, whereas virus can be recovered from the same ganglion by cocultivating explanted ganglion fragments with permissive cultured cells. Furthermore, in situ techniques can be used to demonstrate persistence of the viral genome, which is maintained as a closed circular double-stranded DNA nonintegrated episome located in the nucleus of ganglion cells (Fig. 9.4). In a given ganglion, up to 5%–10% of the sensory neurons may carry the latent genome, and there may be up to 30 copies of the genome per infected cell. Latent HSV can also be detected in second-order sensory neurons located within the CNS and could be a source of reactivation encephalitis.

The molecular mechanisms that determine whether HSV will become latent or complete its full replicative cycle are not well understood, but the metabolic state of the infected sensory neuron is probably a major determinant. The neuron either fails to provide sufficient factors to permit expression of viral genes or produces proteins that inhibit the expression of these genes. No individual viral genes appear to be essential for establishment of latency, since a wide variety of HSV mutants, each

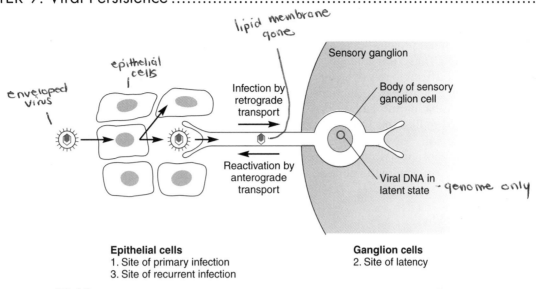

[handwritten annotations on figure: "lipid membrane gone", "epithelial cells", "enveloped virus", "genome only"]

Epithelial cells
1. Site of primary infection
3. Site of recurrent infection

Ganglion cells
2. Site of latency

FIG. 9.9

Life cycle of a persistent latent virus, herpes simplex virus (HSV). (1) HSV infects cells of the skin or mucous membranes, where it causes a productive lytic infection. (2) Virions may enter the terminals of primary sensory neurons, and the nucleocapsid of the virus is passively transported in a retrograde direction to the cell body—located in a sensory ganglion—where it also can undergo a productive lytic cycle. In some sensory neurons the virus becomes latent, and the DNA genome is maintained in the nucleus as an episome. Latent HSV DNA undergoes limited transcription but no proteins are synthesized. (3) The latent genome may be activated, in which case a productive lytic infection is again initiated, and nucleocapsids can be passively transported in an anterograde direction, delivering the virus to the epithelium, where it can once again replicate to produce a "cold sore" or "fever blister." HSV is shown as a nucleocapsid surrounded by an outer envelope. (After Ahmed R, Morrison LA, Knipe DM. Viral persistence. In: Nathanson N, et al., ed. *Viral pathogenesis*. Philadelphia: Lippincott–Raven Publishers, 1997.)

lacking specific viral genes, can establish latency and subsequently be recovered by growth of ganglion explants under conditions permissive for the mutant virus. For instance, ts mutants of HSV that are restricted at the body temperature of the mouse (37°C) can establish latency after peripheral infection and be recovered from sensory ganglia explanted at the permissive temperature (33°C).

During latency most viral genes are not transcribed, with the exception of the latency-associated transcripts (LAT) genes, one small segment of the HSV genome. However, the LAT transcripts, which encode two open reading frames, are never translated. Furthermore, LAT-negative mutants can establish latency, although less efficiently than wild-type virus. The function of the LAT genes, and their relevance to latency, remains enigmatic. Latent HSV persists throughout life ("in contrast to love, HSV is forever") and may be periodically reactivated in some infected hosts whereas others never experience reactivation. During reactivation, the virus initially replicates in the sensory ganglion. Viral nucleocapsids are then passively transported in the anterograde direction to the skin where they can

produce a lytic infection, similar in character to the initial infection. Again, this infection is brought to a close by host immune defenses. Rarely, following reactivation, HSV initiates an acute and—prior to the development of antiviral drugs—often fatal encephalitis. It is not known whether reactivation encephalitis arises from virus transported from the sensory ganglia (first-order sensory neurons) or from virus latent in second-order sensory neurons in the CNS.

A variety of natural or experimental insults can lead to reactivation, including fever, sunburn, stress, hormonal changes, immunosuppression, or trauma to the trigeminal nerve. It is postulated that reactivation of latent HSV is initiated by a physiologic change in the sensory neurons harboring the latent genome. Under this view, reactivating stimuli lead to up-regulation of cellular proteins, such as transcriptases, that directly or indirectly initiate transcription of the HSV IE genes. Alternatively, activation might involve release of a hypothetical inhibitor of HSV replication. The specific cellular factors involved in activation have yet to be defined. In addition, certain viral genes appear to modulate reactivation, particularly the LAT genes and a protein designated ICP0 (infected cell protein 0), as viral mutants lacking these genes are more difficult to reactivate in vivo.

It should be noted that there is an alternative hypothesis that suggests that during latency there is a continual low level of replication limited to a few ganglion cells. Under this view, some activation triggers, such as immunosuppression, could release productive replication from immune control, leading to spread of the virus.

Picornaviruses (Theiler's Virus)

Theiler's murine encephalomyelitis virus (TMEV) is an example of a virus that persists as a smoldering infection in the presence of an apparently brisk and intact immune response. TMEV is an enzootic picornavirus (small RNA virus) of wild mice that spreads in nature as an enteric virus, being excreted in the feces and contracted by ingestion. Wild-type isolates of TMEV fall into two distinct groups, virulent and persistent, based on their biological properties (Table 9.5). After intracerebral injection, the virulent strains replicate to high titer in neurons of the brain and spinal cord, and produce a rapidly progressive encephalomyelitis that is invariably fatal. Intracerebral injection of sublethal doses of virulent strains cause a transient CNS infection, but the virus is always cleared by immune defenses and never persists. The persistent strains of TMEV cause a sublethal encephalitis due to an infection of neurons, similar in character but less widespread than that caused by the virulent strains. In contrast to the virulent strains, the persistent strains of TMEV are never cleared from the CNS and low titers of infec-

tious virus can be isolated without requiring special methods such as cocultivation.

During the persistent phase of CNS infection, there is a dramatic change in the localization of TMEV. The virus disappears from the gray matter (part of the CNS where most neurons are located) and moves into the white matter (part of the CNS where nerve fibers are concentrated), infecting glial cells (supporting cells of the CNS) and microglia (macrophages of the CNS). The infection of oligodendroglia destroys some of the myelinated sheaths that surround nerve fibers, and this demyelination causes neural dysfunction, manifested by a waddling gait and incontinence.

In spite of the marked biological differences between virulent and persistent strains of TMEV, their RNA sequences are 90% identical and their amino acid sequences are 95% identical, illustrating the subtle differences that can exist between persistent and nonpersistent viruses. To elucidate the genetic determinants of virulence and persistence, chimeric viruses have been constructed that combine different domains of the genomes of the two groups of viruses. Full expression of virulence appears to require the capsid structural proteins and the 5′ noncoding domain of the genome. Substitution of individual segments (any of four segments that spanned the whole genome) of the virulent viral genome into the genome of the persistent virus produced chimeric viruses all of which had markedly reduced acute virulence, persisted, and caused demyelination. This suggested that the virulent TMEV strains were potentially capable of persisting and causing demyelination if mice survived the acute phase of infection.

The mechanism(s) of persistence of TMEV has been difficult to elucidate. During persistence, TMEV is confined to glia and macrophages, both of which are quite restrictive and produce low levels of viral capsid proteins and little free infectious virus. This may make infected cells less susceptible targets for the humoral and cellular immune response. Persistence does not appear to be associated with selection of viral variants, such as antigenically altered escape mutants, although a few neutralization escape mutants have been observed. TMEV elicits brisk neutralizing antibody and CTL responses, and abrogation of either component of the immune response delays viral clearance from neurons during the acute phase of infection. However, there is no evidence that persistence is due to the failure of immune surveillance because persistent strains of TMEV induce an active immune response.

TABLE 9.5

A comparison of two Theiler's murine encephalomyelitis virus groups with different biological phenotypes, one of which persists in the CNS

Property	Virulent (GDVII) subgroup	Persistent (TO) subgroup
Relative lethality after IC injection	10,000	1
Acute polioencephalitis	Marked, lethal	Moderate, sublethal
Demyelination	No	Yes
Persistence in the CNS	No	Yes
Plaque size	Large	Small
Temperature sensitive in cell culture	No	Yes

The two groups are named after prototype isolates, the virulent GDVII and the less virulent persistent TO isolate, and there are a number of wild-type isolates of each group.

After Tsunoda I, Fujinami RS. Theiler's murine encephalomyelitis virus. In: Ahmed R, Chen I, eds. *Persistent viral infections*. New York: John Wiley and Sons, 1999.

Morbilliviruses (Measles)

SSPE is an example of a smoldering infection, in this case due to measles virus, that persists in the face of a supernormal immune response. In this example, escape from immune surveillance is associated with the selection of variant viruses that lose the ability to mature and bud properly while maintaining the gene functions for replication of the viral nucleocapsid. Measles virus is a human morbillivirus consisting of an envelope enclosing a nucleocapsid containing a single-stranded RNA genome of negative polarity. The genome encodes six proteins; the nucleocapsid (N), phosphoprotein (P), and large (L)–polymerase–protein are associated with the viral RNA, whereas the hemagglutinin (H) and fusion (F) proteins are inserted into the viral envelope, and the matrix (M) protein binds the viral core to the envelope during virion maturation, which is accomplished by budding through the plasma membrane. SSPE variant viruses exhibit underexpression or defects in one or more of the M, F, and H proteins, inhibiting budding and the production of free infectious virions (Fig. 9.10).

Measles is a ubiquitous virus that spreads to children by the respiratory route and causes a systemic febrile infection with a rash that usually resolves in 1–2 weeks with no serious consequences. However, one rare (approximately 1 case per 100,000 primary infections) complication of measles is SSPE, which occurs unpredictably in apparently normal children following uneventful recovery from acute measles. Several years after measles, these children develop a progressive subacute encephalitis that is invariably fatal in 6–12 months. On brain biopsy or at autopsy, measles antigens can be detected in the brain, and electron microscopy reveals measles nucleocapsids in neurons and glial cells in the brain.

It is very difficult to isolate measles virus from the brain tissue of patients with SSPE, except by special methods involving explantation and cocultivation. Furthermore, viral isolates are usually defective, with most defects localized to the matrix protein or the viral envelope proteins, so that SSPE isolates can be considered to be "maturation-defective" virus variants. As such, they probably spread in vivo by intercellular bridges (Fig. 9.5). During the disease, patients exhibit supernormal virus-specific immune responses, with very high levels of measles antibodies in their blood and spinal fluid. Although these antibodies are capable of neutralizing normal measles virus, they cannot interrupt the spread of SSPE variant viruses, which fail to form free infectious virions.

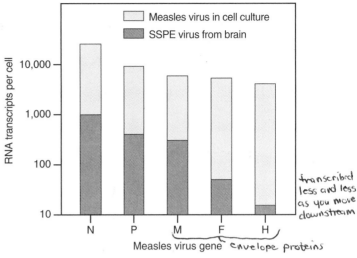

defective in transcribing proteins which assist in budding

transcribed less and less as you move downstream

envelope proteins

FIG. 9.10

Persistent measles virus infection of the brain (subacute sclerosing panencephalitis, SSPE) is associated with reduced and defective virus replication. The figure compares the per-cell expression of five viral RNAs, plotted according to their order of transcription (3' to 5' in this negative strand virus). Viral RNA in a lytic cell culture is compared with that in the brain of a patient with SSPE. In SSPE there is marked reduction in all viral mRNAs. The "gradient" of RNA transcription is much steeper, resulting in a relative paucity of the "downstream" RNAs that encode the M, F, and H that form the viral envelope. This pattern explains the defectiveness of the SSPE virus genome, which produces few if any free infectious virions. (After Cattaneo R, Rebman G, Baczko K, et al. Altered ratios of measles virus transcripts in diseased human brains. *Virology* 1987;160:523–526.)

In summary, it appears that SSPE viruses are defective variants that spread very poorly but can elude neutralizing antibody. As a result, they cause a slowly progressive encephalitis that is eventually fatal.

DISEASES ASSOCIATED WITH PERSISTENT INFECTION

Persistent virus infections are often associated with chronic progressive illnesses that are quite diverse, although individual viruses usually produce a stereotyped and limited spectrum of disease. Some persistent infections invariably cause progressive lethal disease, whereas others often are inapparent for the lifetime of the host. Table 9.6 lists some examples of the different kinds of diseases associated with virus persistence. A number of persistent viruses are oncogenic, causing a wide variety of neoplasms (discussed in Chapters 10 and 11). High-titer persistence is often associated with circulating immune complexes, which eventually produce glomerulonephritis or vasculitis due to complex de-

TABLE 9.6

Diseases associated with persistent viral infections: selected examples

Virus family	Virus	Host	Disease
Oncogenic Viruses			
Retro	MuLV	Mice	Hematopoietic, lymphoreticular neoplasms
Hepadna	HBV	Humans	Hepatocellular carcinoma
Papilloma	HPV	Humans	Cervical carcinoma
Herpes	EBV	Humans	Burkitt's lymphoma
High-Titer Persistence			
Arena	LCMV	Mice	Glomerulonephritis, vasculitis
Parvo	Aleutian disease	Mink	Glomerulonephritis, vasculitis
Latent Infections			
Herpes	HSV	Humans	Cold sores, encephalitis
	CMV	Humans	Pneumonitis, retinitis, encephalitis
	EBV	Humans	Mononucleosis
	VZV	Humans	Herpes zoster
Smoldering Infections			
Morbilli	Measles	Humans	Subacute sclerosing panencephalitis
	CDV	Dogs	Encephalitis, demyelination
Retro	HTLV I	Humans	Tropical spastic paraparesis (HAM)
Polyoma	JC	Humans	Progressive multifocal leucoencephalopathy
Lenti	VMV	Sheep	Interstitial pneumonitis, demyelination
	EIAV	Horses	Episodic hemolytic anemia
	HIV	Humans	AIDS

MuLV, murine leukemia virus; HBV, hepatitis B virus; HPV, human papilloma virus; EBV, Epstein–Barr virus; LCMV, lymphocytic choriomeningitis virus; HSV, herpes simplex virus; CMV, cytomegalovirus; VZV, varicella-zoster virus; CDV, canine distemper virus; HTLV, human T-cell leukemia virus; JC, a human polyoma virus; VMV, visna maedi virus; EIAV, equine infections anemia virus; HIV, human immunodeficiency virus.

position in and around the walls of blood vessels. Latent herpesviruses are periodically activated, replicate, and cause lytic infections in permissive cells. The nature of the consequent disease depends on the permissive cell type and includes destructive lesions of the mucosa or skin (HSV, VZV), pneumonitis (CMV), retinitis (CMV), encephalitis (HSV), and mononucleosis (Epstein–Barr virus). Smoldering infections of the brain cause progressive chronic destructive encephalitis or myelitis (measles virus, CDV, JC polyoma virus, HTLV-1) that may be accompanied by demyelination (Theiler's virus, VMV, CDV, HTLV-1). Other smoldering infections are associated with diverse illnesses, such as AIDS (HIV, SIV, FIV) or episodic acute anemia (equine infectious anemia virus).

FURTHER READING

Reviews and Chapters

Ahmed R, Chen ISY, eds. *Persistent viral infections.* New York: John Wiley and Sons, 1999.

Jakob J, Roos RP. Molecular determinants of Theiler's murine

encephalomyelitis-induced disease. *J Neurovirol* 1996;2: 70–77.

Koup RA. Viral escape from CTL recognition. *J Exp Med* 1994;180:779–782.

Tsunoda I, Fujinami RS. Theiler's murine encephalomyelitis virus. In: Ahmed R, Chen I, eds. *Persistent viral infections.* New York: John Wiley and Sons, 1999.

Walker DL. The viral carrier state in animal cell cultures. *Prog Med Virol* 1964;6:111–148.

Original Contributions

Aebischer T, Moskophidis D, Rohrer UH, et al. In vitro selection of lymphocytic choriomeningitis virus escape mutants by cytotoxic T lymphocytes. *Proc Natl Acad Sci USA* 1991;88:11047–11051.

Ahmed R, Jamieson B, Porter DD. Immune therapy of a persistent and disseminated viral infection. *J Virol* 1987;61: 3920–3929.

Ahmed R, Fields BN. Role of the S4 gene in the establishment of persistent reovirus infections in L cells. *Cell* 1982;28:605–612.

Allen TM, O'Connor DH, Jing P, et al. Tat-specific cytotoxic T lymphocytes select for SIV escape variants during resolution of primary viraemia. *Nature* 2000;407:386–390.

Appay V, Nixon DF, Donahoe SM, et al. HIV-specific CD8$^+$ T cells produce antiviral cytokines but are impaired in cytolytic function. *J Exp Med* 2000;192:63–75.

Billeter MA, Cattaneo R, Spielhofer P, et al. Generation and properties of measles virus mutations typically associated with subacute sclerosing panencephalitis. *Ann NY Acad Sci* 1994;724:367–377.

Bouley DM, Kanagat S, Wire W, Rouse BT. Characterization of herpes simplex virus type 1 infection and herpetic stromal keratitis development in IFNγ knockout mice. *J Immunol* 1995;155:3964–3971.

Cattaneo R, Rebman G, Baczko K, ter Meulen V, Billeter MA. Altered ratios of measles virus transcripts in diseased human brains. *Virology* 1987;160:523–526.

Ciurea A, Hunziker H, Martinic MMA, Et al. CD4+ T-cell-epitope escape mutant virus selected in vivo. *Nat Med* 2001;7: 795–800.

Ciurea A, Klenerman P, Hunziker L, et al. Viral persistence in vivo through selection of neutralizing antibody-escape variants. *Proc Natl Acad Sci USA* 2000;97:2749–2754.

Cole GA, Nathanson N. Potentiation of experimental arbovirus encephalitis by immunosuppressive doses of cyclophosphamide. *Nature* 1968;220:399–401.

Gonzalez-Scarano F, Shope RE, Calisher CH, Nathanson N. Monoclonal antibodies against the G1 and nucleocapsid proteins of La Crosse and Tahyna viruses. In: Calisher CH, ed. *California serogroup viruses*. New York: Alan R Liss, 1983.

Igarishi T, Endo Y, Englund G, et al. Emergence of a highly pathogenic simian/human immunodeficiency virus in a rhesus macaque treated with anti-CD8 mAb during a primary infection with a nonpathogenic virus. *Proc Natl Acad Sci USA* 1999; 96:14049–14054.

Kimata J, Kuller A, Anderson DB, Dailey P, Overbaugh J. Emerging cytopathic and antigenic simian immunodeficiency virus variants influence AIDS progression. *Nature Med* 1999;5:535–541.

Kono Y, Kobayashi K, Fukunaga Y. Antigenic drift of equine infectious anemia virus in chronically infected horses. *Arch Virusforschung* 1973;41:1–10.

Leib DA, Bogard CL, Kosz-Vanenchak M, et al. A deletion mutant of the latency-associated transcript of herpes simplex virus type 1 reactivates from the latent state with reduced frequency. *J Virol* 1989;63:2893–2900.

Liebert UG, Baczko K, Budka H, ter Meulen V. Restricted expression of measles virus proteins in brains from cases of subacute sclerosing panencephalitis. *J Gen Virol* 1986;67: 2435–2444.

Lill NL, Tevethia MJ, Hendrickson WG, Tevethia SS. Cytotoxic T lymphocytes (CTL) against a transforming gene product select for transformed cells with point mutations within sequences encoding CTL recognition epitopes. *J Exp Med* 1994;176:449–457.

Lutley R, Petursson G, Palsson PA, et al. Antigenic drift in visna: virus variation during long-term infection of Icelandic sheep. *J Gen Virol* 1983;64:1433–1440.

Moskophidis D, Zinkernagel RM. Immunobiology of cytotoxic T-cell escape mutants of lymphocytic choriomeningitis virus. *J Virol* 1995;69:2187–2193.

Oldstone MBA, Buchmeier MJ. Restricted expression of viral glycoprotein in cells of persistently infected mice. *Nature* 1982;300:360–362.

Smith PM, Wolcott RM, Chervenak R, Jennings SR. Control of acute cutaneous herpes simplex virus infection: T cell mediated viral clearance is dependent upon interferon-γ (IFNγ). *Virology* 1994;202:76–82.

Stevens JG, Wagner EK, Devi-Rao GB, et al. RNA complementary to a herpes α gene mRNA is prominent in latently infected neurons. *Science* 1987;235:1056–1059.

Takemoto KK, Habel K. Virus-cell relationship in a carrier culture of HeLa cells and Coxsackie A9 virus. *Virology* 1959;7:28–44.

Chapter 10
Viral Oncogenesis: Retroviruses

ONCOGENIC VIRUSES OF HUMANS AND ANIMALS

A large number of viruses, from several families, have oncogenic potential in animals, and some of these have representatives among the viruses that infect humans. Most oncogenic RNA viruses belong to the family Retroviridae (Table 10.1), and the following account will focus on a few of the oncogenic retroviruses that illustrate the mechanisms of transformation and tumorigenesis. Much more extensive information can be found in the books and reviews listed at the end of this chapter.

STRUCTURE AND REPLICATION OF RETROVIRUSES

Retroviruses are enveloped viruses with a core containing a single-stranded RNA genome of positive polarity. The envelope is composed of a lipid bilayer into which is inserted a single glycopeptide that is posttranslationally cleaved into a transmembrane protein noncovalently linked to a surface protein, and these heterodimers are usually arranged in a trimer. The core consists of two copies of the viral genome associated with a nucleocapsid protein, enclosed in a capsid composed of a capsid protein, a matrix protein, and a phosphorylated protein (p12).

A prototypical simple retrovirus has three principal genetic domains. The *gag* (group specific antigen) gene encodes four proteins that compose the core of the virus. The *pol* (polymerase) gene encodes three enzymes: the protease that cleaves the gag polyprotein into its constituent four peptides, the reverse transcriptase that transcribes the RNA genome into its DNA complement, and the integrase that catalyzes the integration of the double-stranded DNA copy into the host genome. The *env* gene encodes the two envelope proteins. In addition there are specific cis-acting sites for critical steps in replication, such as a primer binding site to initiate reverse transcription, a packaging site for incorporation of the genome into virions, and a site for 3' polyadenylation of nascent mRNA molecules. The genomic organization of a simple retrovirus is shown in Fig. 10.1.

Retroviruses are distinguished by their mode of replication, which employs reverse transcription to synthesize a DNA intermediate (the provirus) that integrates into the host genome prior to further steps in viral replication. Unique sequences at the 5' and

TABLE 10.1
Oncogenic RNA viruses of animals and humans: a selected list

Host	Virus class	Example (virus and disease)	Oncogenic mechanism(s)
Human	Retrovirus	HTLV-1* Adult T-cell leukemia	Tax protein transactivates cellular genes
	Togavirus-like	HCV Hepatocellular carcinoma	?
Animal	Retrovirus	BLV Bovine leukemia	Tax protein transactivates cellular genes
		MuLV* Murine leukemia	Insertional mutagenesis
		RSV* Avian sarcoma	Viral oncogene

The viruses described in this chapter are indicated with an asterisk.

BLV, bovine leukemia virus; HCV, hepatitis C virus; HTLV-1, human T-cell leukemia virus type 1; MuLV, murine leukemia virus; RSV, Rous sarcoma virus.

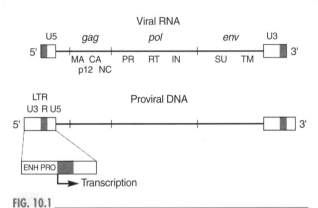

FIG. 10.1 _____

Organization of the genome of a typical simple retrovirus. The RNA genome is about 9 kb long and is bounded at both ends by a noncoding repeat (R) region that encloses the three major coding genes, *gag* (group antigen), *pol* (polymerase), and *env* (envelope). The diagram indicates the position of the major proteins encoded by each of these genes, including the MA (matrix), p12, CA (capsid), NC (nucleocapsid) proteins of *gag*, the PR (protease), RT (reverse transcriptase), IN (integrase) enzymes of *pol*, and the SU (surface) and TM (transmembrane) proteins of *env*. During replication, the RNA genome is copied by the viral reverse transcriptase (RT) into a DNA complement of the genome (provirus). In the process of reverse transcription, the DNA copy acquires a complete long terminal repeat (LTR) sequence at both its ends. The LTR is divided into three domains: U3 (unique 3'), R (repeat), and U5 (unique 5'). For virus replication, the DNA genome is transcribed back into RNA. Transcription begins at the upstream end of the R domain and proceeds through the genome, terminating in a polyadenylation signal at the downstream end of the 3' LTR. The U3 domain contains basal promoters (PRO, such as a TATA box) and upstream enhancers (ENH) that drive transcription. The enhancers are regulatory elements in DNA that bind cellular proteins that act to "open" condensed DNA (DNA associated with histones) so that it is accessible to the polymerase complex.

3' ends of viral RNA contribute to long terminal repeat (LTR) sequences at each end of the provirus that provide the transcriptional control elements for provirus expression. Transcription of the provirus produces both RNA messages and new RNA genomes (Fig. 10.1). This mode of replication has several important consequences. Integration of the DNA intermediate into the genome of the host cell offers the possibility of perpetuating the viral genome through the lifetime of the host and even, under some circumstances, in the germ line of the host. Reverse transcription is not monitored by the cellular enzymes that edit DNA, permitting high rates of mutations.

The integration of viral DNA also provides a possibility to disrupt host genes or to alter their expression by up- or down-regulation via the insertion of promoters or enhancers that are contained in the retroviral LTRs. In addition, integration into cellular DNA facilitates recombination with cellular sequences that can be added to or substituted for segments of the viral genome. Some cellular genes acquired by recombination can confer transforming ability on the virus. If recombinant retroviruses lose segments of their own genome they may become defective, depending for their replication upon nondefective replication-competent "helper" retroviruses that must coinfect the host cell to permit the defective virus to complete its life cycle.

The other essential feature of retroviruses is that in general they are not cytocidal. They mature by budding through the plasma membrane, and this need not compromise the vital functions of the host cell. Also, none of the retroviral proteins appear to induce apoptosis. The innocuous nature of viral replication and the integration of the DNA provirus lead to lifelong associations of virus and cell, which promote certain modes of cellular transformation. In some instances, infection of the host is from the mother, either via the germline or soon after birth, so that the infected host is likely to be tolerant of the neoantigens encoded by the virus, further enhancing the possibility of high-titer persistent infection.

MECHANISMS OF ONCOGENESIS

There are three major classes of oncogenic retroviruses (Table 10.2): the nonacute transforming viruses, the acute transforming viruses, and the trans-acting viruses. There are distinct differences in the genome structure of each group, differences that

		TABLE 10.2		
		Major categories of oncogenic retroviruses		
Category	Occurrence (incubation period)	Mechanism of transformation (clonality)	Replication competence	Examples
Nonacute transforming	In nature, more common (years)	Insertional up-regulation of cellular protooncogenes (clonal)	Competent	MuLV, ALV FeLV
Acute transforming	In nature, uncommon (weeks)	Action of viral oncogenes (polyclonal)	Defective, requires helper virus	ASV, MSV, FeSV
Trans-acting transforming	In nature, uncommon (years)	Action of viral accessory genes (oligoclonal)	Competent	HTLV-1, BLV

Viral oncogenes are derived from cellular protooncogenes but are usually modified by the structural alteration or loss of domains involved in regulation of protooncogene expression.

ALV, avian leukemia virus; ASV, avian sarcoma virus; BLV, bovine leukemia virus; FeLV, feline leukemia virus; FeSV, feline sarcoma virus; HTLV, human T-cell leukemia virus; MuLV, murine leukemia virus; MSV, murine sarcoma virus.

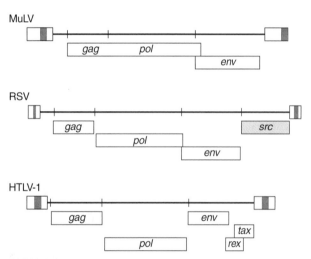

FIG. 10.2 _____

Organization of the genomes of different groups of oncogenic retroviruses. Three different groups of viruses are distinguished in this figure, shown in the form of DNA proviruses. Open rectangles indicate the messages for the principal genes, offset vertically if they are in different reading frames. (*Top*) The nonacute transforming viruses have the organization of simple retroviruses (Fig. 10.1), are replication competent, and transform by promoter or enhancer insertion. (*Middle*) The acute transforming retroviruses usually carry an additional gene (oncogene), which they have acquired by recombination with cellular protooncogene sequences. As a result, these viruses lose viral genes and require a "helper" virus that encodes the missing viral genes to form infectious virions. Some strains of Rous sarcoma virus have expanded the genome to include the *src* oncogene without loss of the replicative genes. (*Bottom*) The trans-acting retroviruses have the complete genomic content of a simple retrovirus and are replication competent, but they also contain accessory genes, the most important of which are the *rex* and *tax* genes; both *rex* and *tax* are required for replication, and tax also confers oncogenic properties on these viruses. (After Coffin JM, Hughes SH, Varmus HE, eds. *Retroviruses*. Cold Spring Harbor, NY: Cold Spring Harbor Laboratory Press, 1997.)

play a central role in their oncogenic activity and other biological properties (Fig. 10.2). This introductory treatment describes only a few examples of these patterns of oncogenesis (see Further Reading).

Certain common themes are present in all transforming retroviruses. Tumor formation is generally a multistep process that involves steps initiated by the virus and other steps that are not virus mediated. Often, tumor induction requires activation of more than one cellular oncogene. In addition to the overexpression of oncogenes, down-regulation of cellular tumor suppressor genes frequently plays a role in tumor induction. Each oncogenic virus tends to produce a characteristic narrow range of tumors, related to the cells that it infects and the cell-specific activity of the enhancers in its LTR.

Nonacute Transforming Retroviruses: Insertional Mutagenesis

The nonacute transforming retroviruses are found in a number of species, particularly mice and chickens, but there are no human representatives of this virus group. The following description is based on the murine leukemia viruses (MuLV), the examples that have been studied in the greatest detail. MuLV are replication-competent simple retroviruses that are limited in their host range to cells that express their specific receptor. The DNA provirus integrates at many sites in the genome of the infected cell and persists for the lifetime of the specific cell type.

Transformation is due to the effect of the LTR of the provirus upon expression of host genes. The effect can be exerted in two distinct ways (Fig. 10.3), although there are several variations in these routes

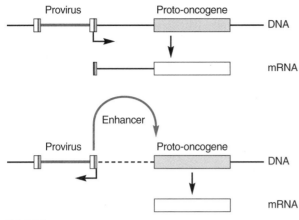

FIG. 10.3

Mechanisms of transformation by nonacute transforming retroviruses. *(Top)* Insertional activation. The DNA provirus is integrated upstream and in the same orientation to a protooncogene, a cellular gene whose product can alter cellular growth by one of many possible mechanisms. Transcription is initiated in a long terminal repeat (LTR) of the provirus and reads through the downstream cellular gene, increasing its rate of transcription. *(Bottom)* Enhancer activation. The DNA provirus is inserted either 3' or 5' (insertion can be more than 100,000 bp away) of a cellular gene that influences growth. For 5' insertion, provirus transcription is in the upstream direction away from the cellular gene. For 3' insertion, provirus transcription is in the same orientation as the protooncogene and downstream from it. The enhancers are regulatory elements in DNA that bind modulatory factors, some of which act to "open" condensed DNA (DNA associated with histones) so that it is accessible to the polymerase complex. The broken arrows indicate the direction of transcription from the viral LTR. (After Fan H. Murine leukemia virus. In: Ahmed R, Chen I, eds. *Persistent viral infections.* New York: John Wiley and Sons, 1999.)

to transformation. If the provirus is integrated upstream from a cellular gene and transcribes toward the cellular gene, then the proviral promoter can initiate transcription that reads through into the cellular gene, increasing the transcription of the cellular gene (*promoter insertion*). If the cellular gene that is up-regulated has an influence on cellular growth, then transformation may result. The resulting tumors are usually clonal but the transformed target cell will depend upon the genomic site of insertion. Alternatively, if the provirus is located near a cellular gene but is oriented to read away from the cellular gene, the enhancer sequences in the provirus may bind cellular factors that "open" condensed DNA and enhance transcription of neighboring cellular genes regardless of their orientation (*enhancer activation*). Again, if the gene influences cellular growth, transformation may result. Enhancer activation is a more common phenomenon than promoter insertion, probably because it can occur at more sites in the cellular genome

Detailed studies have been made of many leukemias and other cancers in mice infected with various nonacute transforming retroviruses. Based on these studies, catalogues have been constructed of the cellular genes that are up-regulated. This work has identified many genes that influence cell growth and division, each of which can produce a transformed phenotype.

The likelihood that an MuLV will integrate in the cellular genome in a position leading to transformation is very low, and many integration events occur for each instance of transformation. The type of tumor that develops reflects the function of the up-regulated gene, and a particular virus can cause different tumors, depending on the chance site of insertion. The potential for tumor induction is directly correlated with the number of integration events, and those MuLV strains that replicate to highest titer in mice tend to be the most leukemogenic. Thus, infection of newborn (compare to adult) animals carries a higher risk of leukemia, partly because MuLV replicates to higher titers in young animals and partly due to immunologic tolerance.

The other critical point is that the development of cancer is usually a multistep process. In other words, cellular transformation by a nonacute retrovirus is not sufficient to cause leukemia. Rather, there must be additional subsequent oncogenic events (described later in this chapter). The combined low probability of insertional mutagenesis of protooncogenes and the requirement for subsequent genetic events explain the clonal nature of the resulting tumors.

In summary, the salient features of the nonacute transforming retroviruses (Table 10.1) are the replication competence of the virus, transformation via insertional mutagenesis and up-regulation of a protooncogene, the ability to cause different types of tumors by inserting into different protooncogenes, the long incubation period, and the clonal nature of the transformed cells.

Acute Transformation by Viral Oncogenes

The acute transforming retroviruses account for only a small proportion of naturally occurring tumors caused by retroviruses, and they have never been isolated from humans. Several properties make them uncommon in nature, such as the requirement for a helper virus and the acute nature of the malignancies that they cause, which reduces the probability of horizontal transmission. However, the study of these viruses has played a very important role in the elucidation of viral oncogenesis.

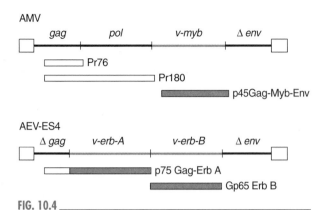

FIG. 10.4

The genomic organization and major proteins encoded by two representative acute transforming retroviruses. The genome is shown above and the mRNAs representing the proteins are shown below, with the oncogenes and oncoproteins in blue. In some cases the oncoproteins are fused with viral proteins, and in all instances they are modified from their cellular counterparts (not indicated in this figure). AMV, avian myeloblastosis virus; AEV, avian erythroblastosis virus; erb, erythroblastosis; myb, myeloblastosis; Δ env, truncated env gene. (After Coffin JM, Hughes SH, Varmus HE, eds. *Retroviruses*. Cold Spring Harbor, NY: Cold Spring Harbor Laboratory Press, 1997.)

Most of the acute transforming viruses have two defining characteristics.

First, their genomes contain a viral oncogene (v-*onc*), which is a gene that possesses a specific transforming activity, often at a high level. It appears that most, perhaps all, viral oncogenes are derived from protooncogenes (c-*onc*), normal genes that can in-fluence cell growth. These are often the very same protooncogenes whose up-regulation by nonacute oncogenic retroviruses leads to transformation. Protooncogenes are so named because they are thought to be the ancestors of viral oncogenes that were first "captured" by recombination between retroviruses and cellular sequences, and subsequently evolved by mutation.

Second, acute transforming viruses are—with some exceptions—defective due to the loss of part of their genome during a putative ancestral recombination event. Usually, the oncogene is located in the *env*, *pol*, and/or *gag* region of the genome, so that the virus cannot encode the surface, transmembrane, or some gag proteins (Fig. 10.2). These proteins are required to produce infectious virions, and their absence renders such viruses defective and able to grow only in the presence of a replication-competent nontransforming retrovirus that encodes the *env* (or other missing) proteins in which the transforming virus genome is packaged. Figure 10.4 shows the genomic organization of a few representative acute transforming viruses.

Study of many acute transforming virus isolates has led to a catalogue of the oncogenes that they carry, which has produced an explosion of information regarding the normal control of cell growth and division. Table 10.3 summarizes some of the more common viral oncogenes.

Acute oncogenic retroviruses transform cells in

TABLE 10.3

Some common oncogenes associated with acute transforming retroviruses

Category of oncogene (*example*)	Function of c-*onc*	Alteration in v-*onc* relative to c-*onc*	Action of v-*onc*
Protein tyrosine kinase (*src*)	Tyrosine phosphorylation activates intracellular signaling proteins	Loss of C terminal tyrosine or Loss of receptor binding domain	Increased tyrosine phosphorylation causes sustained intracellular signaling
Serine-threonine kinase (*mos*)	Phosphorylation activates cell cycle proteins	Deletion of N-terminal regulatory sequences	Increased phosphorylation activates cell cycle
G-protein (*ras*)	Cycles between GTP/GDP forms regulating intracellular signaling	Change in amino acid at codon 12	High GTP/GDP ratio causes sustained activation signals
Transcription factor (*myb*)	Interactions with transcription complex to enhance or reduce transcription	Deletion of regulatory domains, or increase of mRNA or protein accumulation	Increase transcription of growth factors or cell cycle genes or reduction of antioncogene expression

c-*onc*, cellular protooncogene; *mos*, Moloney mouse sarcoma; *myb*, myeloblastosis; *ras*, rat sarcoma; *src*, sarcoma; v-*onc*, viral oncogene.

culture rapidly and are—like most oncogenic retroviruses—noncytocidal. The rapid transforming effect reflects the fact that the oncogenes encoded by these viruses have evolved from their ancestral cellular genes in such a manner that they are relatively unresponsive to the checks and balances that control normal cellular growth factors. Oncogenes may escape control because they are transcribed and translated at a very high rate under the influence of the viral LTR with their potent promoters and enhancers. Also, viral oncoproteins are often modified from their cellular counterparts by the loss or change in domains involved in down-regulation by other cellular proteins. Based on their rapid transforming activity, these viruses may be quantified by "focus forming" assays that are similar in principle to the plaque assays used to count lytic viruses.

In vivo, the acute transforming retroviruses cause tumors in weeks rather than months to years. This reflects their high transforming activity, such that further mutations may not be necessary for the production of tumors. Because cells are transformed rapidly and with a high probability, the tumors that result are usually polyclonal. Another consequence of the aggressive transforming activity is that some acute viruses may cause tumors in the absence of helper viruses, that is, without replicating in the host.

In summary, the salient features of the acute transforming retroviruses (Table 10.1) are the defectiveness of the virus requiring a helper virus for propagation, transformation by a viral oncogene, short incubation period, specificity of tumor induction, and the polyclonal nature of the transformed cells.

Trans-Activation by Viral Accessory Genes

Human T-cell leukemia virus type 1 (HTLV-1) and bovine leukemia virus compose a relatively rare group of retroviruses whose oncogenic activity is due to a mechanism distinct from that used by the nonacute and acute transforming retroviruses. These viruses encode several accessory genes—particularly the *tax* gene—that are not only essential for their replication but also play a role in their transforming activity. However, the mechanism of oncogenesis is complex because it appears that *tax* is necessary but not sufficient for oncogenesis, and subsequent nonviral genetic events are required to produce tumors. HTLV-1, the best studied example, is described later in this chapter.

Common Themes

Certain common themes are present in all of the transforming retroviruses. Tumor formation is generally a multistep process that involves steps initiated by the virus and subsequent steps that are not virus mediated. Often, tumor induction requires activation of more than one cellular oncogene. In addition to the overexpression of oncogenes, down-regulation of cellular tumor suppressor genes frequently plays a role in tumor induction. Each oncogenic virus tends to produce a characteristic narrow range of tumors, related to the cells that it infects and the cell-specific activity of its LTR.

ONCOGENIC RETROVIRUSES: EXAMPLES

Murine Leukemia Viruses, Prototypic Nonacute Transforming Retroviruses

MuLV were originally isolated from laboratory or wild mice and have been studied in more detail than any other group of oncogenic retroviruses. MuLV can be classified into several groups, including ecotropic, xenotropic, amphotropic, and polytropic or mink cell focus-forming viruses (MCFV) (Table 10.4). Ecotropic viruses infect only murine cells, xenotropic viruses infect cells from rats and other species but not mice, and amphotropic viruses infect cells from both mice and other species. MCFV do not exist as exogenous viruses; rather, they arise from endogenous sequences present in the mouse genome that can recombine with exogenous replication-competent MuLV. Each of these four classes of MuLV is distinguished by having a different cellular receptor, and cells infected with one class of viruses resist superinfection with other viruses of the same class because of receptor blockade (Fig. 4.5). Some receptors have been isolated whereas others are yet to be defined.

Mice, like other higher vertebrates, have endogenous retroviral sequences in their genomes. Presumably, these sequences originated when a retrovirus infected a germ cell, and the provirus was then incorporated into the germline. Most laboratory mice have about 50 copies of retrovirus sequences scattered throughout their genome, mainly representing xenotropic and polytropic MCFV with a few copies of ecotropic viruses, all derived from the MuLV family of retroviruses. Endogenous retroviral sequences usually are defective, containing only part of the viral genome, which frequently includes the *env* gene. Although the endogenous sequences rarely give rise to replication-competent viruses, some of them can recombine with each other or with exogenous replication-competent murine retroviruses. Recombination is thought to occur when RNA transcripts of

TABLE 10.4					
Major classes of Murine leukemia viruses					
Class	Species infected	Cellular receptor	Exogenous	Endogenous (no.)	Source
Ecotropic	Mice	Transporter of cationic amino acids	Yes	Yes (0–3)	Laboratory mice
Xenotropic	Rats, other	Not known	No	Yes (~25)	Laboratory mice
Amphotropic	Murine, other	Phosphate transporter	Yes	No	Wild mice
Polytropic (MCF)	Murine, other	Not known	No	Yes (~25)	Laboratory mice

Each class utilizes a different cellular receptor, causing interference between individual viruses within each class but not between classes. Exogenous viruses exist as replication competent viruses that are transmitted horizontally in mice, whereas endogenous viruses exist as germline sequences (the number of copies per mouse genome is indicated) that are usually not replication competent but may recombine with exogenous viruses during the course of infection. Ecotropic viruses fall into two groups based on sharing of antigenic determinants: the FMR (Friend, Maloney, Rauscher) group and the Gross group (Gross and AKR viruses). Polytropic viruses are also called MCFV because they form transformed foci in mink cells. Except for the xenotropic group, members of all groups are oncogenic, producing tumors of T lymphocytes and—in a few instances—of other hematopoietic cells.

endogenous viruses and exogenous viruses are co-packaged in a virion. During the next round of replication, reverse transcriptase can use both the exogenous and the endogenous RNA during DNA synthesis. Viral recombinants play a significant role in the production of leukemia following infection with certain of the nonacute oncogenic retroviruses.

Leukemogenesis by MuLV is a complex multi-step process. One major step is activation of protooncogenes by insertional mutagenesis (Fig. 10.3). From the analysis of many leukemic mice, a host of different protooncogenes have been shown to be activated in individual animals. Many of these are related to the viral oncogenes associated with the acute retro-

viruses (Table 10.3) and include proteins involved in transcriptional regulation, tumor suppressors such as p53, cyclins, and cytokine receptors. It has long been recognized that different MuLV tend to cause different types of leukemia depending on the transformed cell (lymphocytes, erythrocytes, or other hematopoietic cell types). One major determinant of cellular specificity is the LTR of different viruses, each of which is most active in certain types of cells. These specificities have been mapped to the enhancer sequences in some instances, based on experiments showing that engineered chimeric viruses exhibit the specificity of the virus that donated the enhancer sequence (Table 10.5).

Polytropic MCFV also play an important role in

TABLE 10.5					
The LTR of retroviruses can influence oncogenicity					
			NUMBER OF MICE WITH INDICATED DIAGNOSIS		
MuLV strain	Genome backbone	Enhancer U3 region	Erythroleukemia	Lymphoblastic leukemia	Myelogenous leukemia
Moloney	Moloney	Moloney	0	15	0
Friend	Friend	Friend	50	0	0
Moloney/Friend	Moloney	Friend	53	1	3
Friend/Moloney	Friend	Moloney	1	19	0

The LTR of MuLV is an important determinant of the type of neoplasm produced by this group of oncogenic retroviruses. Moloney MuLV produces T-cell lymphomas whereas the Friend MuLV causes erythroleukemias. In this experiment, the two virus strains were compared with chimeric viruses in which the U3 (unique 3′) region of the genome of one virus was substituted into the backbone of the other virus. After infection of mice, the type of neoplasm cosegregated with the viral origin of the U3 region. The U3 domain of the viral LTR contains enhancer sequences and the tumor specificity was attributed to this part of the genome. Presumably the U3 region binds cellular proteins that differ between cells of lymphopoietic and erythropoietic lineages.

LTR, long terminal repeat; MuLV, murine leukemia virus.

From Chatis PA, Holland CA, Silver JE, et al. A 3′ end fragment encompassing the transcriptional enhancers of nondefective Friend virus confers erythrogenicity on Moloney leukemia virus. *J Virol* 1984;52:248–254.

leukemogenesis induced by ecotropic MuLV. Recombinant ecotropic viruses containing sequences (often in the *env* region) derived from MCFV appear concomitant with the development of leukemia and their sequences are frequently found in leukemic clones. It is thought that MCFV recombinants may potentiate leukemogenesis because they use a different receptor, which permits dual infection of individual cells, thereby increasing viral load and raising the probability of insertional mutagenesis. In addition, certain MCFV envelopes can bind to the erythropoietin receptor releasing the receptor from dependence on erythropoietin or interleukin-2 (IL-2) so that it is constitutively up-regulated, creating autocrine stimulation.

Based on a variety of complex experiments, a conjectural scheme of leukemogenesis has been developed (Fig. 10.5). Initial infection of the bone marrow with MuLV induces transformation of multiple cell lines, leading to hyperplasia of the spleen. MCFV recombinant viruses are produced and play a significant role in these early events, in part by increasing vi-

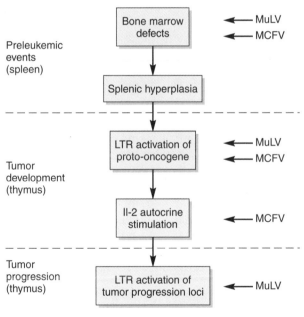

FIG. 10.5

Multistep nature of leukemogenesis in mice infected with Moloney murine leukemia virus (M-MuLV). MuLV acts by insertional mutagenesis to up-regulate many different protooncogenes, which can act at several points in the development of leukemic clones. In addition, a second virus, the polytropic mink cell focus-forming virus, is produced by the recombination of endogenous retroviral sequences with the replication competent MuLV. Mink cell focus-forming virus increases viral load and up-regulates selected receptors for growth factors, complementing the activity of MuLV. (After Fan H. Murine leukemia virus. In: Ahmed R, Chen I, eds. *Persistent viral infections*. New York: John Wiley and Sons, 1999.)

ral load in infected cells. Some of these hyperplastic cell lines are further transformed into tumor cells by insertional mutagenesis that activates additional protooncogenes or inactivates tumor suppressor genes, enhanced by autocrine stimulation via cytokine receptors such as the erythropoietin receptor. Further activation of protooncogenes leads to tumor progression.

Avian Leukosis-Sarcoma Viruses, Prototypic Acute Transforming Retroviruses

The avian leukosis-sarcoma viruses (ALSV) are a group of oncogenic retroviruses that include both acute and nonacute transforming viruses. The avian leukosis virus (ALV) group mainly causes leukemias (leukosis), tumors of hematopoietic cells (lymphocytes, myelocytes, erythrocytes, or their precursors), whereas the avian sarcoma virus (ASV) group mainly causes fibrosarcomas. The ASV group includes Rous sarcoma virus (RSV) strains—all of which carry the *src* gene—and other avian sarcoma viruses that carry viral oncogenes other than *src*, such as *fps, ros,* or *yes.* The ALSV were the first retroviruses isolated and have served as prototypes to work out many of the basic biological and molecular features of retroviruses and establish new paradigms in virology and oncology. RSV was one of the first viruses shown to be a filterable agent (Peyton Rous, 1910), capable of causing a transmissible sarcoma in chickens, and investigations of RSV led to development and proof of the hypothesis of reverse transcription (Temin, 1970), and to the discovery of the relationship between viral and cellular oncogenes (Bishop and Varmus, 1976; see Sidebar 1.4). This account will focus on RSV.

Replication-Competent and Defective Viruses A number of isolates of RSV, from chickens with sarcomas, have been characterized. All of these viruses carry the src oncogene (also called v-src), are acutely transforming in chick embryo cultures, and induce sarcomas under appropriate experimental conditions. Some of these viruses are replication competent because they have an expanded genome that accommodates a full complement of retroviral genes in addition to the src gene (Fig. 10.2).

Other isolates of RSV are defective in their *env* genes and can only replicate in cells that are co-infected with a replication competent helper virus that encodes a functional *env* gene. The helper viruses are all members of the ALV group and stocks of defective RSV must be propagated together with an ALV isolate. ALV, although they cannot transform fibroblasts, can cause leukosis in chickens. ALV isolates differ in their envelopes and can be classified into subgroups depending on the receptor specificity of the envelope (see below). Since different ALV can be used to complement a given defective RSV isolate, that isolate can exhibit different subgroup identities depending on the ALV virus that serves as helper. Historically, when it was discovered that some stocks of RSV contained an associated virus that was required for propagation of the sarcoma virus, the term Rous-associated virus (RAV) was introduced; subsequently RAV was recognized to be ALV that acted as a helper for the defective transforming RSV.

Endogenous Viruses Low multiplicities of defective strains of RSV produce foci that do not contain infectious virus (called nonproducer foci) because they do not carry the helper ALV. However, it was found that such foci could release retrovirus particles (seen in the electron microscope). These particles were produced by certain avian embryo cultures that carried endogenous virus (ev) sequences coding for infectious ALV or ALV envelope. The endogenous ALV—when expressed—acted as helper for defective strains of RSV.

About 30 different ev loci have been identified in various strains of chickens and a typical breed will carry about 5 loci. The ev sequences range from those encoding a complete infectious ALV to those that do not express any detectable viral proteins. The complete endogenous viruses belong to subgroup E (Table 10.6). Endogenous chicken viruses do not appear to be oncogenic (lack of oncogenicity has been mapped to the LTR) nor are they essential for the growth and development of chickens, as ev-negative strains of chickens can be derived and are normal.

Cellular Receptors Different isolates of RSV use different receptors, and receptor utilization depends on the receptor specificity of the envelope of the virus (replication-competent RSV) or that of the helper virus (defective RSV). The receptor binding domain of the viral envelope determines three interlinked properties: interference pattern, host range, and antigenic signature. These properties can be used to classify RSV into different subgroups. An interference classification of RSV is shown in Table 10.6 and reflects the fact that when an ALV infects a cell culture, sufficient envelope protein (viral attachment protein) is produced to saturate the cellular receptors, which interferes with the ability of another virus that uses the same receptors to superinfect the culture (Fig. 4.5).

TABLE 10.6
Classification of Rous sarcoma virus into subgroups depending on interference patterns

Subgroup Interfering ALV	SUBGROUP OF CHALLENGE RSV				
	A	B	C	D	E
A	I				
B		I		±I	I
C			I		
D		±I		I	I
E					I

The interference assay involves infection of a culture with an ALV, followed by an interval of several days during which sufficient virus is produced to saturate the receptors used by that particular isolate. The ALV does not cause any cytopathic effect, and the infected culture can then be used to titrate a strain of RSV using a focus-forming assay. The RSV stock is simultaneously titrated on an uninfected companion cell culture. If prior infection with ALV "interferes" by reducing the titer of the RSV isolate, it is classified in the same subgroup. Although there are some instances of cross interference, the subgroups can be distinguished by reduced interference, one-way interference, and differences in host range and neutralization.

ALV, avian leukosis virus; RSV, Rous sarcoma virus; I, interference.

Adapted from Payne LN. Biology of avian retroviruses. In: Levy JA, ed. *The retroviridae*. New York: Plenum Press, Vol. 1. 1992:299–404.

Embryo cells derived from different strains of chickens (as well as cells from other avian species) vary in which receptors they express, and a panel of such cells, can be used to classify different RSV isolates. These patterns provide another independent criterion for the subgrouping of RSV, and they also explain the genetically determined pattern of susceptibility of different lines of chickens to various RSV. Finally, antisera raised against each subgroup of RSV will neutralize homologous and heterologous viruses of the same subgroup because neutralization reflects binding of antibodies to domains on the viral envelope that are involved in binding to the cellular receptor. Again, this provides another probe to determine relationships between the viral attachment proteins of different subgroups.

Human T-Cell Leukemia Virus Type 1, a Trans-Acting Oncogenic Retrovirus

HTLV-1 is an exogenous virus of humans that causes adult T-cell leukemia (ATL) as well as a neurologic disease (HTLV-associated myelopathy) that is not relevant to the present discussion. HTLV-1 is transmitted from person to person in several ways, via breast-feeding, sexual contact, or blood (transfusions, blood products, and contaminated needles). HTLV-1 and ATL have an unusual geographic pattern, being endemic in a few selected areas, such as Japan and the Caribbean islands, but relatively rare in most regions of the world. HTLV-1 infections are usually asymptomatic and only about 1%–5% of infected persons ever develop the disease after an incubation period that can be as long as 20–30 years. Acute ATL is characterized by circulating malignant CD4$^+$ T cells, which infiltrate skin and viscera, usually causing death within a year.

The mechanism of oncogenesis is only partially understood for HTLV-1 and bovine leukemia virus, a genetically related virus that produces B-cell leukemia in cows. There is no evidence of up-regulation of a cellular protooncogene, and these viruses do not carry established oncogenes like the acutely transforming retroviruses, suggesting that they utilize a different mechanism of oncogenesis. The *tax* gene of HTLV-1, which trans-activates the viral LTR (similar to the action of *tat* in HIV) and is required for viral replication, probably plays an essential role in transformation.

Several observations are relevant to the mechanism of oncogenesis. (a) HTLV-1 infects CD4$^+$ T lymphocytes, using the CD4 molecule as a viral receptor. The provirus is integrated randomly into the genome of the host cell, followed by viral replication that requires the action of *tax* and *rex* gene products, which are analogues of the *tat* and *rev* genes of HIV. The virus is noncytocidal and persists in infected cultures but remains highly cell associated, so that it is most readily transmitted by cocultivation of infected with uninfected cells. (b) HTLV-1 can immortalize primary T-cell cultures, producing CD4$^+$ cell lines that can grow continuously in the absence of IL-2 (exogenous IL-2 is required to maintain cultures of primary T cells). However, transformation occurs in only a small fraction of infected T cells, and the cultures evolve slowly through a polyclonal to a monoclonal phase, suggesting that additional events are required to produce the transformed phenotype. (c) Tax trans-activates a number of cellular genes, including those encoding IL-2 and the IL-2 receptor. If Tax established an autocrine loop for IL-2, it could explain the IL-2 independence of HTLV-1–transformed cells. (d) Tax transforms immortalized rodent cell lines, converting them to cells that will form colonies in agar and produce tumors in immunosuppressed mice. (e) ATL is oligoclonal and cell lines cloned from patients contain HTLV-1 proviruses, but these are often highly deleted sequences that are incapable of replicating. Also, there is little transcription or translation of the *tax* gene in ATL cell lines.

These observations have been used to construct a working hypothesis regarding the oncogenic activity of HTLV-1 (Fig. 10.6). Initially, the virus infects CD4$^+$ T lymphocytes, and the viral genome is inte-

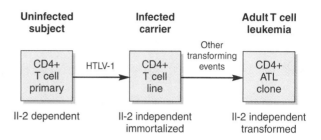

FIG. 10.6

Diagram of putative transforming action of human T-cell leukemia virus type 1 (HTLV-1). It is postulated that adult T-cell leukemia, the cancer associated with HTLV-1, is produced by a multistep sequence of events. The first event is infection of CD4$^+$ T lymphocytes by HTLV-1, which leads to integration of the provirus in the genome of the host cell. The action of the HTLV-1 *tax* gene immortalizes clones of T lymphocytes. In a small proportion (less than 5%) of infected carriers, these cells undergo subsequent genetic events that convert them into leukemogenic clones. (After Kashanchi F, Pise-Masison C, Brady JN. Human T cell leukemia virus. In: Ahmed R, Chen I, eds. *Persistent viral infections*. New York: John Wiley and Sons, 1999.)

grated as a provirus. The tax protein—through the trans-activation of selected cellular genes—immortalizes T lymphocytes, which carry the viral genome for the life of the human host. In a small proportion of viral carriers, one or more additional low-probability genetic events—such as chromosomal recombination—transform an occasional T-cell line into an oncogenic clone. Initially, these clones produce smoldering ATL, which develops into acute ATL if there is an outgrowth of a highly aggressive cell clone. During the genetic events that create highly oncogenic T-cell clones, the HTLV-1 provirus may be altered, "inactivating" the *tax* gene, which is no longer required to maintain the transformed phenotype.

BIOLOGY OF CANCER

Cellular Specificity of Tumors Induced by Oncogenic Retroviruses

Each retrovirus tends to produce a narrow range of characteristic neoplasms in its natural host. This specificity is determined by a number of variables. First, most viruses utilize one or a very few cellular surface molecules as their receptors. Thus, HTLV-1 uses CD4 as its receptor and causes transformation of the corresponding subset of T lymphocytes. Second, for retroviruses where the LTR plays a role in transformation, the activity of promoters and enhancers within the LTR is often greater in those cell types that are typical targets of each virus (Table 10.5). Finally, the cellular environment may be a determinant in the further conversion of virus-transformed cells into malignant clones. Thus, T lymphocytes transformed by MuLV apparently migrate to the thymus where they undergo additional genetic changes that confer tumorigenic potential, thereby causing T-cell leukemias or thymomas.

Multifactorial Induction of Cancer

Understanding of the molecular biology of cancer has evolved rapidly, to the point where a comprehensive set of mechanisms have been implicated in the induction of most neoplasms (Sidebar 10.1). All of these mechanisms involve the molecular biology of the cell and may be grouped into three major categories: lack of need for external growth signals and insensitivity to antigrowth signals; unlimited growth potential and evasion of apoptosis; tumor metastasis and sustained angiogenesis. This comprehensive view has important implications. First, it explains the multistep nature of oncogenesis because the multiple genetic changes required could only occur over a period of time. Second, it explains why an oncogenic virus may be necessary but not sufficient to produce neoplasia. The virus, through a variety of molecular actions, can perturb some of the cellular determinants that protect the host against neoplasia. The routes to tumorigenesis are undoubtedly diverse, both in the order of the changes that occur and in the number of mutations involved. In some instances, a single mutation may activate or inactivate an essential change in cellular phenotype, whereas in other instances several mutations may be required; conversely, some mutations may alter several critical phenotypes. By the same token, an oncogenic virus may be associated with one or several of the mutations involved in the emergence of a given neoplasm.

FURTHER READING

Reviews and Chapters

Bergers G, Hanahan D, Coussens LM. Angiogenesis and apoptosis are cellular parameters of neoplastic progression in transgenic mouse models of tumorigenesis. *Int J Dev Biol* 1998;42:995–1002.

Bryan TM, Cech TR. Telomerase and the maintenance of chromosome ends. *Curr Opin Cell Biol* 1999;11:318–324.

Coffin JM, Hughes SH, Varmus HE, eds. *Retroviruses.* Cold Spring Harbor, NY: Cold Spring Harbor Laboratory Press, 1997.

Fan H. Murine leukemia virus. In: Ahmed R, Chen I, eds. *Persistent viral infections.* New York: John Wiley and Sons, 1999.

Hanahan D, Weinberg RA. The hallmarks of cancer. *Cell* 2000;100:57–70.

Kashanchi F, Pise-Masison C, Brady JN. Human T cell leukemia virus. In: Ahmed R, Chen I, eds. *Persistent viral infections.* New York: John Wiley and Sons, 1999.

Levine AJ. p53, the gatekeeper for growth and division. *Cell* 1997;88:323–331.

Payne LN. Biology of avian retroviruses. In: Levy JA, ed. *The retroviridae,* Vol. 1. New York: Plenum Press, 1992:299–404.

Robinson HL, Rein A, Speck NA. Avian and murine retroviruses. In: Nathanson N, et al., ed. *Viral pathogenesis.* Philadelphia: Lippincott–Raven Publishers, 1997:629–656.

Classic Papers

Astrin SM, Robinson HL, Crittenden LB, et al. Ten genetic loci in the chicken that contain structural genes for endogenous avian leukosis viruses. *Cold Spring Harbor Symp Quant Biol* 1980;44:1105–1109.

Baltimore D. RNA-dependent DNA polymerases in virions of RNA tumor viruses. *Nature* 1970;226:1209–1211.

Hayward WS, Neel BG, Astrin SM. Activation of a cellular onc gene by promoter insertion in ALV-induced lymphoid leukosis. *Nature* 1981;290:475–480.

Hahn WC, Counter CM, Lundberg AS, et al. Creation of human tumor cells with defined genetic elements. *Nature* 1999; 400:464–468.

Hartley JW, Rowe WP. Naturally occurring murine leukemia viruses in wild mice: characterization of a new "amphotropic" class. *J Virol* 1976;19:19–25.

Poisz BJ, Ruscetti FW, Gazdar AF, et al. Detection and isolation of type C retrovirus particles from fresh and cultured lymphocytes of a patient with cutaneous T-cell lymphoma. *Proc Natl Acad Sci USA* 1980;77:7415–7419.

Rous P. Transmission of a malignant new growth by means of a cell-free filtrate. *JAMA* 1911;56:198.

Rubin H. A virus in chick embryos which induces resistance in vitro to infection with Rous sarcoma virus. *Proc Natl Acad Sci USA* 1960;46:1105–1119.

Temin HM, Rubin H. Characteristics of an assay for Rous sarcoma virus and Rous sarcoma cells in tissue culture. *Virology* 1958;6:669–688.

Temin HM, Mizutani S. RNA-dependent DNA polymerase in virions of Rous sarcoma virus. *Nature* 1970;226:1211–1214.

Original Contributions

Chatis PA, Holland CA, Silver JE, et al. A 3′ end fragment encompassing the transcriptional enhancers of nondefective Friend virus confers erythrogenicity on Moloney leukemia virus. *J Virol* 1984;52:248–254.

Gazzolo I, Duc Dodon M. Direct activation of resting T lymphocytes by human T-lymphotropic virus type I. *Nature* 1987;326:714–717.

Grassman R, Dengler C, Muller-Fleckenstein I, et al. Transformation to continuous growth of primary human T lymphocytes by human T-cell leukemia virus type I X-region genes transduced by a *Herpesvirus saimiri* vector. *Proc Natl Acad Sci USA* 1989;86:3351–3355.

Grossman WJ, Kimata JT, Wong FH, Zutter M, Ley TJ, Ratner L. Development of leukemia in mice transgenic for the *tax* gene of human T-cell leukemia virus type I. *Proc Natl Acad Sci USA* 1995;92:1057–1061.

Laimins LA, Fruss P, Pozzatti R, Koury G. Characterization of enhancer elements in the long terminal repeat of Moloney murine sarcoma virus. *J Virol* 1984;49:183–189.

Li JP, Baltimore D. Mechanism of leukemogenesis induced by mink cell focus-forming murine leukemia viruses. *J Virol* 1991;65:2408–2414.

Nerenberg M, Hinrichs SH, Reynolds RK, et al. The *tat* gene of human T-lymphotropic virus type 1 induces mesenchymal tumors in transgenic mice. *Science* 1987;237:1324–1329.

Pozzati R, Vogel J, Jay G. The human T-lymphotropic virus type I *tax* gene cooperates with the *ras* oncogene to induce neoplastic transformation of cells. *Mol Cell Biol* 1990;10:413–423.

Short MK, Okenquist SA, Lenz J. Correlation of leukemogenic potential of murine retroviruses with transcriptional tissue preference of the viral long terminal repeats. *J Virol* 1987;61:1067–1072.

Stehelin D, Varmus HE, Bishop JM, Vogt PK. DNA related to the transforming gene(s) of avian sarcoma viruses is present in normal avian DNA. *Nature* 1976;260:170–173.

Vogt PK, Ishizaki R. Patterns of viral interference in the avian leukosis and sarcoma complex. *Virology* 1966;30:368–374.

Chapter 11
Viral Oncogenesis: DNA Viruses

Many different families of DNA viruses have members that are oncogenic either in their natural hosts or in experimental animals (Table 11.1). The small oncogenic DNA viruses (adenoviruses, polyoma viruses, papillomaviruses) utilize mechanisms to induce neoplasms different from those used by the large oncogenic DNA viruses (such as some herpesviruses and poxviruses).

Most of the small oncogenic DNA viruses can immortalize or transform cells in culture, although this potential is usually restricted to selected cell types and specific conditions. This capability is attributed to virus-encoded proteins that are sometimes dubbed oncoproteins. However, unlike the RNA viral oncogenes, they are not derived from cellular counterparts but serve essential functions for virus propagation, often by inducing S-phase genes and extending the life of the infected cell to facilitate and enhance the yield of new infectious virions. These viral oncoproteins function by blocking the action of normal cellular factors, including tumor suppressor proteins.

Relatively few large DNA viruses are oncogenic; for the most part they do not encode identifiable "oncogene," nor do they act primarily by blocking cellular tumor suppressor proteins. In many instances, these viruses immortalize cells by interacting with extracellular growth factors and the intracellular signaling systems that respond to growth factors, resulting in perpetual cellular replication. In addition, they may encode homologs of cellular proteins to activate the cell cycle or interfere with intracellular signaling that leads to apoptosis.

MECHANISMS OF ONCOGENESIS

Tumor Suppressor Genes

Most of the protooncogenes described in Chapter 10 function in normal cells to promote cell growth. In addition, cells also encode a number of genes whose products act as negative regulators of cell growth. These genes are often called tumor suppressor genes because they are mutated or deleted at a high frequency in certain cancers. Together, the positive and negative regulators of cell growth constitute a complex system of checks and balances. The small DNA viruses produce proteins that bind and inactivate these tumor suppressor proteins (Table 11.2), thereby activating S-phase genes. Consequently, some of these viruses can occasionally promote tumor formation.

TABLE 11.1

Oncogenic DNA viruses of animals and humans: a selected list

Host	Virus class	Example (virus and *disease*)	Natural tumors in host of origin	Oncogenic mechanism(s)
Human	Adeno	Adenovirus*	No Experimental only	Encodes transforming protein(s) Inactivates tumor suppressors
	Papilloma	HPV* *cervical carcinoma*	Yes	Encodes transforming protein(s) Inactivates tumor suppressors
	Hepadna	HBV *Hepatocellular carcinoma*	Yes	X protein inhibits DNA repair
	Herpes	EBV* *Burkitt's lymphoma*	Yes	Immortalizes cells
		HHV-8* *Kaposi's sarcoma*	Yes	Encodes growth-altering proteins
Animal	Polyoma	MPV (mouse)	No	Encodes transforming protein(s)
		SV40 (monkey)	Experimental only	Inactivates tumor suppressors
	Papilloma	CRPV (rabbit) *Papilloma*	Yes	Encodes transforming protein(s) Inactivates tumor suppressors
	Hepadna	WHV (woodchuck) *Hepatocellular carcinoma*	Yes	X protein inhibits DNA repair
	Herpes	MDV (chicken) *Marek's disease*	Yes	Immortalizes cells Encodes transforming protein
	Pox	Myxoma (rabbit) *Papilloma*	Yes	Encodes growth-altering proteins

The viruses described in this chapter are indicated with an asterisk (*).

CRPV, cottontail rabbit papilloma virus; EBV, Epstein–Barr virus; HBV, hepatitis B virus; HHV-8, human herpesvirus 8; HPV, human papillomavirus; MDV, Marek's disease virus; MPV, mouse polyoma virus; SV40, simian virus 40; WHV, woodchuck hepatitis virus.

pRb and p53 Proteins

Two of the most important tumor suppressor genes are the retinoblastoma susceptibility (*Rb*) gene and the *p53* gene. The pRb protein controls the transition from S to G_1 phase of the cell cycle (Fig. 11.1). In the active, hypophosphorylated form, the Rb protein associates with the family of E2F transcription fac-

TABLE 11.2

DNA tumor viruses encoding proteins that inactivate cellular antioncogenes

Virus	pRb CELLULAR ANTI-ONCOPROTEIN	p53 CELLULAR ANTI-ONCOPROTEIN
	Viral protein (below) that complexes with cellular anti-oncoprotein (above)	
SV40	Large T antigen	Large T antigen
Human papillomavirus	E7	E6
Adenovirus	E1A	E1B

pRb, retinoblastoma protein.

After Brooks GF, Butel JS, Morse SA. Tumor viruses and oncogenes. *Jawetz, Melnick, & Adelberg's Medical microbiology*, 21st ed. Stamford: Appleton and Lange, 1998.

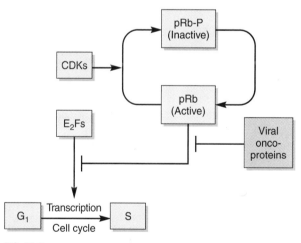

FIG. 11.1

The action of the retinoblastoma protein (pRb) and the effect of viral oncoproteins that bind it. pRb cycles between an active (unphosphorylated) and an inactive (phosphorylated) form, catalyzed by cyclin-dependent kinases. The active form (pRb) binds several transcription factors (E2Fs) that are required to recruit cells from G_1 to S phase of the cycle, and sequestration of these factors acts as a brake on cell division. Several viral oncoproteins (E7 of HPV; E1A of adenovirus; large T antigen of simian virus 40) bind and sequester pRb, abrogating its braking action and promoting cellular proliferation and tumorigenesis.

tors, sequestering them and abrogating their function, which is to promote the transition of cells from G₁ to S phase. By this action, pRb acts as a brake on the cell cycle. Cells are moved into and through the cell cycle by the action of different cyclin-dependent kinases (CDK), proteins that are activated by complexing with cyclins. One of the functions of some of the CDK is to inactivate the pRb protein by phosphorylation. Highly phosphorylated pRb no longer binds E2Fs and the responsive genes are then de-repressed, removing the braking action of pRb on the cell cycle.

p53 is a key tumor suppressor protein. Alteration or inactivation of p53 is a common component of the multistep process that leads to oncogenesis (Fig. 11.2). More than 50% of a variety of human tumors exhibit mutations in the p53 gene. The importance of p53 is further substantiated by the observa-

tion that knockout mice lacking functional p53 develop normally but are prone to tumor development.

p53 acts to repair or delete abnormal cells that would have a deleterious effect on the organism. The concentration of p53 increases in response to DNA damage and induces growth inhibitory genes, DNA repair genes, or genes that promote apoptosis (programmed cell death). p53 acts as a transcription factor, inducing expression of a number of proteins, including p21. p21 binds and inhibits certain CDK, thereby providing a brake on the cell cycle. Alternatively, extensive DNA damage can result in p53-mediated apoptosis by inducing proteins such as Bax, which triggers the caspase cascade, a sequence of protein activations leading to proteolysis of selected proteins essential for viability of the cell.

Both growth arrest and apoptosis initiated by p53 are blocked by a number of oncoproteins encoded by DNA viruses. Some viral oncoproteins interfere with effector domains of p53 whereas others enhance its ubiquitination and degradation. Viral oncoproteins may also act downstream of p53 by binding and inactivating Bax or possibly by elevating cyclins that then neutralize p21 activity.

There is interaction between the pRb and p53 pathways. The inactivation of pRb by aberrant growth signals can activate p14ARF, an E2F1-responsive gene. This protein stabilizes p53, leading to apoptosis and circumventing pRb inactivation.

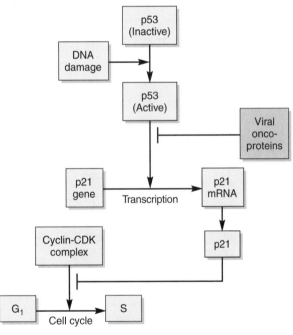

FIG. 11.2 _____

The action of the p53 protein and the effect of viral oncoproteins that block its action. p53 accumulates and is activated by damage to cellular DNA or aberrant growth signals, and functions to protect multicellular organisms against the growth or survival of such damaged cells. One action of p53, illustrated in this figure, is to cause growth arrest by initiating the transcription of a protein, p21, which binds to a complex of cyclin and cyclin-dependent kinase). This complex activates the cell cycle by moving cells from G₁ to S phase. When p21 binds this complex it is inactivated, providing a brake on the cell cycle. DNA viral oncoproteins (human papillomavirus E6, adenovirus E1B, SV40 large T antigen) bind p53 and interfere with p53-mediated growth arrest, thus removing a critical brake on the cell cycle and contributing to the transformation or immortalization of cells.

SELECTED EXAMPLES OF DNA TUMOR VIRUSES

To illustrate the mechanisms of oncogenesis by DNA tumor viruses, four examples will be discussed (Table 11.1). Three of these, human papillomaviruses (HPV), Epstein–Barr virus (EBV), and Kaposi's sarcoma (KS) herpesvirus (HHV-8), are significant causes of cancer in humans, whereas the fourth, human adenovirus, has served as an important model for the study of oncogenes and has been used as a vector for human gene therapy. These four viruses also illustrate the differences between the small DNA viruses (HPV and adenovirus) and the large DNA viruses (EBV, HHV-8).

Papillomaviruses: HPV and Cervical Carcinoma

HPVs are a large family of ubiquitous small DNA viruses that are distinguished by their tropism for epithelial cells at different body sites (Table 11.3). Based on molecular differences (>10% sequence diversity), more than 100 different types of HPV have been identified by cloning and sequencing.

TABLE 11.3

Skin lesions in transgenic mice bearing the E6 or E7 oncogenes (high copy number) of human papilloma virus 16 under control of the human keratin type 14 promoter

EVIDENCE OF EPIDERMAL HYPERPLASIA

Transgenes (number of mice)	Cataracts (%)	Thickened ears or snout (%)	Skin tumors (%)
K14E7 (18)	89	89	0
K14E6 (253)	100	100	13 (total) 9 (malignant)

After Herber R, Liem A, Pitot H, Lambert PF. Squamous epithelial hyperplasia and carcinoma in mice transgenic for the human papillomavirus type 16 E7 oncogene. *J Virol* 1996;70:1873–1881; and Song S, Pitot HC, Lambert PF. The human papilloma virus type 16 E6 gene alone is sufficient to induce carcinomas in transgenic animals. *J Virol* 1999;73:5887–5893.

One of the attributes of HPV is the striking differences in pathogenicity of these different types. HPV can be sorted into 5–10 distinct groups based on their localization to different cutaneous and mucosal surfaces of the body. All of these viruses cause benign epithelial overgrowths (variously called warts, papillomas, and condylomas). Under normal circumstances, immune surveillance keeps the viruses under control such that infections are either latent or regress spontaneously. However, some infections can persist and some virus types can also cause, at a low frequency, malignant tumors, notably cervical and penile cancers (Table 11.4).

Initially the virus infects basal cells in the epidermis through a wound in the epithelium. As the wound heals and the epithelium is reestablished, the cells at the lower epidermal strata divide and some of the daughter cells move toward the surface of the skin while undergoing differentiation. The virus undergoes sequential steps in its replicative cycle in the differentiating epithelium; progeny virions are found exclusively in some of the superficial dying and dead cells (Fig. 11.3). This dependence on squamous differentiation has made it impossible to grow and study HPV in conventional cell cultures, impeding research for many years. However, HPVs will replicate in special keratinocyte rafts or organotypic cultures.

HPV Genome and Its Expression The HPV genome is divided into three regions: early (E) and late (L) genes and a noncoding transcription regulatory region that overlaps the origin of replication. Early genes encode mainly factors required for transcription and replication of the viral genome, whereas the late genes encode the virion structural proteins.

The transcription profile of HPV mRNA and physical states of the HPV DNA differ in benign papillomas and condylomas (warts) from that seen in high-grade dysplasias and carcinomas. In papillomas, condylomas, and low-grade dysplasias, viral DNA is detected as nonintegrated nuclear episomes, together with varying levels of mRNA representing different viral genes, of which the E4 and E5 messages are most abundant. There is great heterogeneity in the amounts of viral DNA and RNA in different cells within any papillomatous lesion or among lesions from different patients. Viral capsid

TABLE 11.4

Some of the more common types of human papillomaviruses and the diseases they cause

HPV type	Anatomic site	Skin or mucosal disease	Risk of cancer
1, 4	Sole, palm	Plantar warts	None
2, 57	Skin, genital mucosa	Common warts	None
3, 10	Skin, genital mucosa	Flat warts	None
6, 11	Anogenital area, larynx	Warts	<<1%
16, 18, 31	Cervix, anogenital area, esophagus	Condylomas, dysplasias, carcinomas	1–3%
5, 8, 47	Skin, esophagus	Epidermodysplasia verruciformis	30–40%

After Chow LT, Broker TR. Small DNA tumor viruses. In: Nathanson N, et al., eds. *Viral pathogenesis*. Philadelphia: Lippincott–Raven Publishers, 1997;267–301.

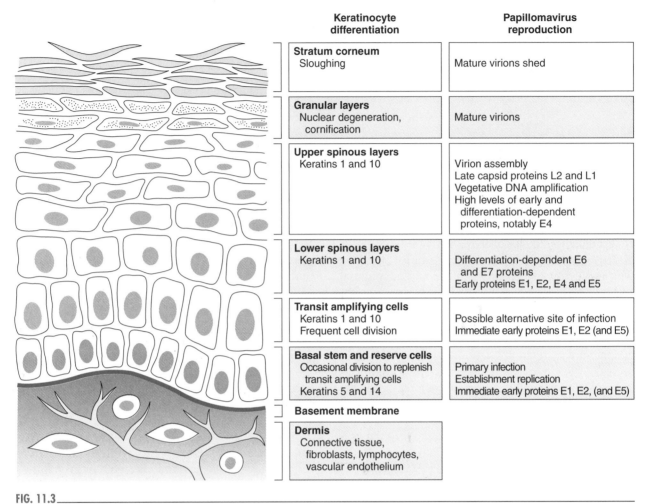

Keratinocyte differentiation	Papillomavirus reproduction
Stratum corneum Sloughing	Mature virions shed
Granular layers Nuclear degeneration, cornification	Mature virions
Upper spinous layers Keratins 1 and 10	Virion assembly Late capsid proteins L2 and L1 Vegetative DNA amplification High levels of early and differentiation-dependent proteins, notably E4
Lower spinous layers Keratins 1 and 10	Differentiation-dependent E6 and E7 proteins Early proteins E1, E2, E4 and E5
Transit amplifying cells Keratins 1 and 10 Frequent cell division	Possible alternative site of infection Immediate early proteins E1, E2 (and E5)
Basal stem and reserve cells Occasional division to replenish transit amplifying cells Keratins 5 and 14	Primary infection Establishment replication Immediate early proteins E1, E2, (and E5)
Basement membrane	
Dermis Connective tissue, fibroblasts, lymphocytes, vascular endothelium	

FIG. 11.3

Sequence of transcriptional events in the synthesis of papillomavirus in different layers of the epidermis. This diagram shows the steps in generation and maturation of keratinocytes and the corresponding steps in the sequential cascade of viral replication. (After Chow LT, Broker TR. Small DNA tumor viruses. In: Nathanson N, et al. *Viral pathogenesis*. Philadelphia: Lippincott–Raven Publishers, 1997:267–301, with permission.)

proteins may be visualized in some of the differentiated cells.

In contrast, in carcinomas in situ and invasive cancers, infectious virus is no longer produced. Rather, viral DNA is often integrated into the host chromosomes, accompanied by some loss of viral sequences. The viral DNA copy number is significantly reduced relative to that in productive infections, as it no longer can amplify. Furthermore, integration disrupts the viral transcription unit, and most HPV genes are not expressed. However, the E6 and E7 mRNA are present at elevated levels in these cancerous lesions and are not detected in the adjacent, histologically normal tissues. These observations, together with the properties of the E6 and E7

proteins, constitute compelling evidence that HPV can cause cancers in humans.

The HPV receptor appears to be expressed in many cell types and species, suggesting that the exquisite tissue specificity cannot be attributed to receptor distribution. The intimate relationship between productive infection and epithelial differentiation probably reflects the requirement for certain combinations of cellular transcription factors that are only supplied by epithelial cells at sequential stages of squamous differentiation (Fig. 11.3). By using the keratinocyte raft culture system, the activities of the promoters responsible for the expression of the viral oncogenes have been investigated. They are regulated by cis elements that bind transcription factors to activate transcription in the differentiated cells, by cis elements that repress transcription in the undifferentiated basal cells, and by histone deacetylases that remodel chromatin and down-regulate transcription in the basal cells.

In addition to virus-encoded proteins, viral DNA replication also depends on the host cells to supply cellular replication proteins, such as DNA polymerases, topoisomerases, and enzymes that synthesize the deoxyribonucleoside triphosphate substrates. Most of these cellular proteins are normally present in S-phase cells but not in postmitotic, differentiated cells. This explains the need for viral oncoproteins that reactivate S-phase genes in these cells. It also accounts for the oncogenicity of certain HPV types when these viral oncogenes are inappropriately up-regulated in the normally quiescent stem cells. The transforming ability of HPV can be mapped to three of the early proteins—E5, E6, and E7—all of which are also required for productive virus replication.

E5 Protein The E5 protein appears to act by its influence on signaling through the epidermal growth factor pathway. The epidermal growth factor receptor is constantly recycled (like other receptors) by internalization into endosomes from which some molecules return to the cell surface while others are degraded. E5 associates with the epidermal growth factor receptor and increases the proportion of molecules that are returned to the cell surface, thereby enhancing the concentration of receptor on the cell surface. The result is expansion of the infected cell population.

E6 and E7 Proteins The E6 and E7 genes are required for productive infection with HPV (Fig. 11.3). E7 binds to and inactivates the retinoblastoma (pRb) and related proteins, thereby overcoming the cell cycle brake (Fig. 11.1). For certain HPV types, such as the oncogenic HPV-18, expression of the E7 protein in differentiated keratinocytes is sufficient to promote S-phase reentry in a fraction of differentiated keratinocytes, thus providing a milieu conducive for viral DNA replication.

The exact role of E6 in differentiated cells is not completely understood. It cannot promote S-phase entry in postmitotic differentiated cells, but it may postpone programmed cell death and lengthen the duration of viral reproductive phase. One of the better understood properties of E6 is its ability to inactivate the p53 protein by targeting p53 to the ubiqui-

tin-mediated proteosome pathway (Fig. 11.2). In addition, E6 activates telomerase, an enzyme necessary to maintain the telomere lengths of chromosomes, critical for escaping senescence and sustaining long-term cell proliferation.

The oncogenicity of the E6 and E7 genes, when expressed in undifferentiated epithelial cells, has been demonstrated in two experimental systems: immortalization of primary keratinocytes in culture and tumor production in transgenic mice. E6 and E7 can extend the life span of primary cells, leading eventually to immortalization, but immortalization also requires mutations in yet-to-be identified cellular genes. Furthermore, immortalized cells are not transformed because they are not tumorigenic in nude mice, and tumor formation requires additional cellular mutations. In HPV-immortalized cells, mutations can occur as a consequence of excessive cell cycling in the absence of a functional p53 to safeguard the integrity of the genome. Because these cells are already deficient in functional p53 protein, few HPV-caused cancers have mutated p53, at least until late stages when the genome is highly unstable. In contrast, cancers that are not associated with HPVs often exhibit mutations in the p53 gene.

In animal model systems, the effects of E6 and E7 depend on the promoters that drive them, which determine the cell type in which the viral oncogenes are expressed, the level of transcription, and the nature of the tumors that are induced. Transgenic systems that mimic HPV-associated premalignant lesions and cancers, such as a keratin 14 promoter driving E6 or E7 from HPV-16, produce dysplasias or cancers of epithelial tissues because this promoter is active in basal and stem cells. On the other hand, when expressed from a differentiation-dependent promoter, such as the keratin 1 promoter, E6 and E7 of HPV-18 cause only papillomas.

Constitutive expression of E6 and E7 has different tumorigenic effects and the two proteins act synergistically in vivo (Fig. 11.4). Transgenic mice expressing E7 develop epidermal hyperplasia, suggesting that E7 plays a role in the early stages of carcinogenesis, whereas E6 transgenic mice develop epidermal hyperplasia that progresses to malignant tumors in some animals, suggesting that E6 plays a role in the both early and late stages of carcinogenesis (Table 11.3). E6 and E7 proteins from highly oncogenic HPV types are more potent than the same proteins derived from HPV types that cause only warts, and this distinction probably reflects subtle differences in the activity of the proteins of different

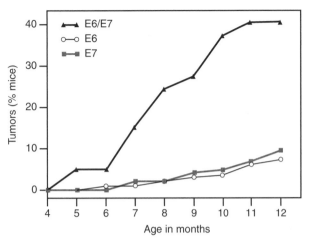

FIG. 11.4

The synergistic effect of *E6* and *E7*, two oncogenes of human papillomavirus. In this experiment, three groups of mice were followed for the frequency of skin tumors: mice transgenic for the *E6* gene, for the *E7* gene, and for both genes. Expression of the oncogenes is controlled by the keratin 14 promoter, which is active in the basal cells and hair follicles where stem cells reside. Normal (nontransgenic) mice would show a very low incidence of skin tumors (not shown). (After Song S, Liem A, Miller JA, Lambert PF. Human papillomavirus type 16 E6 and E7 contribute differently to carcinogenesis. *Virology* 2000;267:141–150.)

HPV types. For example, E6 from benign and oncogenic types of HPV binds to the amino terminus of p53 but only E6 from the oncogenic types also binds to the cellular E6-AP protein. E6-AP in complex with E6 functions as a ubiquitin ligase for p53 to promote its ubiquitination and degradation.

Immunologic Factors   The immune response to HPV antigens plays a major role in the natural history of the benign and malignant lesions caused by these viruses. Observations in humans, although they are anecdotal, are consistent with the ability of the immune response to control HPV-induced lesions. (a) Cutaneous warts are a disease of early childhood. They disappear in most subjects, suggesting the development of protective immunity. In adults, benign warts frequently regress "spontaneously" in normal patients. Likewise, genital wart disease is commonly seen in the years after commencement of sexual activity and then usually regresses. (b) Patients who are immunosuppressed due to AIDS, treatment with immunosuppressive regimens, or anticancer chemotherapy are at increased risk of HPV-associated lesions, including cervical carcinoma, and lesions are larger, more aggressive, and less likely to spontaneously regress than in immunologically unimpaired subjects. This increase of

HPV lesions in immunosuppressed patients is most likely due to reactivation of latent infections, rather than acquisition of new infections. When immunosuppression is discontinued, HPV lesions often regress. (c) Patients with dys- or agammaglobulinemia do not experience excessive HPV-associated lesions, suggesting that cellular immunity plays a more important role in the control of HPV disease.

The cottontail rabbit papillomavirus is the best animal model for HPV disease; it has been used to study the role of immunity in prevention and regression of papillomavirus lesions. A number of studies indicate that immunization of rabbits provides partial or complete protection against subsequent intracutaneous challenge with cottontail rabbit papillomavirus (Fig. 11.5). In immunized rabbits papillomas either fail to develop, are smaller, or regress more frequently than in non-immunized control animals. Both early (E) and late (L) proteins can provide protection, but it is likely that the most important antigens are E1, E2, E6, and E7 because only the early proteins are expressed in the basal stem cell and lower spinous layers.

Adenovirus: Oncogenesis in an Experimental System
Adenoviruses are ubiquitous human viruses, the majority of which infect the respiratory tract and cause

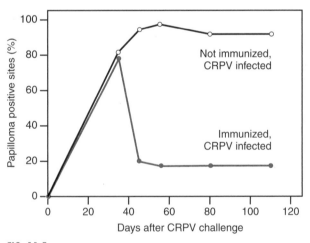

FIG. 11.5

Immune control of papillomavirus-induced papillomas demonstrated with the best animal model for human papillomavirus–induced papillomas, infection of rabbits with the cottontail rabbit papillomavirus (CRPV). In this experiment, rabbits were immunized with E1 and E2 proteins and challenged 2 weeks later with CRPV applied to lightly scarified skin. The graph shows the subsequent growth of papillomas in control and immunized animals. (After Selvakumar R, Borenstein LA, Lin Y-L, et al. Immunization with nonstructural proteins E1 and E2 of cottontail rabbit papillomavirus stimulates regression of virus-induced papillomas. *J Virol* 1995;69:602–605.)

respiratory illness. Infection is self-limiting due to immune surveillance. Although they have not been associated with tumors or cancer in their human hosts, adenoviruses can produce tumors in experimentally infected rodents. In nature, various types of adenoviruses infect epithelial tissues, including respiratory epithelium, conjunctiva, and intestinal epithelium. As with HPV, adenoviruses also require the S-phase cellular milieu to complete their replication, and they encode proteins that promote S-phase reentry in differentiated cells. In permissive human cells, virus is subsequently produced and released.

Human adenovirus can also infect cultured rodent cells. As these cells are not permissive and therefore do not undergo lytic productive infection, they can be transformed. The transformed cells contain integrated viral genomes, express some of the early proteins (encoded by the E1A and E1B region), exhibit the hallmarks of tumor cells, and are no longer subject to the growth restrictions of normal cells. Although all human adenoviruses transform cultured rodent cells, only some are tumorigenic in experimental rodents. In vitro cell culture systems and in vivo rodent models have been exploited to analyze adenovirus transformation.

Cellular transformation by adenoviruses is accomplished by two regions of the genome, E1A and E1B, both of which encode proteins that are expressed soon after infection. The ability to cause tumor development in animals also requires another action of adenoviruses, the ability to down-regulate the expression of major histocompatibility complex (MHC) class I proteins on the cell surface, which is also a function of E1A (described in Chapter 4).

E1A and E1B   The E1A proteins have several domains that bind different cellular proteins. The CR1 (conserved region 1) and CR2 domains bind the retinoblastoma protein (pRb) and prevent its action. As discussed already, pRb controls the transition from G_1 to S phase (Fig. 11.1), and removing its braking action activates E2Fs and the cell cycle. In addition, the CR1 domain of E1A binds a group of proteins, such as p300, which also act as a regulator of the cell cycle. p300 and related proteins cooperate with p53 to induce transcription of p21, the CDK inhibitor that blocks the cell cycle, and the sequestration of p300 removes another constraint on the cycle (Fig. 11.6).

E1B synergizes the action of E1A (Fig. 11.7). When E1A alone is introduced into cells, the cells are immortalized at low frequency, if at all. This is because the expression of E1A alone not only trig-

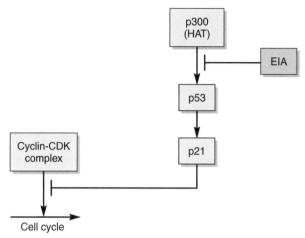

FIG. 11.6 _____

Adenovirus E1A acts as an oncoprotein by interfering with several brakes on the cell cycle, one of which is illustrated in this cartoon. p300, a histone acetyltransferase (HAT) protein, promotes the transcription of p53 by altering chromatin from a condensed to an open form. p53, in turn, acts as a transcription factor for p21, a protein that binds and sequesters cyclin/cyclin-dependent kinase (CDK) complexes. Cyclin-CDK complexes activate the cell cycle, and p21 acts as a brake on the cycle. EIA sequesters p300 and retards the downstream activation of p21, removing the p21 brake on the cell cycle.

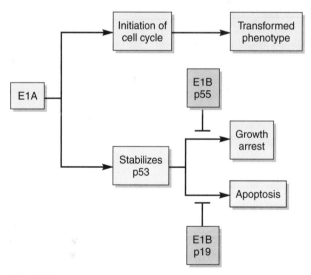

FIG. 11.7 _____

The oncogenic activity of adenovirus is associated with the E1A and E1B proteins, both of which are encoded by early (E) genes. These two proteins work in a complementary manner to transform cells. E1A drives resting cells into cell cycle and thence into a transformed phenotype. The same gene product also stabilizes p53, an anti-oncoprotein, and p53 initiates growth arrest and apoptosis. Thus cells transformed with E1A alone undergo transient transformation but the transformed foci quickly die. Adenovirus E1B encodes a p55 and a p19 protein; p55 blocks growth arrest and p19 blocks the apoptosis pathway. Thus, cells cotransformed with both *E1A* and *E1B* genes undergo stable transformation.

gers S-phase entry but also elevates p53 levels, which induces apoptosis. E1B encodes two proteins, p55 and p19, which block the action of p53 in different ways. p55 binds p53 and blocks the up-regulation of p21 (Fig. 11.2). The apoptosis action of p53 is mediated by the induction of "cell death" proteins, such as Bax, that trigger the caspase pathway leading to apoptosis; p19 binds and sequesters Bax, thereby preventing apoptosis.

In mice injected with human adenovirus, many cell types and tissues are infected, including liver, muscles, and hematopoietic cells. Defective adenovirus (missing E1A and/or E1B) has been used as a vehicle for human gene transfer. Its wide range of tissue tropism is an advantage because many cell types or tissues can be transduced, but its potential widespread cytotoxicity is also a limitation. In addition, because adenovirus engineered to delete E1B promotes apoptosis, it is being tested as an oncolytic agent via intratumoral injection of solid tumors.

Herpesviruses: Epstein–Barr Virus

EBV is a member of the gamma subfamily of herpesviruses, characterized by their ability to infect lymphoid cells. Biologically, EBV differs widely from adenoviruses in its oncogenic activity because it immortalizes rather than transforms its target cells and is a less potent carcinogen than the small oncogenic DNA viruses.

The great majority of humans are persistently but silently infected with EBV. Initial infections occurring in childhood are usually asymptomatic, whereas an estimated third of infections in adolescents or young adults are associated with infectious mononucleosis (a transient self-limited systemic illness with sore throat, malaise, and fever). In a small minority of infected persons, EBV is associated with one of several neoplasms—Burkitt's lymphoma, nasopharyngeal carcinoma, or Hodgkin's disease—that usually occur long after initial infection. By itself EBV is not capable of causing cancer, but it acts as a cocarcinogen in a multistep process.

B Lymphocytes EBV targets B lymphocytes by virtue of its receptor, the CD21 molecule, that is expressed mainly on B cells. The gp350 molecule, the major glycoprotein of EBV, is the virus attachment protein. However, entry is a complex process and involves at least one coreceptor, the class II MHC molecule that binds gp42, another viral glycoprotein. Upon entry into B lymphocytes, there is partial expression of the viral genome, persistence in a latent state, and immortalization of the infected cell.

Peripheral blood lymphocytes obtained from seropositive donors, depleted of T cells and cultured appropriately, usually yield immortalized lymphoblastoid cell lines. However, these immortalized cell lines are actually generated during the process of ex vivo culture. Cell culture triggers the lytic cycle in some persistently infected B cells, and the virus released from these cells recruits uninfected B cells that initiate the lymphoblastoid cell lines.

Infected B cells account for the persistence of the virus in humans, but they rarely release virus and are probably not responsible for viral shedding. In infected humans, therapeutic whole-body ablation of the lymphoid population (which leaves the epithelial cells intact) eradicates EBV infection. This suggests that the virus depends for its maintenance mainly on long-lived transformed B cells, which are present at a frequency of 10^{-5} to 10^{-6} among peripheral blood B cells.

Epithelial Cells In addition to B cells, EBV infects epithelial cells in the oral cavity. Within the epithelium, EBV is only partially expressed in the basal cells but completes its replicative cycle in the more superficial differentiated cells in the upper spinous and granular layers, in a manner reminiscent of HPV (Fig. 11.3). Productive infection of epithelial cells is probably responsible for shedding of infectious EBV, which can be isolated from throat swabs in many seropositive carriers. Likely this accounts for spread of the virus from host to host, but is not a pathway to oncogenesis. Figure 11.8 summarizes the likely pathogenesis of EBV infections.

Latent Infection and Cellular Immortalization Following viral entry into human B lymphocytes, the nucleocapsid is released into the cytosol, and the linear double-stranded DNA genome is released. The genome, with unspecified viral proteins, passes through a pore in the nuclear membrane. Once in the nucleus, the genome remains in a linear form until the host cell has passed through the next cell division, when the genome is circularized. During latency, the genome is maintained as a circularized replicating episome, which replicates before or during cell division so that there may be several copies per infected cell.

During initial infection of cultured B lymphocytes, a large number of EBV genes are expressed, and several of the expressed proteins are required for immortalization (Table 11.5). However, cultured immortalized B-cell clones express mainly two EBV proteins, EBNA-1 and LMP1. EBNA-1 is a

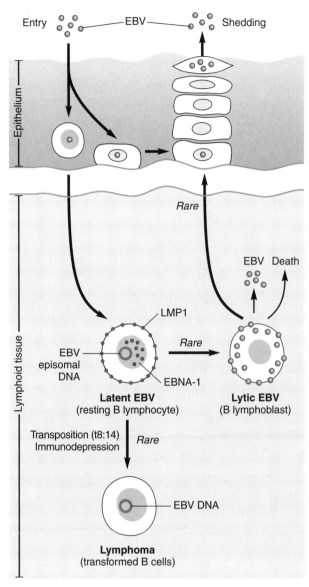

FIG. 11.8

Diagrammatic summary of possible pathogenesis of EBV and Burkitt's lymphoma. The virus initially infects epithelial cells and B lymphocytes. It persists in resting B cells in which it is rarely activated into a latent infection, whereas it causes both latent and lytic infection in epithelial cells from which virus is shed into the oral cavity. During latent persistence, several EBV genes are expressed, particularly EBNA-1 (EB nuclear antigen) and LMP1 (lymphoblast membrane protein), which immortalizes B-cell clones. In a few infected persons, persistently infected B cells undergo further transformation into tumor cells, producing a variety of lymphomas, such as Burkitt's lymphoma, which is frequently associated with a chromosomal transposition (t8:14) that up-regulates the c-*myc* oncogene. In addition, immunodepression markedly increases the risk of overt neoplasia, acting as another factor in the multistep progression to cancer.

nuclear protein required to maintain and replicate the viral genome as an unintegrated episome. LMP1, an integral plasma membrane protein, has a pleiotropic action, including constitutive activation of some B-cell growth factor receptors, such as TRAF (TNF receptor–associated factor) 1 and 3, which regulate B-cell proliferation. LMP1 has been considered an oncogene because transgenic mice expressing this protein develop B-cell lymphomas. However, in humans, EBV-associated tumors appear to involve a multifactorial process.

Immune Control of EBV   As mentioned above, a high proportion of normal subjects are carriers of EBV, and lymphoblastoid cells lines can be isolated from their peripheral blood in many instances. When human EBV carriers are immunosuppressed, as occurs in transplantation recipients or in AIDS patients, they are subject to lymphoproliferative disorders, such as immunoblastic lymphoma, and the proliferating cells are EBNA (EBV nuclear antigen) positive. It is assumed that these disorders represent EBV-infected lymphoid cells that have escaped immune surveillance. This implies that in normal EBV carriers, cellular immune responses control the proliferation of EBV-immortalized lymphoblastoid clones.

Burkitt's Lymphoma   First described by Dennis Burkitt as a fatal lymphoma of the head and neck in African children, three categories of Burkitt's lymphoma are now recognized. The originally described endemic form is particularly common in West Africa and coastal New Guinea; there is a worldwide rarer "sporadic" form; and a third form is seen as a complication of AIDS. One salient feature of Burkitt's lymphoma is that EBV DNA can be detected in tumor cells in more than 95% of endemic cases and in about half of the sporadic and AIDS-associated cases. In individual patients the EBV DNA sequences are uniform, suggesting that each tumor represents the clonal expansion of a single transformed cell.

The second characteristic feature of Burkitt's lymphoma is that the lymphoma cells exhibit a chromosomal translocation (t8:14) between chromosome 8 and (most frequently) chromosome 14, resulting in the relocation of c-*myc*, a protooncogene on chromosome 8, downstream from the enhancer regions for immunoglobulin heavy or light chains. Also, mutations in c-*myc* are common. In proliferating B-lymphoblastoid cells, these genetic alterations lead to constitutive overexpression of c-*myc*, which acts as a

TABLE 11.5		
EBV genes essential for B-lymphocyte immortalization		
EBV gene	Localization	Function (all are essential for immortalization)
EBNA-1	Nuclear	Required for the maintenance of the EBV episome
EBNA-2	Nuclear	Transactivator of cellular genes and LMP1
EBNA-3A, 3C	Nuclear	Transcription regulator of cellular genes
LMP1	Plasma membrane	Pleiotropic functions, including the constitutive activation of cellular receptors TRAF 1, TRAF 3

EBV immortalizes B lymphocytes by establishing a persistent latent infection. During latent infection a large number of EBV genes are expressed, and some of these, listed above, are essential for immortalization, based on experiments with EBV mutant viruses.

EBNA, Epstein–Barr nuclear antigen; LMP, lymphoblast membrane protein; TRAF, TNF receptor association factor.

After Hutt-Fletcher L. Epstein–Barr virus. In: Ahmed R, Chen I, eds. *Persistent viral infections*. New York: John Wiley and Sons, 1999.

transcription factor important in the cell cycle, driving B lymphocytes to unrestrained proliferation.

A likely third factor in the multistep process leading to cancer formation is loss of immune surveillance, which could increase the proliferation of lymphoblastoid cell lines both before and after chromosomal translocation. Endemic Burkitt's lymphoma is coextensive with *Plasmodium falciparum* malaria, and it is postulated that chronic malaria would reduce immune competence; consistent with this hypothesis is the increased risk of Burkitt's lymphoma in patients with AIDS.

In summary, EBV immortalizes B lymphocytes through the action of several of its gene products that convert infected cells to a constitutive state of proliferation. EBV infection may initiate Burkitt's lymphoma but only as part of a multistep process (Fig. 11.8) that includes chromosomal translocation of the c-*myc* protooncogene and immunosuppression.

Herpesviruses: HHV-8 and Kaposi's Sarcoma

KS is a neoplasm of the skin and viscera that is associated with HHV-8 (human herpesvirus 8, also called Kaposi's sarcoma herpesvirus). Epidemiologic and pathologic evidence strongly suggest that HHV-8 plays a causal role in KS, but the exact mechanisms remain to be elucidated. The discovery of HHV-8 is an instructive example of the use of molecular methods to unravel the etiology of an infectious disease and illustrates the complexities of viral oncogenesis.

History and Background KS was described more than 100 years ago as a relatively uncommon sarcoma of the skin in older men in Eastern Europe and the Mediterranean region. In the 1960s, prior to the emergence of AIDS, a more aggressive form of KS was reported in East African children. In the 1980s,

KS achieved new prominence as one of the diseases associated with AIDS. In the United States, KS is mainly seen in gay men with AIDS, and more than 95% of cases in the United States are in this group. The incidence of KS in U.S. gay men is more than 10,000-fold greater than in the general population and more than 10-fold greater than in other AIDS patients, such as injecting drug users and blood recipients. These observations led to a number of hypotheses regarding the etiology of KS that would explain its enigmatic epidemiologic patterns (discussed below).

The histopathology of KS is unusual for sarcomas because the lesions consist of many different cell types, of which the dominant one is the spindle cell, thought to represent a transformed endothelial cell. In addition, tumors are infiltrated with inflammatory cells and exhibit large numbers of newly formed vascular elements. Cultured spindle cell lines secrete proinflammatory and angiogenic factors, suggesting that the spindle cell is responsible for the characteristic cellular composition of KS tumors. Thus, an etiologic hypothesis should also explain the origin and nature of the spindle cells.

Etiology One hypothesis suggested that KS is caused by a previously undetected infectious agent. Searching for footprints of such a putative agent, Chang and colleagues used the methods of differential representation analysis (Sidebar 11.1) to identify DNA sequences specific for KS tumor tissue. Several DNA fragments were identified, shown to bear homology with sequences in known human and primate herpesviruses, and used as probes to sequence the complete genome of a previously undescribed herpesvirus, since named HHV-8.

The discovery of HHV-8 as a putative causal agent of KS led to serological surveys that explained

■ ■ ■
SIDEBAR 11.1

"To search for foreign DNA sequences belonging to an infectious agent in AIDS-KS, we used representational difference analysis (RDA) to identify and characterize unique DNA sequences in KS tissue . . . .

"The initial round of amplification-hybridization from KS and excess normal-tissue DNA . . . resulted in . . . four bands at approximately 380, 450, 540, and 680 base pairs . . . [which] . . . became discrete after a third round of amplication-hybridization. The four KS-associated bands (designated KS330Bam, KS390Bam, KS480Bam, and KS631Bam) . . . were gel purified. KS390Bam and KS480Bam Southern (DNA) hybridized nonspecifically to both KS and non-KS human tissues and were not further characterized. The remaining two RDA bands, KS330Bam and KS631Bam, were cloned and sequenced. K330Bam is 51% identical by amino acid homology to a portion of the ORF26 open reading frame encoding the capsid protein VP23 of herpesvirus saimiri . . . [and is] . . . also 39% identical to the amino acid sequnce encoded by the corresponding BDLF1 ORF of Epstein–Barr virus. KS631Bam has homology to the tegument protein (ORF75) of herpesvirus saimiri and to the tegument protein of EBV (ORF BNRF1). To determine the specificity of KS330Bam . . . for AIDS-KS, these sequences were hybridized to Southern blots of DNA extracted from cryopreserved tissue obtained from patients. . . .

Patients	Tissue type	Positive by KS330Bam DNA hybridization
AIDS	KS lesions	20/27 (74%)
AIDS	Lymphomas and lymph nodes	6/39 (15%)
Non-AIDS	Lymphomas and lymph nodes	0/36 (0%)

"Although these sequences suggest the presence of a new human herpesvirus in KS lesions, a causal link between these sequences and AIDS-KS cannot be established by our retrospective case control study." ■

Chang Y, Cesarman E, Pessin MS, et al. Identification of herpesvirus-like DNA sequences in AIDS-associated Kaposi's sarcoma. *Science* 1994;266:1865–1869.

its enigmatic epidemiologic patterns (Table 11.6). The great majority of patients with KS are infected with HHV-8, consistent with the hypothesis that HHV-8 plays a critical role in the etiology of KS. The incidence of HHV-8 is much higher in gay men than in drug injectors or recipients of blood or blood products, explaining why KS is much more common in homosexual men than in individuals from other AIDS risk groups. The much higher incidence of KS in HIV-positive gay men with AIDS than in HIV-positive gay men without AIDS is consistent with a role of immunosuppression as a cofactor in the development of KS. This is supported by the observation that effective treatment of AIDS with highly active antiretroviral therapy leads to arrest or remission of KS lesions. Finally, prospective studies of gay men showed that subjects who were infected with both HIV and HHV-8 (in either order) developed KS at a high incidence, whereas those infected with only HHV-8 or only HIV did not develop KS (Fig 11.9).

Pathogenesis of KS The role of HHV-8 in the pathogenesis of KS has been under intense investigation since the virus was isolated in 1994 and still awaits definitive explanation. Meanwhile, some important clues have been unearthed. HHV-8 is a gammaherpesvirus that bears considerable similarity to EBV. In many HHV-8–seropositive subjects, HHV-8 can be detected in peripheral blood B lymphocytes where it probably persists. Furthermore, HHV-8 is associated with a variety of unusual lymphoproliferative neoplasms, such as primary effusion lymphoma, suggesting that it may transform B cells. Some cell lines derived from these unusual lymphomas carry HHV-8 as a persistent infection in which the viral DNA exists as an episome. Most of the cells in such cultures are latently infected and express a limited number of HHV-8 (latent) transcripts, whereas productive replication with expression of the whole genome and generation of infectious virus is seen in only a few cells.

Many spindle cells in the KS lesion are positive for HHV-8 DNA; most of these express transcripts typical of latent infection, whereas a few express whole-genome transcripts. HHV-8 encodes a number of genes that could play a role in cell transformation or tumorigenesis. These include proteins that resemble cyclins, chemokines, cytokines, G-protein-coupled receptors, as well as proteins similar to bcl-2 that can block apoptosis. In addition, the major latency-associated antigen, LANA-1, modulates the transcription of pRb and p53. It appears likely that these and other HHV-8–encoded proteins play a role in the cellular proliferation of endothelial and B cells that is characteristic of HHV-8–associated neoplasms. An in vitro model has been developed using a line of immortalized human dermal microvascular cells that, when exposed to HHV-8, become latently infected and acquire a spindle morphology together

TABLE 11.6		
Human herpesvirus 8 as a putative causal agent of kaposi's sarcoma		
	PREVALENCE OF HHV-8 ANTIBODY	
Subjects	HIV negative	HIV positive
Kaposi's sarcoma		
Endemic KS (Greece)	17/18 (94%)	
AIDS cases		84/103 (82%)
Without Kaposi's sarcoma		
Homosexual men	8/65 (12%)	10/33 (30%)
Injecting drug users	0/25 (0%)	0/38 (0%)
Hemophiliacs		0/26 (0%)

Prevalence of HHV-8 infection and Kaposi's sarcoma (KS) shows that most patients with KS are infected with HHV-8, consistent with the hypothesis that HHV-8 plays an essential role in the etiology of KS. Regardless of HIV status, HHV-8 is much more prevalent in homosexual men than other risk groups, such as injecting drug users and patients with hemophilia. This could explain why AIDS-associated KS is mainly seen in homosexual men compared with AIDS patients from other risk groups. The serologic test for HHV-8 infection used in this study was probably suboptimal for sensitivity.

HHV, human herpesvirus; KS, Kaposi's sarcoma; HIV, human immunodeficiency virus.

Modified from Simpson GR, Schulz TF, Whitby D, et al. Prevalence of Kaposi's sarcoma associated herpesvirus infection measured by antibody to recombinant capsid protein and latent immunofluoresscence antigen. *Lancet* 1996;349:1133–1138.

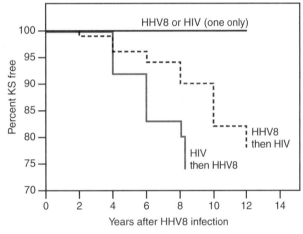

FIG. 11.9

The cumulative incidence of Kaposi's sarcoma (KS) in a group of gay men infected with human herpesvirus type 8 (HHV-8) and with HIV. Infection with HIV followed with HHV-8 infection carried a higher risk than HHV-8 infection followed by HIV infection. Subjects infected with HHV-8 only or with HIV only did not develop KS. (After Jacobson SP, Jenkins FJ, Springer G, et al. Interaction of human immunodeficiency virus type 1 and human herpesvirus type 8 infections on the incidence of Kaposi's sarcoma. *J Infect Dis* 2000;181:1940–1949.)

with the stigmata of transformation, such as loss of contact inhibition and anchorage-independent growth. This model reproduces many of the characteristics of KS in humans and strongly supports the etiologic role of HHV-8 in this disease.

In summary, Kaposi's sarcoma illustrates the multifactorial nature of virus-induced neoplasia. In this instance, it appears that HHV-8, a newly de-scribed gammaherpesvirus, is the etiologic agent that infects and transforms endothelial cells, converting them to the spindle cells characteristic of KS lesions. Transformation is probably mediated by multiple viral genes that encode proteins that upregulate the cell cycle, constitutively activate growth factor pathways, and block apoptosis. The transformed cells have limited tumorigenic capacity and are normally controlled by immune surveillance. However, immunosuppression, associated with AIDS, holoendemic malaria, or aging, markedly increases the risk of overt KS disease.

REPRISE

The oncogenic DNA viruses reinforce some general themes regarding virus-induced tumorigenesis. First, in all instances, the viruses encode proteins that perturb normal controls on cellular proliferation, differentiation, and death. The oncogenic DNA viruses can initiate the process of oncogenesis under conditions where their transforming genes inactivate the effects of cellular proteins, such as p53 and pRb, which act as brakes on the cell cycle. Second, cancer is a multifactorial, multistep process, and additional mutations in host genes are usually necessary for carcinogenesis. Third, in many instances the appearance and aggressiveness of tumors are modulated by immune responses against both viral gene products and tumor antigens, indicating the potential for preventive and therapeutic vaccines.

FURTHER READING

Reviews and Chapters

Antman K, Chang Y. Kaposi's sarcoma. *N Engl J Med* 2000;342:1027–1038.

Bergers G, Hanahan D, Coussens LM. Angiogenesis and apoptosis are cellular parameters of neoplastic progression in transgenic mouse models of tumorigenesis. *Int J Dev Biol* 1998;42: 995–1002.

Brooks GF, Butel JS, Morse SA. Tumor viruses and oncogenes. In: *Jawetz, Melnick, and Adelberg's Medical microbiology,* 21st ed. Norwalk, CT: Appleton & Lange, 1998;543–565.

Bryan TM, Cech TR. Telomerase and the maintenance of chromosome ends. *Curr Opin Cell Biol* 1999;11:318–324.

Chow LT, Broker TR. In vitro experimental systems for HPV: epithelial raft cultures for viral reproduction and pathogenesis and for genetic analyses of viral proteins and regulatory sequences. *Clin Dermatol* 1997;15:217–227.

Chow LT, Broker TR. Small DNA tumor viruses. In: Nathanson N, et al., eds. *Viral pathogenesis.* Philadelphia: Lippincott-Raven Publishers, 1997:267–301.

Cohen JI. Epstein–Barr virus infection. *N Engl J Med* 2000;343:481–492.

de Villiers EM. Papillomavirus and HPV typing. *Clin Dermatol* 1997;15:199–206.

Dyson, N. The regulation of E2F by pRb-family proteins. *Genes Dev* 1998;12:2245–2262.

Ganem D. KHSV and Kaposi's sarcoma: the end of the beginning? *Cell* 1997;91:157–160.

Hanahan D, Weinberg RA. The hallmarks of cancer. *Cell* 2000;100:57–70.

Hutt-Fletcher L. Epstein–Barr virus. In: Ahmed R, Chen I, eds. *Persistent viral infections.* New York: John Wiley and Sons, 1999;243–268.

Levine AJ. p53, the gatekeeper for growth and division. *Cell* 1997;88:323–331.

Ricciardi RP. Adenovirus transformation and tumorigenicity. In: Prem S, ed. *Adenoviruses: basic biology to gene therapy.* Austin, TX: RG Landes, 1999.

Sherr CJ, Roberts JM. CDK inhibitors: positive and negative regulators of G_1-phase progression. *Genes Dev* 1999;13: 1501–1512.

Vogelstein B, Lane D, Levine AJ. Surfing the p53 network. *Nature* 2000;408:307–310.

Weinberg RA. The retinoblastoma protein and cell cycle control. *Cell* 1995;81:323–330.

White E. Regulation of p53-dependent apoptosis by E1A and E1B. *Curr Top Microbiol Immunol* 1995;199:34–58.

Classic Papers

Burkitt D. A sarcoma involving the jaws in African children. *Br J Surg* 1958;45:218–223.

Chang Y, Cesarman E, Pessin MS, et al. Identification of herpesvirus-like DNA sequences in AIDS-associated Kaposi's sarcoma. *Science* 1994;266:1865–1869.

Epstein MA, Achong BG, Barr YM. Virus particles in cultured lymphoblasts from a Burkitt's lymphoma. *Lancet* 1964; 1:252–253.

Gallimore PH, Sharp PA, Sambrook J. Viral DNA in transformed cells. II. A study of the sequences of adenovirus 2 DNA in nine lines of transformed rat cells using specific fragments of the viral genome. *J Mol Biol* 1974;89:49–72.

Nishikura K, Ar-Rushidi A, Erikson J, et al. Differential expression of the normal and of the translocated human c-*myc* oncogene in B cells. *Proc Natl Acad Sci USA* 1983; 80:4822–4826.

Rous P, Beard JW. The progression to carcinoma of virus-induced rabbit papilloma (Shope). *J Exp Med* 1935;62:523–548.

Shope RE, Hurst EW. Infectious papillomatosis of rabbits: with a note on the histopathology. *J Exp Med* 1933;58:607–624.

Trentin JJ, Yabe Y, Taylor G. The quest for human cancer viruses. *Science* 1962;137:835–849.

Whyte P, Buchkovich KJ, Horowitz JM, et al. Association between an oncogene and an anti-oncogene: the adenovirus E1A proteins bind to the retinoblastoma gene product. *Nature* 1988;334:124–129.

Original Contributions

Ai W, Toussaint E, Roman A, CCAAT displacement protein binds to and negatively regulates human papillomavirus type 6 E6, E7, and E1 promoters. *J Virol* 1999;73:4220–4229.

Ballestas ME, Chjatis PA, Kaye KM. Efficient persistence of extrachromosomal HSHV DNA mediated by latency-associated nuclear antigen. *Science* 1999;284:641–644.

Bates S, Phillips AC, Clark PA, et al. Vousden KH. p14ARF links the tumour suppressors RB and p53. *Nature* 1998;395:124–125.

Beral V, Peterman TA, Berkelman RL, Jaffe HW. Kaposi's sarcoma among persons with AIDS: a sexually transmitted infection? *Lancet* 1990;335:123–128.

Brehm A, Miska EA, McCance DJ, et al. Retinoblastoma protein recruits histone deacetylase to repress transcription. *Nature* 1998;391:597–601.

Chakravarti D, Ogryzko V, Kao H-Y, et al. A viral mechanism for inhibition of p300 and PCAF acetyltransferase activity. *Cell* 1999;96:393–403.

Chang J, Renne R, Dittmer D, Ganem D. Inflammatory cytokines and the reactivation of Kaposi's sarcoma–associated herpesvirus lytic replication. *Virology* 2000;266:17–25.

Cheng EH-Y, Nicholas J, Bellows DS, et al. A Bcl-2 homolog encoded by Kaposi sarcoma–associated virus, human herpesvirus 8, inhibits apoptosis but does not heterodimerize with Bax or Bak. *Proc Natl Acad Sci* 1997;94:690–694.

Dahiya A, Gavin MR, Luo RX, Dean DC. Role of the LXCXE binding site in Rb function. *Mol Cell Biol* 2000;20:6799–6805.

Cheng S, Schmidt-Grimminger DC, Murant T, et al. Differentiation-dependent up-regulation of the human papillomavirus E7 gene reactivates cellular DNA replication in suprabasal differentiated keratinocytes. *Genes Dev* 1995;9:2335–2349.

Dollard SC, Wilson JL, Demeter LM, et al. Production of human papillomavirus and modulation of the infectious program in epithelial raft cultures. *Genes Dev* 1992;6:1131–1142.

Flores ER, Allen-Hoffman BL, Lee D, Lambert PF. The human papillomavirus type 16 E7 oncogene is required for the productive stage of the viral life cycle. *J Virol* 2000;74:6622–6631.

Francis DA, Schmid SI, Howley PM. Repression of the integrated papilloma virus E6/E7 promoter is required for growth suppression of cervical cancer cells. *J Virol* 2000;74:2679–2686.

Greenhalgh DA, Wang XJ, Rothnagel JA, et al. Transgenic mice expressing targeted HPV-18 E6 and E7 oncogenes in the epidermis develop verrucous lesions and spontaneous, rasHa-activated papillomas. *Cell Growth Diff* 1994;5:667–675.

Han R, Cladel NM, Reed CA, et al. Protection of rabbits from viral challenge by gene gun-based intracutaneous vaccination with a combination of cottontail rabbit papillomavirus E1, E2, E6, and E7 genes. *J Virol* 1999;73:7039–7043.

Han J, Sabbatini P, Perez D, et al. The E1B 19K protein blocks apoptosis by interacting with and inhibiting the p53-inducible and death-promoting Bax protein. *Genes Dev* 1996;10:461–477.

He Z, Wlazlo AP, Kowalczyk DW, et al. Viral recombinant vaccines to the E6 and E7 antigens of HPV-16. *Virology* 2000;270:146–161.

Herber R, Liem A, Pitot H, Lambert PF. Squamous epithelial hyperplasia and carcinoma in mice transgenic for the human papillomavirus type 16 E7 oncogene. *J Virol* 1996;70:1873–1881.

Jacobson SP, Jenkins FJ, Springer G, et al. Interaction of human immunodeficiency virus type 1 and human herpesvirus type 8 infections on the incidence of Kaposi's sarcoma. *J Infect Dis* 2000;181:1940–1949.

Kiyono T, Foster SA, Koop JI, et al. Both Rb/p16INK4a inactivation and telomerase activity are required to immortalize human epithelial cells [see comments]. *Nature* 1998;396:84–88.

Kowalik TF, DeGregori J, Leone G, et al. E2F1-specific induction of apoptosis and p53 accumulation, which is blocked by Mdm2. *Cell Growth Diff* 1998;9:113–118.

Jian Y, Van Tine BA, Chien WM, et al. Concordant induction of cyclin E and p21cip1 in differentiated keratinocytes by the HPV E7 protein inhibits cellular and viral DNA synthesis. *Cell Growth Diff* 1999;10:101–111.

Li X, Coffino P. High-risk human papillomavirus E6 protein has two distinct binding sites with p53, of which only one determines degradation. *J Virol* 1996;70:4509–4516.

Lin BY, Ma T, Liu JS, et al. HeLa cells are phenotypically limiting in cyclinE/cdk2 for efficient human papillomavirus DNA replication. *J Biol Chem* 2000;275:6167–6174.

Liu Y, Chen JJ, Gao Q, et al. Multiple functions of human papillomavirus type 16 E6 contribute to the immortalization of mammary epithelial cells. *J Virol* 1999;73:7279–7307.

Luo RX, Postigo AA, Dean DC. Rb interacts with histone deacetylase to repress transcription. *Cell* 1998;92:463–473.

Ma T, Zou N, Lin BY, et al. Interaction between cyclin-dependent kinases and human papillomavirus replication initiation protein E1 is required for efficient viral replication. *Proc Natl Acad Sci USA* 1999;96:382–387.

Martin MED, Berk AJ. Adenovirus E1B 55K represses p53 activation in vitro. *J Virol* 1998;72:3146–3154.

Meyers C, Frattini MG, Hudson JB, Laimins LA. Biosynthesis of human papillomavirus from a continuous cell line upon epithelial differentiation. *Science* 1992;257:971–973.

Moses AV, Fish KN, Ruhl RE, et al. Long-term infection and transformation of dermal microvascular endothelial cells by human herpesvirus 8. *J Virol* 1999;73:6892–6902.

Parker JN, Zhao W, Askins KJ, Broker TR, Chow LT. Mutational analyses of differentiation-dependent human papillomavirus type-18 enhancer elements in epithelial raft cultures of neonatal foreskin keratinocytes. *Cell Growth Diff* 1997;8:751–762.

Radkov SA, Kellam P, Boshoff C. The latent nuclear antigen of Kaposi sarcoma–associated herpesvirus targets the retinoblastoma-E2F pathway and with the oncogene Hras transforms primary rat cells. *Nature Med* 2000;6:1121–1127.

Roth J, Konig C, Wienzek S, et al. Inactivation of p53 but not p73 by adenovirus type 5 E1B 55-kilodalton and E4 34-kilodalton oncoproteins. *J Virol* 1998;72:8510–8516.

Russo JJ, Bohenzky RA, Chien MC. Nucleotide sequence of the Kaposi sarcoma–associated herpesvirus (HHV-8). *Proc Natl Acad Sci USA* 1996;93:14862–14867.

Sabbatini P, Lin J, Levine AJ, White E. Essential role for p53-mediated transcription in E1A-induced apoptosis. *Genes Dev* 1995;9:2184–2192.

Scheffner M, Huibregtse JM, Vierstra RD, Howley PM. The HPV-16 E6 and E6-AP complex functions as a ubiquitin-protein ligase in the ubiquitination of p53. *Cell* 1993;75:495–505.

Selvakumar R, Borenstein LA, Lin Y-L, et al. Immunization with nonstructural proteins E1 and E2 of cottontail rabbit papillomavirus stimulates regression of virus-induced papillomas. *J Virol* 1995;69:602–605.

Sibilla M, Fleischmann A, Behrens A, et al. The EGF receptor provides an essential survival signal for SOS-dependent skin tumor development. *Cell* 2000;102:211–220.

Smith K, Brown CC, Spindler KR. The role of mouse adenovirus type 1 early region 1A in acute and persistent infections in mice. *J Virol* 1998;72:5699–5706.

Song S, Pitot HC, Lambert PF. The human papilloma virus type 16 E6 gene alone is sufficient to induce carcinomas in transgenic animals. *J Virol* 1999;73:5887–5893.

Song S, Liem A, Miller JA, Lambert PF. Human papillomavirus type 16 E6 and E7 contribute differently to carcinogenesis. *Virology* 2000;267:141–150.

Steenbergen RDM, Walboomers JMM, Meijer CJLM, et al. Transition of human papillomavirus type 16 and 18 transfected human foreskin keratinocytes towards immortality: activation of telomerase and allele losses at 3p, 10p, 11q and/or 18q. *Oncogene* 1996;13:1249–1257.

Stoppler MC, Ching K, Stoppler H, et al. Natural variants of the human papillomavirus type 16 E6 protein differ in their ability to alter keratinocyte differentiation and to induce p53 degradation. *J Virol* 1996;70:6987–6993.

Straight SW, Hinkle PM, Jewers RJ, McCance DJ. The E5 oncoprotein of human papillomavirus type 16 transforms fibrob-lasts and effects the down regulation of the epidermal growth factor receptor in keratinocytes. *J Virol* 1993;67:4521–4532.

Teodoro JG, Branton PE. Regulation of p53-dependent apoptosis, transcriptional repression, and cell transformation by phosphorylation of the 55-kilodalton E1B protein of human adenovirus type 5. *Virology* 1997;90:595–606.

Thomas JT, Hubert WG, Ruesch MN, Laimins LA. Human papillomavirus type 31 oncoproteins E6 and E7 are required for the maintenance of episomes during the viral life cycle in normal human keratinocytes. *Proc Natl Acad Sci USA* 1999;96:8449–8454.

Tierney RJ, Steven N, Young LS, Rickinson AB. Epstein–Barr virus latency in blood mononuclear cells: analysis of viral gene transcription during primary infection and in the carrier state. *J Virol* 1994;68:7374–7385.

Wienzek S, Roth J, Dobbelstein M. E1B 55-kilodalton oncoproteins of adenovirus types 5 and 12 inactivate and relocalize p53, but not p51 or p73, and cooperate with E4 orf6 proteins to destabilize p53. *J Virol* 2000;74:193–202.

Yao QY, Ogan P, Rown M, et al. Epstein–Barr virus–infected cells persist in the circulation of the acyclovir-treated virus carriers. *Int J Cancer* 1989;43:67–71.

Zhao W, Noya F, Chen WY, et al. Trichostatin A up-regulates HPV-11 URR-E6 promoter activity in undifferentiated primary human keratinocytes. *J Virol* 1999;73:5026–5033.

Chapter 12
Host Susceptibility to Viral Diseases

A central theme of this book is that the outcome of a viral infection depends on both the parasite and the host. Just as a particular virus can vary in its virulence, so a given host species can vary in its response to a single virus. This diversity is apparent during an outbreak of smallpox or poliomyelitis, when individual infections range from inapparent to fatal. However, potential differences in the route of infection, as well as in the strain and dose of virus, could account for the observed variance in illness and mortality in natural infections, so that the role of host determinants is ambiguous. By contrast, experimental comparison of the responses of inbred strains of animals to a standardized virus inoculum provides convincing evidence of host variation in the outcome of infection.

This chapter describes host genes that have a striking influence on the course of viral infection. Most experimental studies employ mice because of the availability of a vast number of inbred genetically defined strains. It might be predicted that immune response diversity could explain genetic variation but, in most mouse models, nonimmune determinants are operative.

Similar genetic determinants undoubtedly exist in other species although the data are sparse. For instance, domestic rabbits are considerably more susceptible to the oncogenic effects of cottontail rabbit papillomavirus (CRPV) than are cottontail rabbits themselves, but there are no tools to map the responsible genetic loci. By implication, these genetic determinants also operate in humans, although it has been difficult to test this hypothesis. However, recent studies of HIV/AIDS have revealed a few such genetic loci, which undoubtedly represent only the tip of the iceberg.

In addition to genetic determinants, other host factors, such as age and nutritional status, can play a significant role in viral infection. These variables are described at the end of this chapter.

GENETIC DETERMINANTS OF SUSCEPTIBILITY IN MICE

Overview

There are many inbred strains of mice that can be readily tested for their susceptibility or resistance to viruses. If any single virus is compared in enough mouse strains, some variability in disease or mortality will be seen, as illustrated in Table 12.1. From studies of these mouse models, a number of generalizations can be made.

TABLE 12.1

Difference in susceptibility to a given virus among mouse strains

Mouse Strain	H-2 Haplotype	Female dead/total (% mortality)	Male dead/total (% mortality)
A.WY/SnJ	a	60/62 (97%)	50/51 (98%)
A.SW/SnJ	s	34/37 (92%)	33/35 (94%)
C57BL/6J	b	50/66 (73%)	16/19 (73%)
B10.A	a	22/35 (63%)	8/16 (50%)
A/J	a	13/22 (59%)	Not done
C57BL/10ScN	b	47/93 (50%)	35/42 (73%)
DBA/2J	d	4/40 (10%)	4/9 (44%)
BALB/c	d	1/52 (2%)	21/50 (42%)
CBA/J	k	0/56 (0%)	2/34 (6%)
SJL/J	s	0/74 (0%)	0/62 (0%)

[handwritten annotations: "has to do with immune response" next to haplotype column; circled 'a' entries; "3 strains with same haplotype have different % mortality ∴ susceptibility"; "∴ haplotype does not determine susceptibility in this system or example"; "may be some correlation in this system"]

In this example, mice were inoculated with rabies virus, $10^{7.7}$ mouse IC LD_{50} by the intraperitoneal route, and followed for 21 days to determine mortality. Mouse strains differed markedly in their susceptibility, and this did not correlate with the H-2 (MHC) haplotype. Also, there was some suggestion that female mice were less susceptible than males.

After Lodmell DL. Genetic control of resistance to street rabies virus in mice. *J Exp Med* 1983;157:451–460.

- In many instances, differences in susceptibility are determined by a single genetic locus, as shown by testing F1 and F2 crosses. Either resistance or susceptibility can be dominant.

- Determinants identified in animal experiments may or may not be observable in cultured cells derived from resistant and susceptible hosts.

- Most genetic determinants do not map to the major histocompatibility complex (MHC) locus (H-2 in the mouse) and so are due to nonimmune mechanisms.

- A few genetic determinants map to the MHC locus (H-2 in the mouse).

- Most genetic determinants affect the response to a single virus or virus family. Therefore, each determinant is unique, which can be documented by mapping loci to specific chromosomes.

- For murine retroviruses, some genetic loci represent endogenous viral sequences that have been incorporated into the host genome (discussed in Chapter 10).

A representative list of genetic loci that determine susceptibility is shown in Table 12.2. A few of these will be described to illustrate the effects upon viral infection and cellular mechanisms.

Nonimmunologic Determinants

The Mx Protein, Interferon, and Influenza Virus The Mx protein is a classical example of a genetic determinant of disease susceptibility that has been studied exhaustively, although much remains to be eluci-

dated. It was observed in the 1960s that most strains of inbred mice are susceptible to influenza and related orthomyxoviruses, but the A2G mouse strain is resistant (Fig. 12.1), based on the severity of pneumonia. Furthermore, there was a correlation between susceptibility to disease and level of virus replication in the lung. When F1 and F2 hybrids between the two strains of mice were tested, it was found that susceptibility segregated as a single genetic locus and that resistance was dominant (Table 12.3). This genetic locus (designated as *Mx* for myxovirus) was shown to have no influence on susceptibility to most other virus families.

Macrophages from susceptible and resistant mice show the Mx phenotype when tested soon after culture, but 2 weeks later the resistant cells have become susceptible. The loss of resistance was found to correlate with a loss of production of interferon by macrophages from A2G-resistant mice. It was then shown that the Mx phenotype (resistance to influenza virus) was dependent on interferon since treatment with anti-interferon antiserum abrogated resistance to influenza virus (Table 12.3). However, the Mx locus did not code for interferon itself because both susceptible and resistant mice produced similar interferon responses to viral infection.

Interferon produces its effects indirectly by inducing the expression of a plethora of cellular genes (Fig. 5.1), and one of these is the Mx gene. The murine Mx gene has been cloned and shown to have two alleles, Mx1 and Mx2. Resistant mice possess the Mx1 allele, which encodes a protein that confers resistance to influenza virus, whereas susceptible

non-immunologic determinants of susceptibility

TABLE 12.2

Representative mouse genetic loci that influence the outcome of viral infections

Virus (Family)	Disease or effect	Locus designation (no. of genes)	Dominant trait	Maps to H-2 (Chromosome)
	Nonimmune			
Influenza (Orthomyxoviridae)	Pneumonia	*Mx* (1)	Resistance	No (C16)
Flaviviridae	Encephalitis	*Flv* (1)	Resistance	No (C5)
Rabies (Rhabdoviridae)	Encephalitis	(1)	Resistance	No
MHV (Coronaviridae)	Hepatitis (MHV-3)	*Musfbp* (1)	Susceptibility	No
	Death	*Hv-1* (1)	Susceptibility	No
	Infection	*Hv-2* (1)	Susceptibility	No (C7)
MuLV (Retroviridae)	Susceptibility to infection	*Fv-1* (1)	Resistance	No (C4)
Herpes simplex (Herpesviridae)	Encephalitis	(2+)	Resistance	No
Ectromelia (MP) (Poxviridae)	Mousepox	Rmp-1 (1)	Resistance	No
	Immune			
LCMV (Arenaviridae)	Immune response	(?)	?	Yes (C17)
MuLV (Retroviridae)	Erythroleukemia	Rfv-1 (1)	Codominant	Yes (C17)
	Erythroleukemia	Rfv-2 (1)	Codominant	Yes (C17)
	Erythroleukemia	Rgv-1 (1)	Codominant	Yes (C17)
MPV (Polyomaviridae)	Tumor induction	(1)	Resistance	Yes (C17)

MHV, mouse hepatitis virus; MuLV, murine leukemia virus; MP, mousepox; LCMV, lymphocytic choriomeningitis virus; MPV, mouse polyoma virus.

Adapted from Brinton M. Host susceptibility to viral disease. In: Nathanson N, et al., ed. *Viral pathogenesis*. Philadelphia: Lippincott-Raven Publishers, 1997.

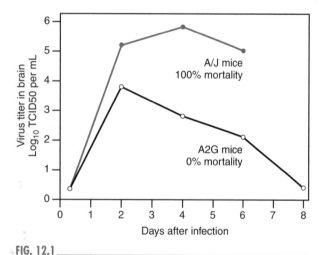

FIG. 12.1

Inbred strains of mice may exhibit dramatic differences in disease susceptibility to specific viruses. In this example, A/J mice—like most other strains—are susceptible, whereas A2G mice are resistant. Both strains of mice were injected intracerebrally with 70 LD$_{50}$ (as determined in susceptible mice) of a neurotropic strain (WSN) of a type A influenza virus, and observed for symptoms and assayed for replication in the brain. The susceptible A/J mice supported a higher level of replication and died, whereas resistant A2G mice survived. (After Fiske RA, Klein PA. Effect of immunosuppression on the genetic resistance of A2G mice to neurovirulent influenza virus. *Infect Immun* 1975;11:576–586.)

mice possess the Mx2 allele that encodes a related—often truncated—protein that does not induce influenza virus resistance. The Mx1 protein appears to interfere with processing of virus mRNA, perhaps by blocking intracellular transportation or transcription of the viral messages, but the details remain to be elucidated (Fig. 12.2). Also, Mx acts as a guanosine triphosphatase (GTPase), but it is not clear if this function is necessary for its anti-influenza virus activity.

Viral Replication, the Flv Gene, and Flaviviruses Flaviviruses, named after yellow fever virus, cause encephalitis when injected intracerebrally in mice. Most strains of mice are susceptible to flaviviruses, but a few strains, such as the PRI (Princeton Rockefeller Institute) strain, are resistant. When crosses were made between susceptible and resistant mice, the trait segregated as a single autosomal gene and resistance was dominant. The locus controlling flavivirus susceptibility was named *Flv* and has been mapped to mouse chromosome 5.

A congenic strain was developed by repeatedly back-crossing resistant F1 hybrids with susceptible C3H mice. The resistant congenic strain (C3H/RV)

TABLE 12.3

Determination of inheritance of susceptibility or resistance among inbred strains of mice

Mouse strain	Genotype	Treatment	Mortality dead/total	Log10 virus titer
A/J	r/r	None	4/4	6.0
A2G	R/R	None	0/4	3.7
F1	R/r	None	0/4	3.5
A/J	r/r	AIF	4/4	6.0
A2G	R/R	AIF	4/4	6.3
F1	R/r	AIF	4/4	6.5

The progeny of genetic crosses between inbred strains of mice can be used to determine the inheritance of susceptibility or resistance. In this example, influenza A virus was used to test A/J mice (susceptible) and A2G mice (resistant), and the results with F1 hybrids (A/J X A2G) showed that resistance was dominant. In addition, treatment of resistant animals with anti-interferon antiserum (AIF) abrogated resistance, indicating that interferon played a role in the resistance phenotype. Virus titers were determined on blood samples obtained 2 days after infection and are expressed as EID$_{50}$ (50% egg infectious doses).

After Haller O, Arnheiter H, Lindenmann J, Gresser I. Host gene influences sensitivity to interferon action selectively for influenza virus. *Nature* 1980;283:660–662.

was compared with the susceptible C3H/He parental strain following infection with West Nile virus (WNV, a virulent flavivirus) (Table 12.4, Fig. 12.3). After intraperitoneal injection of WNV, 100% of susceptible animals—but no resistant animals—died. After intracerebral injection of WNV, virus replicates much less rapidly in the brains of resistant animals but sufficiently to kill some of them after an extended incubation. Resistance is not absolute and can be overcome by intracerebral inoculation of a virulent flavivirus, immunosuppression, or infection of newborn animals. However, in contrast to the Mx gene, interferon is not involved in expression of the resistance phenotype.

Cell cultures derived from the two strains of animals also differ in the efficiency with which they replicate flaviviruses and have been used to probe the cellular mechanism. WNV could be readily maintained by serial passage in susceptible cells but disappeared after three passages in resistant cells. Flaviviruses are positive-stranded RNA viruses and are transcribed into minus copies, which are used as the templates for full-length genomes that are incorporated into nascent virions. Transcription of flavivirus RNA is less efficient in resistant cells, presumably because the *Flv* gene encodes a cellular protein that is involved in transcription. The nature of the Flv protein and its mechanism of action await further study.

Inhibition of Viral Replication: The Fv-1 Gene and Murine Leukemia Viruses In the course of studies of the murine leukemia viruses (MuLV), it was discovered

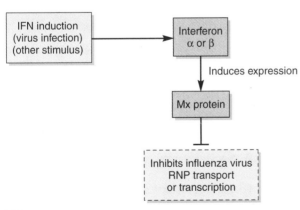

FIG. 12.2

If a genetic determinant of host susceptibility can be mapped to a single locus, it should encode a specific host protein. One of the best studied examples is the Mx protein, a cellular protein whose normal function is unknown. The *Mx* gene can be induced by interferon and establishes a state of resistance against group A influenza viruses. The Mx protein appears to act by blocking either intracellular transport of influenza virus mRNA or its transcription.

TABLE 12.4

Differences in susceptibility to flaviviruses among mice

Mouse population	Dead/total	Percent mortality
Inbred strains		
C3H/He	38/38	100
C3H/RV	0/34	0
Wild outbred mice		
Maryland	2/10	20
Soledad, California	0/5	0
La Puenta, California	0/5	0
Devonshire, California	0/5	0

Different strains of mice differ in their susceptibility to flaviviruses. In this example, a few inbred and wild mouse populations were tested for their susceptibility to an attenuated yellow fever virus after intracerebral injection of 0.03 mL into adult animals.

After Darnell MB, Koprowski H, Lagerspetz K. Genetically determined resistance to infection with group B arboviruses. I. Distribution of the resistance gene among various mouse populations and characteristics of gene expression in vivo. *J Infect Dis* 1974;129:240–247.

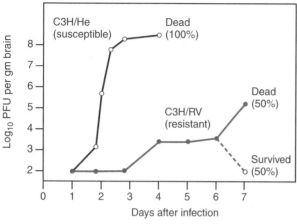

FIG. 12.3

Effect of a genetic locus (Flv) that influences susceptibility to fla-viviruses. Two strains of mice are compared: C3H/He, an in-bred strain that is susceptible, and C3H/RV, a congenic strain that is identical except for a 50-centimorgan region from the resistant strain that contains the resistance allele at the *Flv* locus. Adult mice were injected intracerebrally with $10^{5.5}$ plaque-forming units of West Nile virus. All resistant animals exhibited lower virus titers, but some eventually died while others survived. (After Brinton M. Host susceptibility to viral diseases. In: Nathanson N, et al., eds. *Viral pathogenesis.* Philadelphia: Lippincott–Raven Publishers, 1997.)

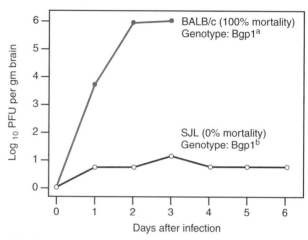

FIG. 12.4

Pathogenesis of a viral infection in susceptible and resistant in-bred strains of mice, where susceptibility is determined by cellular receptors. Susceptible BALB/c mice are compared with resistant SJL mice after intracerebral injection of $10^{2.85}$ plaque-forming units into adult animals. (After Ohtsuka N, Taguchi F. Mouse susceptibility to mouse hepatitis virus infection is linked to viral receptor genotype. *J Virol* 1997;71: 8860–8863.)

that mouse cells could be divided into B type (named after BALB/c strain of mice) and N type (named after NIH strain of mice). MuLV isolates were either N-tropic or B-tropic depending on which cells supported their replication. Cross-breeding of mice indicates that cellular susceptibility is determined by a single locus, named the Fv-1 locus, with two alleles, Fv-1$^{n/n}$ or Fv-1$^{n/b}$. Fv-1$^{n/b}$ cells are resistant to infection with both N- and B-tropic viruses. The Fv-1 locus encodes endogenous retroviral sequences that control a step in viral replication that occurs after entry and involves the CA (capsid) protein encoded by the viral *gag* gene. However, the nature of the interaction between the CA protein and the Fv-1 gene product remains to be determined.

Receptors: Mouse Hepatitis Virus Differences in the susceptibility of inbred strains of mice to mouse hep-atitis virus (MHV, a coronavirus) was one of the first described examples of the genetic determinants of viral infection. Several distinct genetic loci have been identified in mice (Table 12.2), and discussion is limited to the Hv-2 locus, mapped to chromosome 7, which influences the replication of MHV. Susceptible BALB/c mice can be infected and—depending on age, route, and dose—killed by a number of different isolates of MHV, whereas Swiss Jackson Laboratory (SJL) mice appear resistant to infection

(Fig. 12.4). Based on cross-breeding of resistant and susceptible strains, susceptibility is determined by a single autosomal dominant genetic locus.

A search for a putative receptor for MHV has identified a murine biliary glycoprotein (Bgp1), which is classified as a member of the immunoglob-ulin superfamily based on its structure as a trans-membrane glycoprotein with four distinct globular domains. The Bgp1 gene from different strains of mice has two allelic forms (Bgp1^a and Bgp1^b), and these alleles cosegregate with the Hv-2 locus. Evidence that the Hv-2 locus encodes a viral receptor is provided by studies of F1 and F2 crosses of the resistant and susceptible strains, which show that there is a correspondence between Bgp 1 alleles and responses to MHV infection (Table 12.5).

Transforming Retroviruses and Their Receptors Different avian retroviruses (avian leukosis-sarcoma viruses, ALSVs) utilize different cellular membrane proteins as their receptors (Fig. 4.5 and Table 10.6), and this forms the basis for genetic differences between inbred strains of chickens in their susceptibility to a panel of ALSV (discussed in Chapter 10).

Immunologic Determinants

Lymphocytic Choriomeningitis Virus Lymphocytic choriomeningitis virus (LCMV) is a virus whose persistence and disease induction is intimately en-twined with its ability to induce an immune response

H-2 (animals) = MHC (humans)

TABLE 12.5

Virus susceptibility as determined by host-encoded cellular receptors

Mouse strain	Bgp1 alleles	Mortality dead/total
BALB/c	a/a	33/33
SJL	b/b	0/27
F1 (BALB X SJL)	a/b	13/13
F2 (F1 X F1)	a/a	8/8
	a/b	23/23
	b/b	0/14

bad guy is a allele

In the above example, two inbred strains of mice (BALB/C and SJL) encode different variants of the MHVR, designated Bgp1[a] and Bgp1[b]. These experiments show that mice expressing Bgp1[a] are susceptible and illustrate segregation of the MHV susceptibility as a single genetic locus, where susceptibility is dominant. The genotypes of individual mice were determined by single-strand conformation polymorphism, which exploits the different migration on polyacrylamide gels of PCR fragments of DNA of the Bgp1[a] and Bgp1[b] genotypes. Adult mice were infected with $10^{7.85}$ of the JHM sp-4 strain of MHV, and mortality was determined 2 weeks later.

Bgp, biliary glycoprotein; MHVR, mouse hepatitis virus receptor; PCR, polymerase chain reaction.

After Ohtsuka N, Taguchi F. Mouse susceptibility to mouse hepatitis virus infection is linked to viral receptor genotype. *J Virol* 1997;71:8860–8863.

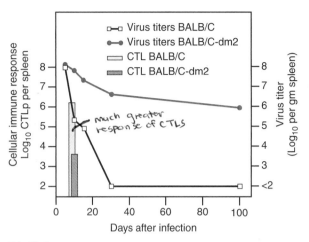

much greater response of CTLS

FIG. 12.5

Some genetic determinants of host response to infection are associated with immune responses and map to the major histocompatibility complex (MHC). In this example, two inbred strains of mice differing only in one region of the H-2 region (mouse MHC) are compared after infection with 10^2 plaque-forming units of the docile strain of lymphocytic choriomeningitis virus (LCMV). BALB/c and BALB/c-dm2 mice are H-2[d] (d haplotype), but the dm2 mice are mutants lacking the L region of the H-2 locus. The dm2 mice are poor immunologic responders to LCMV and clear their virus more slowly. (After Moskophidis D, Lechner F, Hengartner H, Zinkernagel RM. MHC class I and non-MHC-linked capacity for generating an anti-viral CTL response determines susceptibility to CTL exhaustion and establishment of virus persistence in mice. *J Immunol* 1994;152:4976–4983.)

mediated by CD8[+] cytotoxic T lymphocytes (CTL). Not surprisingly, immmunogenetic differences between mouse strains have been associated with variations in host response to LCMV infection. A comparison of two mouse strains that vary in their cellular immune responses to LCMV show that animals mounting the weakest immune responses clear their infections more slowly (Fig. 12.5).

Murine Leukemia Virus Several MHC-associated genes influence the outcome of infection with MuLVs (see Chapter 10 for background information). These include the *Rfv-1* and *Rfv-2* genes that affect recovery

from Friend (F) MuLV-induced erythroleukemia, and the *Rgv-1* gene that affects recovery from Gross (G) MuLV-induced leukemia. Although the murine leukemia virus model is a somewhat contrived one, based on laboratory derived retroviruses and inbred strains of mice, it illustrates genetic determinants that act through the immune response. Figure 12.6 shows an example of host differences in the development of

TABLE 12.6

Differences in host susceptibility for virus

human

HIV-1 antibody	Parameter	Total	CCR5 +/+	CCR5 +/−	CCR5 −/−
Positive	Number	1,343	1,148	195	0
	Percent	100	85	15	0
Negative	Number	612	508	87	17
	Percent	100	83	14	3

CCR5 – don't have coreceptor for HIV

no individual

In the above example, the chemokine receptor CCR5 acts as a coreceptor for most wild-type HIV isolates. A naturally occurring mutation (Δ32) in the CCR5 gene abrogates the expression of CCR5 and homozygous carriers of the mutation are resistant to HIV-1. Subjects at high risk of HIV exposure were divided into those who were HIV infected and those who were not, and tested for their CCR5 status (CCR5 +/+, +/−, or −/−). There was a striking absence of CCR5[−/−] persons in the HIV-infected group and an excess of CCR5[−/−] persons (above the 1% expectancy) in the HIV-uninfected group.

After Dean M, Carrington M, Winkler C, et al. Genetic restriction of HIV-1 infection and progression to AIDS by a deletion allele of the CKR5 structural gene. *Science* 1996;273:1856–1862.

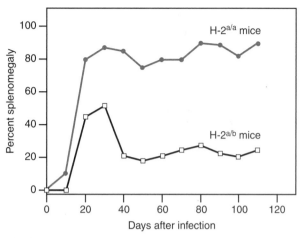

FIG. 12.6

An example of host genetic determinants that map to the major histocompatibility complex. Two congenic strains of mice, which differ only at the H-2 locus (H-2$^{a/a}$or H-2 $^{b/b}$), were injected with 15 focus forming units of Friend virus strain of murine leukemia virus and followed for the development of splenomegaly or death, which are signs of erythroleukemia caused by the virus. The H-2$^{b/b}$ mice were resistant, controlled their leukemia, and survived, whereas the H-2 $^{a/a}$ mice did not control their leukemia and died. Strains of mice with recombinations between the a and b haplotypes were used to show that there were two genetic loci involved, the Rfv-1 locus in the D-L region of H-2 and the class II locus in the K I-A region of H-2. It appears that the Rfv-1 acts via the CD8 response to viral antigens and the K I-A locus influences the activity of virus-specific CD4 helper cells. (After Miyazawa M, Nishio J, Wehrly K, Chesebro B. Influence of MHC genes on spontaneous recovery from Friend retrovirus-induced leukemia. *J Immunol* 1992;148: 644–647.)

leukemia induced by F-MuLV that map to discrete regions within the H-2 (MHC).

GENETIC DETERMINANTS IN HUMANS: HIV/AIDS

The intense investigation of HIV has provided evidence of host determinants in a human viral disease.

Resistance to HIV Infection

Epidemiologic studies indicate that there is significant variation in human susceptibility to HIV infection. Thus, some persons fail to develop the markers of infection (serum antibody and evidence of HIV RNA or DNA in blood) in spite of repeated sexual exposure to infected contacts. There appear to be at least two different mechanisms for resistance to infection: absence of the viral coreceptor and local immunity.

As described in Chapter 3, the entry of HIV into susceptible human cells involves a primary receptor (CD4) and a coreceptor (CCR5 or CXCR4, both chemokine receptors). Some humans have a natural mutation in the CCR5 gene that abrogates expression of the CCR5 protein (the Δ32 mutation), and individuals who are homozygous for Δ32 are resistant to infection with the majority of wild-type strains of HIV-1, since most of these utilize CCR5 but not CXCR4 as a coreceptor. Comparison of primary blood mononuclear cells (PBMC) from CCR5$^{+/+}$ and CCR5$^{-/-}$(Δ32) homozygous subjects shows that T lymphocytes from the latter are resistant to infection with most primary HIV isolates. In the Caucasian population of the United States, the relative incidence of CCR5 is approximately as follows: $-/-$ individuals, 1 %; $+/+$ individuals, 81%; and $+/-$ individuals, 18%. When an HIV at-risk cohort was divided into infected and uninfected groups, the HIV-infected group contained no individuals who were homozygous for the Δ32 mutation, whereas the HIV-negative group had an excess (above expectancy) of individuals with the Δ32 mutation (Table 12.6).

Among subjects at very high risk of HIV infection because of their lifestyle (e.g., commercial sex workers), a very small number (less than 5%) remain uninfected. Such subjects are called "exposed uninfected" (EU) or "exposed seronegative" (ESN). There is some evidence that these persons have developed local immunity in the genital tract because they may demonstrate HIV-specific CTL activity in their peripheral blood lymphocytes or IgA antibody in their genital secretions. It remains to be determined whether a host factor or the circumstance of exposure explains ESN status.

Resistance to AIDS

Among HIV-infected persons, there are marked differences in the rate of progression to AIDS, which can vary from about 1 year to more than 20 years. Many factors influence incubation period, including the pathogenicity of the virus, the dose and route of infection, and the exposure to opportunistic infections. The potential role of host determinants in this variability has yet to be completely analyzed, but one factor is CCR5. Thus, patients who are heterozygous for the Δ32 mutation (CCR5$^{+/-}$) progress to AIDS more slowly than do those homozygous for the coreceptor (CCR5$^{+/+}$). Age is another factor because, in the absence of antiretroviral therapy, median survival time (HIV infection to death) is less than 2 years in infants and more than 10 years in adults.

It is likely that there are other significant host vari-

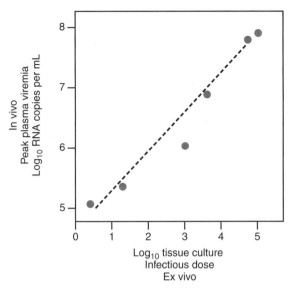

FIG. 12.7

Host variability in SIV infection of rhesus monkeys, which is an excellent model for HIV infection of humans. Using a standardized virus inoculum (cloned pathogenic SIVsmE543-3), monkeys were compared both in vivo and ex vivo for the course of infection. This figure shows data from six animals that were selected for their variable susceptibility, indicating that there was a consistent relationship between the peak titer of virus in PBMCs (ex vivo) and peak viremia (in vivo). (After Goldstein S, Brown CR, Dehghani H, et al. Intrinsic susceptibility of rhesus macaque peripheral CD4+ T cells to simian immunodeficiency virus in vitro is predictive of in vivo viral replication. *J Virol* 2000;74:9388–9395.)

TABLE 12.7		
Immunogenetic determinants influencing the course of HIV infection		
	FREQUENCY OF B57 HLA HAPLOTYPE	
Status of HIV-infected patients	B57/total	Percent B57
Nonprogressors	11/13	85
Progressors	19/200	10

Immunogenetic determinants may influence the course of HIV infection. In the above example, two groups of HIV-infected persons—long-term nonprogressors and progressors—were compared as to their HLA haplotypes. A single HLA haplotype, B57, was overrepresented in the nonprogressor group compared with the progressors, where the B57 haplotype occurred at a frequency similar to that seen in the Caucasian population of the United States. This implies that persons with B57 produce a more protective immune response to HIV than do individuals with other haplotypes, but the exact mechanism remains to be established.

HIV, human immunodeficiency virus; HLA, human leukocyte antigen.

After Migueles SA, Sabbaghian MS, Shupert WL, et al. HLA B*5701 is highly associated with restriction of virus replication in a subgroup of HIV-infected long term nonprogressors. *Proc Natl Acad Sci USA* 2000;97:2709–2714.

ables in HIV infection, a hypothesis suggested by observations on simian immunodeficiency virus (SIV). SIV produces AIDS in monkeys and is an excellent model for HIV-induced AIDS in humans. There are dramatic differences between individual monkeys in their viremia levels, and these correlate with incubation period to clinical AIDS. Furthermore, in vivo susceptibility is mirrored in the levels of SIV replication in PBMC (Fig. 12.7). Since these differences occur in monkeys of similar age that are homozygous for CCR5, additional host factors must be operative.

Immunologic Determinants in AIDS

In addition to the host factors described above, there is evidence that there are human leukocyte antigen (HLA)–associated determinants that influence the progression of HIV infections. Among HIV-infected patients, those that remain normal for the longest period are often called long-term nonprogressors. One comparison of nonprogressors with progressors shows an overrepresentation of a single HLA haplotype (B57), suggesting that their immunogenetic determinants can influence susceptibility to AIDS (Table 12.7).

PHYSIOLOGIC FACTORS

Age

Infancy The effects of infancy have been studied in detail in experimental murine models, as the maturation process, from birth to adulthood, is completed in about 2 months and it is easy to obtain rodents of precisely documented age. Several generalizations can be made (Sidebar 12.1).

Most viruses are more lethal in newborn than in adult animals, which usually reflects differences in the extent of replication (Fig. 12.8). Such distinctions are apparent within the first 24–48 hours of infection, implying that they are not due to acquired immunity. However, they may reflect innate host defenses, such as the induction of interferon. Age-specific reduction in susceptibility is often associated with differentiation of cells. For instance, if multipotent chromaffin cells are treated with nerve growth factor, which causes them to differentiate, there is a marked reduction in the production of LCMV, and this is reversed when the factor is removed. Likewise, if young rats are transplanted with embryonic brain tissue and challenged with Japanese encephalitis virus, replication is restricted to the transplant, reflecting the higher susceptibility of embryonic versus infantile neurons.

In some instances, age influences intracellular responses to viral infection. For instance, bcl-2, an antiapoptosis protein, protects mouse neurons against the lethal consequences of infection with Sindbis, an alphavirus; as mice mature, bcl-2 expression increases, and is thought to explain age-specific development of resistance against Sindbis mortality. Selected Sindbis virus mutants can overcome age-specific resistance and this property has been mapped to a single amino acid in one of the envelope proteins.

Another example is the parvoviruses, small DNA viruses that can only replicate in dividing cells. Proliferation of specific cell groups is an essential part of normal development. External granule cells of the rodent cerebellum divide through the first postnatal week, after which they migrate to form the internal granule cell layer. If rats are infected with rat virus (a murine parvovirus) shortly after birth, the virus destroys the granule cells leading to severe atrophy of the cerebellum, whereas infection of weanling animals (age 3 weeks) is inapparent.

In a few instances, maturation increases susceptibility to specific viruses. For instance, intracerebral injection of a rodent-adapted poliovirus strain paralyzed older mice more rapidly and at a greater frequency than newborn mice. The difference was due to the failure of virus to travel along neuronal pathways from brain to spinal cord in newborn mice, as both newborn and 4-week-old mice were equally susceptible to intraspinal poliovirus injection. It was hypothesized that maturation of the fast axonal transport system—which could move poliovirus nucleocapsids from brain to cord—might account for this paradoxical finding.

Newborn mice are immunologically immature and, depending on the specific virus, evolve to reach

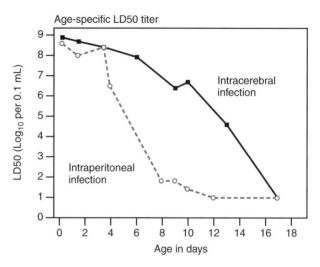

Age-specific LD50 titer

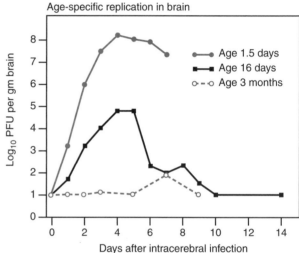

Age-specific replication in brain

FIG. 12.8 _____

Newborn animals are usually much more susceptible to viral infection and disease than are their adult counterparts. In this example, West Nile virus (WNV), a flavivirus, was used to infect Wistar rats. (*Top*) Age-specific titration by intracerebral and intraperitoneal routes. Although highly susceptible at birth, rats developed resistance to disease by age 2 weeks. (*Bottom*) Susceptibility reflects the extent of virus replication, as shown by the replication of WNV in the brain following intracerebral injection of 1- to 2-day-old rats (3 PFU inoculum) or older ($10^{5.1}$ PFU). (After El Dadah AH, Nathanson N, Sarsitis R. Pathogenesis of West Nile virus encephalitis in mice and rats. *Am J Epidemiol* 1967;86:765–790.)

an "adult" immune response at 1–3 months of age. For viruses that produce an immune-mediated illness, newborn animals may be "tolerized" by the same virus inoculum that induces disease in adult animals (see Chapter 7). By the same token, persistent viruses are more likely to initiate a lifelong infection in newborn than in adult mice, which usually correlates with tolerance following infection of newborn animals.

TABLE 12.8

Increased susceptibility of human infants to viral disease as exemplified by measles

	ANNUAL MORTALITY (ALL CAUSES) PER 100		
Age group	1835–1845	1846	Excess in 1846
<1	10.8	30.0	19.2
1–9	0.5	0.5	—
10–19	0.5	0.5	—
20–29	0.5	0.7	0.2
30–39	0.8	2.1	1.3
40–49	1.1	2.7	1.6
50–59	0.9	4.4	3.5
60–69	2.0	7.7	5.7
70–79	6.5	13.1	6.6
80–100	16.8	26.0	9.2

Table shows age-specific differences in mortality from the measles epidemic of 1846 in the Faroe Islands, compared with average mortality for 1835–1845. The excess mortality for 1846 provides a crude estimate of measles-specific mortality during the epidemic, which involved at least 75% of the population. The data demonstrate a dramatic increase in infant mortality and also suggest that there was an increase in mortality in the older age groups.

Data from Panum PL. *Observations made during the epidemic of measles on the Faroe Islands in the year 1846.* New York: American Public Health Association, 1940.

Human infants are also more susceptible to diseases induced by many viruses, as shown in Table 12.8. Since infants acquire passive antibody from their mothers, both across the placenta and via breast feeding, they are often protected during the first 3–24 months of life. If a human population has not been exposed to a given virus, both mothers and their infants lack immunity, and the influence of age on susceptibility is revealed. Such a situation occurred in the Faroe Islands, isolated in the northern Atlantic Ocean, in 1846, during an epidemic of measles. Since the last recorded outbreak of measles in the Faroe Islands had occurred about 75 years previously, most of the population was nonimmune. Under these circumstances, there was a particularly high mortality in infants, illustrating their biological vulnerability due to the absence of immune protection. This example also provides a teleological explanation for the evolutionary development of passive transfer of immunity from mother to infant.

For a number of other virus-related diseases, it has been shown or inferred that infants are more susceptible to the disease induced by the virus, such as Western equine encephalitis, HIV, herpes simplex virus, and smallpox viruses. On the other hand, there are a few viruses in which infection of infants induces immune tolerance and a persistent infection, in contrast to primary infection of adults where the immune response both clears the virus and causes disease. The most prominent example of this paradigm is hepatitis B virus, as described in Chapter 7. Figure 7.10 illustrates the tolerizing effect of rubella virus infection in utero compared with immunizing infection in childhood.

Old Age A few viruses appear to cause more severe disease in older persons, and Figure 12.9 illustrates this phenomenon for St. Louis encephalitis virus. It has been speculated that in older persons

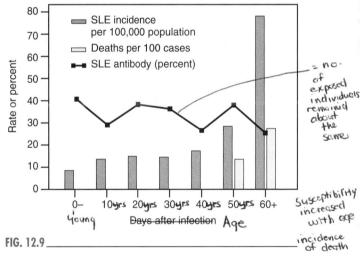

FIG. 12.9

Older age increases disease susceptibility for some viruses, as shown for St. Louis encephalitis (SLE), a flavivirus. There was an outbreak of SLE in a nonimmune population in Houston, Texas, in 1964. Both the rate of clinical encephalitis and the case/fatality rate were distinctly higher in persons older than 60 years. A serosurvey conducted 6 months after the epidemic found that the incidence of infection was similar for all age groups, implying that the age-specific incidence profile reflected an increased host susceptibility in older persons. (After Luby JP, Miller G, Gardner P, et al. The epidemiology of St. Louis encephalitis in Houston, Texas. *Am J Epidemiol* 1967;86: 584–597; and Henderson BE, Pigford CA, Work T, Wende RD. Serological survey for St. Louis encephalitis, and other group B arbovirus antibodies in residents of Houston, Texas. *Am J Epidemiol* 1970;91:87–98.)

localized gaps in the blood–brain barrier increase the risk that virus can pass from blood to the central nervous system, the target system for encephalitic viruses. Also, immune responses wane in old age and may contribute to increased severity of certain viral diseases. For instance, herpes zoster infection (a reactivation of varicella-zoster infection acquired in childhood) is much more common above age 70, and measles appears to be more severe in older persons (Table 12.8).

Other Determinants of Susceptibility

There are limited quantitative data on other host determinants of infection. Severe undernutrition can alter (often enhance) the severity of viral disease, both in animals and in humans, and this may be due to reduction in immune responses. There are gender-specific differences in the outcome of certain viral infections as suggested by the data in Table 12.1. Also, pregnancy has an adverse effect on the prognosis of some viral infections such as hepatitis B, hepatitis E, and poliomyelitis.

REPRISE

Viruses depend on host-encoded proteins for each step of their replication cycle. Furthermore, the course of infection is conditioned by innate host defenses (such as interferon) as well as by acquired immune responses, both of which involve a large number of genetic loci. Since many genes have multiple alleles, it might be predicted that individual host alleles would have a differential affect on the course of a viral infection. It is difficult to identify host genetic determinants in an outbred free-living population, but the availability of a large number of inbred genetically uniform strains of mice has permitted the identification of about 25 host genes that influence susceptibility to viral infection. Most of these determinants are specific for a single virus or virus family, and many involve a step in the viral replication cycle. A smaller number of determinants map within the MHC complex and involve the immune response. Genetic determinants similar to those identified in mice probably operate in many species but relatively few have been documented.

Some of the relevant genes have been sequenced and the normal function of their products determined. However, the intimate mechanisms whereby different allelic variants affect viral infection are not completely understood, and there are a great many relevant genes that have yet to be described. New developments in genomics and molecular cell biology will lead to major advances in understanding host susceptibility to viral infection.

In addition to genetic determinants, many other host variables, such as age, nutritional status, and other physiologic variables, influence the outcome of a viral infection. Analysis of relevant mechanisms is very incomplete and represents a fertile area for future investigation.

FURTHER READING

Reviews and Chapters

Brinton M. Host susceptibility to viral diseases. In: Nathanson N, et al., eds. *Viral pathogenesis*. Philadelphia: Lippincott–Raven Publishers, 1997.

Chesebro B, Miyazawa M, Britt WJ. Host genetic control of spontaneous and induced immunity to Friend murine retrovirus infection. *Annu Rev Immunol* 1990;8:477–499.

Christenson ND, Han R, Kreider JW. Cottontail rabbit papillomavirus. In: Ahmed R, Chen ISY, eds. *Persistent virus infections*. New York: John Wiley & Sons, 1999;485–502.

Rosenberg N, Jolicoeur P. Retroviral pathogenesis. In: Coffin JM, Hughes SH, Varmus H, eds. *Retroviruses*. Cold Spring Harbor, NY: Cold Spring Harbor Laboratory Press, 1996.

Skamene E, ed. *Genetic control of host resistance to infection and malignancy*. New York: Alan R. Liss, 1985.

Classic Papers

Bang F, Warwick A. Mouse macrophages as host cells for the mouse hepatitis virus and the genetic basis of their susceptibility. *Proc Natl Acad Sci USA* 1960;46:1065–1075.

Dean M, Carrington M, Winkler C, et al. Genetic restriction of HIV-1 infection and progression to AIDS by a deletion allele of the CKR5 structural gene. *Science* 1996;273:1856–1862.

Johnson RT, McFarland HF, Levy SE. Age-dependent resistance to viral encephalitis: studies of infections due to Sindbis virus in mice. *J Infect Dis* 1972;125:257–262.

Lindenmann J, Lance CA, Hobson D. The resistance of A2G mice to myxoviruses. *J Immunol* 1963;90:942–951.

Panum PL. *Observations made during the epidemic of measles on the Faroe Islands in the year 1846*. Translation reprinted by the American Public Health Association, New York, 1940.

Pincus T, Rowe WP, Lilly F. A major genetic locus affecting resistance to infection with murine leukemia viruses. II. Apparent identity to a major locus described for resistance to Friend murine leukemia virus. *J Exp Med* 1971;133:1234–1241.

Original Contributions

Best S, Le Tissier P, Towers G, Stoye JP. Positional cloning of the mouse retrovirus restriction gene Fv1. *Nature* 1996; 382:826–830.

Chandra RK. Nutrition, immunity and infection: from basic knowledge of dietary manipulation of immune responses to practical application of ameliorating suffering and improving survival. *Proc Natl Acad Sci USA* 1996;93:14304–14307.

Cole GA, Nathanson N, and Rivet H. Viral hemorrhagic encephalopathy of rats. II. Pathogenesis of hemorrhagic lesions. *Am J Epidemiol* 1970;91:339–350.

El Dadah AH, Nathanson N, Sarsitis R. Pathogenesis of West Nile virus encephalitis in mice and rats. *Am J Epidemiol* 1967;86:765–790.

Fiske RA, Klein PA. Effect of immunosuppression on the genetic resistance of A2G mice to neurovirulent influenza virus. *Infect Immun* 1975;11:576–586.

Freund R, Dubensky T, Bronson R, et al. Polyoma tumorigenesis in mice: evidence for dominant resistance and dominant susceptibility genes of the host. *Virology* 1992;191:724–731.

Goldstein S, Brown CR, Dehghani H, et al. Intrinsic susceptibility of rhesus macaque peripheral CD4$^+$ T cells to simian immunodeficiency virus in vitro is predictive of in vivo viral replication. *J Virol* 2000;74:9388–9395.

Groschel D, Koprowski H. Development of a virus-resistant inbred mouse strain for the study of innate resistance to arbo B viruses. *Arch Virusforschung* 1965;17:379–391.

Haller O, Arnheiter H, Lindenmann J, Gresser I. Host gene influences sensitivity to interferon action selectively for influenza virus. *Nature* 1980;283:660–662.

Henderson BE, Pigford CA, Work T, Wende RD. Serological survey for St. Louis encephalitis, and other group B arbovirus antibodies in residents of Houston, Texas. *Am J Epidemiol* 1970;91:87–98.

Jubelt B, Narayan O, Johnson RT. Pathogenesis of human poliovirus infection in mice. II. Age-dependency of paralysis. *J Neuropathol Exp Neurol* 1980;39:149–159.

Kaslow RA, Carrington M, Apple R, et al. Influence of combinations of human major histocompatibility complex genes on the course of HIV-1 infection. *Nature Med* 1996;2:405–411.

Levine B, Huang Q Isaacs JT, Reed JC, et al. Conversion of lytic to persistent alphavirus infection by the bcl-2 cellular oncogene. *Nature* 1993;361:739–742.

Liu R. Paxton WA, Choe S, et al. Homozygous defect in HIV-1 coreceptor accounts for resistance of some multiply-exposed individuals to HIV-1 infection. *Cell* 1996;86:367–377.

Lodmell D. Genetic control of resistance of street rabies virus in mice. *J Exp Med* 1983;157:451–460.

Luby JP, Miller G, Gardner P, et al. The epidemiology of St. Louis encephalitis in Houston, Texas. *Am J Epidemiol* 1967;86:584–597.

Migueles SA, Sabbaghian MS, Shupert WL, et al. HLA B*5701 is highly associated with restriction of virus replication in a subgroup of HIV-infected long term nonprogressors. *Proc Natl Acad Sci USA* 2000;97:2709–2714.

Miyazawa M, Nishio J, Wehrly K, Chesebro B. Influence of MHC genes on spontaneous recovery from Friend retrovirus–induced leukemia. *J Immunol* 1992;148:644–647.

Morely D. The severe measles of West Africa. *Proc R Soc Med* 1969;57:846–849.

Moskophidis D, Lechner F, Hengartner H, Zinkernagel RM. MHC class I and non-MHC-linked capacity for generating an anti-viral CTL response determines susceptibility to CTL exhaustion and establishment of virus persistence in mice. *J Immunol* 1994;152:4976–4983.

Ogata A, Nagashima K, Hall WW, et al. Japanese encephalitis virus neurotropism is dependent on the degree of neuronal maturity. *J Virol* 1991;65:880–886.

Ohtsuka N, Taguchi F. Mouse susceptibility to mouse hepatitis virus infection is linked to viral receptor genotype. *J Virol* 1997;71:8860–8863.

Tucker PC, Strauss EG, Kuhn RJ, et al. Viral determinants of age-dependent virulence of Sindbis virus for mice. *J Virol* 1993;67:4605–4610.

Woodruff J. The influence of quantitated post-weaning undernutrition on coxsackievirus B3 infection of adult mice. *J Infect Dis* 1970;121:164–181.

Zelus BD, Wessner DR, Williams RK, et al. Purified soluble recombinant mouse hepatitis virus receptor, Bgp1^b, and Bgp2 murine coronavirus receptors differ in mouse hepatitis virus binding and neutralizing activities. *J Virol* 1998;72:7237–7244.

Chapter 13
HIV and AIDS

At the turn of the millennium, human immunodeficiency virus (HIV) is probably the single most significant human viral infection. Furthermore, as a complex and enigmatic example of viral pathogenesis, it has been the subject of more intensive study than any other viral disease. This has led to a dynamic analysis of virus–host interactions, a pioneering achievement that supplements established approaches to pathogenesis. For these several reasons, a full chapter has been devoted to HIV and acquired immunodeficiency syndrome (AIDS).

This chapter first describes virus replication, cellular responses, and the sequence of viral, immunologic, and disease events that follow HIV infection. It then turns to the dynamic aspects of infection, including the turnover of virus and viral latency, the turnover of cellular populations, and the evolution of viral phenotypes. Finally, data are synthesized to explain the pathophysiology of immunodeficiency, opportunistic infections, and HIV-associated neoplasms. Prophylactic vaccines for HIV are discussed in Chapter 14.

LENTIVIRUSES

HIV is a lentivirus. In common with all retroviruses, lentiviruses have three major genetic loci (*gag, pol,*

env) that encode the core proteins, the reverse transcriptase and integrase, and the envelope proteins, respectively (Fig. 10.1). The lentiviruses are distinguished from other retroviruses by several characteristics: (a) They possess six unique accessory genes that encode nonstructural proteins. Two of these genes (*tat* and *rev*) are required for replication in cell cultures, whereas four (*vpr, vpu, vif, nef*) are not necessary for replication in cell culture systems. However, at least some (such as *nef*) are required for full virulence in vivo (Fig. 13.1). (b) The viral attachment protein (the surface envelope protein, known as SU or gp120) of the primate lentiviruses binds to the CD4 molecule that is found on the CD4 subset of T lymphocytes and on monocytoid cells (macrophages, microglia, and dendritic cells), and this determines their cellular host range. (c) They are capable of replicating in nondividing (as well as dividing) cells, in contrast to other retroviruses that replicate only in dividing cells, due to the ability of the preintegration complex of reverse transcribed viral DNA and proteins to cross the nuclear envelope. (d) They cause lifelong infections that are associated with a number of chronic diseases, including AIDS, but they do not encode oncogenes. (e) They are strictly exogenous viruses and host genomes do not include copies of their sequences.

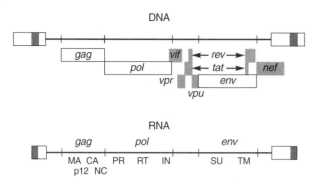

DNA

RNA

FIG. 13.1

Organization of the HIV-1 genome. The RNA genome is about 9 kb long and is bounded at both ends by a noncoding repeat (R) region that encloses the three major coding genes, *gag* (group antigen), *pol* (polymerase), and *env* (envelope), which are transcribed in three different reading frames. The diagram indicates the position of the major proteins encoded by each of these genes, including the MA (matrix), p12, CA (capsid), NC (nucleocapsid) proteins of *gag*, the PR (protease), RT (reverse transcriptase), IN (integrase) enzymes of *pol*, and the SU (surface) and TM (transmembrane) proteins of *env*. HIV-1 also encodes six nonstructural accessory proteins (tat, rev, nef, vpr, vpu, vif) whose open reading frames are shown.

The lentivirus family includes members that infect horses, cows, sheep, goats, cats, and nonhuman primates. The simian immunodeficiency viruses (SIV) are particularly relevant because some of them cause AIDS in nonhuman primates, and thus they constitute useful animal models that have contributed important insights that could not be gleaned from human studies alone. There are two major groups of human lentiviruses, HIV-1 and HIV-2, both of which represent viruses that crossed the species barrier from nonhuman primates. HIV-1 was most likely derived from chimpanzee viruses (SIVcpz) and HIV-2 from sooty mangabey viruses (SIVsm), both primate species that are found only in Africa. HIV-1 can be classified in three subgroups: M (main), N (new), and O (outlier), and the M subgroup (which is most common) can be subdivided into ten genotypes (A–J).

Many SIV strains cause lifelong persistent nonpathogenic infections in the species in which they are enzootic. However if an SIV strain is transmitted to a nonenzootic simian species it may cause AIDS. Thus, there are several laboratory-passaged strains of SIV, probably derived originally from SIVsm, which are pathogenic in rhesus monkeys (Asiatic monkeys never exposed to SIVsm from Africa). Likewise, HIV-1 is about 100% lethal in humans but is relatively nonpathogenic when used for experimental infection of chimpanzees, the species from which it originated. These phenomena probably carry important implications regarding the mecha-

nisms by which HIV causes AIDS (discussed below).

VIRUS–CELL INTERACTIONS

Receptors and Viral Tropism

The entry of HIV into permissive host cells is a multistep process (Fig. 13.2) that involves a primary receptor and a coreceptor. In all cases, the primary receptor is CD4, an immunoglobulin superfamily molecule expressed on two major cell types, the CD4 subset of T lymphocytes and cells of the monocyte lineage. The coreceptor is one of several members of the chemokine family of molecules, particularly CCR5 expressed on T lymphocytes and macrophages, and CXCR4 expressed on T lymphocytes and T-cell lines (TCL). (Complicating this picture, it has recently been reported that CXCR4 is expressed on macrophages but at low levels as an oligomer that cannot functionally associate with CD4.)

Initially, the viral attachment protein (SU protein) binds to CD4, which triggers a conformational change in the SU protein that leads to binding to the coreceptor. Binding to the coreceptor triggers a (second) conformational change in the transmembrane (TM) envelope proteins that is often described as releasing a spring, resulting in the approximation of the N-terminal domain of the TM protein to the plasma membrane of the cell. A hydrophobic fusion peptide at the N terminal domain of the TM protein inserts into the plasma membrane, leading to fusion between the viral envelope and the cellular membrane. (Recently it has been suggested that an acid-mediated step may play a role in the final fusion process.)

HIV-1 virus strains vary in their ability to bind to coreceptors and this affects their cellular host range (Fig. 13.3). Viruses may be roughly classified into three groups, (a) wild-type isolates from patients that utilize mainly CCR5 (often called "macrophage tropic"), (b) wild-type isolates that utilize both CCR5 and CXCR4 (often called "dual tropic" or "pantropic"), and (c) strains that have been passed in the laboratory in TCL (often called "TCL tropic") that preferentially utilize CXCR4. Importantly, all HIV-1 strains can replicate in primary T-lymphocytes cultures; the distinction is that macrophage-tropic isolates replicate poorly in TCL and TCL-adapted viruses replicate poorly in primary cultures of macrophages. The capacity of HIV strains to utilize one or both coreceptors influences their in vivo tropism and pathogenicity (discussed below).

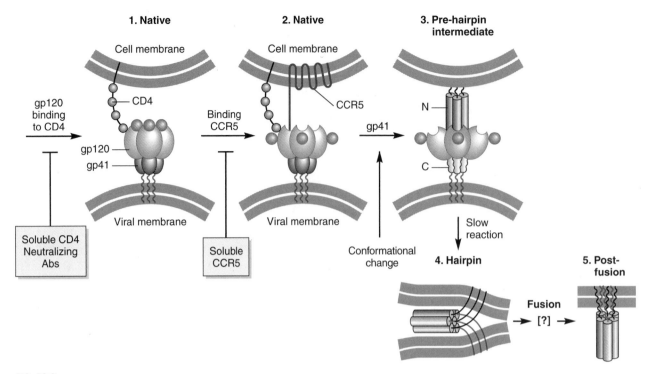

FIG. 13.2

Entry of HIV showing a hypothetical reconstruction of stepwise conformational changes in the SU (gp120) and TM (gp41) proteins. (a) The gp120 protein binds to the CD4 receptor. (b) Binding to CD4 triggers a conformational change that leads to binding of gp120 to the coreceptor (CCR5). (c) Binding to the coreceptor leads to a second major conformational change, this one in the TM (gp41) protein, which unfolds to expose and insert the fusion sequence at its N terminus into the plasma membrane of the cell, producing the prehairpin intermediate. (d) In a third conformational change, helices in the N and C domains of gp41 associate, producing the hairpin configuration, which brings the viral envelope and the plasma membrane into close approximation. (e) The two membranes fuse, leaving gp41 on the external surface. (After Chan DC , Kim PS. HIV entry and its inhibition. *Cell* 1998;93:681–684, and Doms R, personal communication, 2001.)

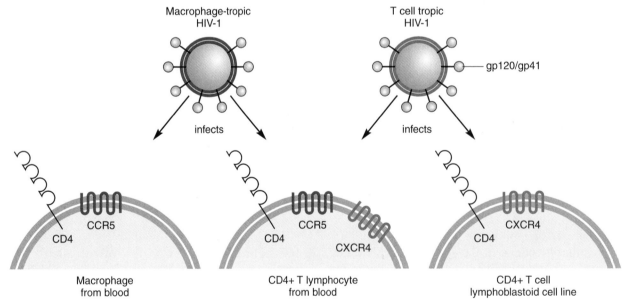

FIG. 13.3

Cells that are permissive for the entry and replication of HIV-1 carry two receptors on their surface. The primary receptor is the CD4 molecule, a protein expressed on the surface of certain subsets of T lymphocytes (so-called CD4$^+$ cells) and on macrophages. In addition, there is a coreceptor, either CCR5 or CXCR4, which is a member of a large family of molecules that serve as chemokine receptors on the surface of lymphoid cells. Some HIV-1 isolates are macrophage tropic because they use the chemokine CCR5 (R5 viruses); macrophage-tropic viruses do not infect T-cell lines (TCL). Other isolates are TCL tropic because they infect cells that express the chemokine CXCR4 (X4 viruses); TCL-tropic viruses do not infect macrophages. Both kinds of isolates can replicate in peripheral blood mononuclear cells, lymphocytes freshly cultured from peripheral blood, because these cells express both coreceptors. Some HIV-1 isolates (not shown in the figure) are "dual-tropic" (R5X4 viruses) because they can utilize both coreceptors and can infect all of the three cell types shown. (Recent studies indicate that CXCR4 is actually expressed on macrophages but is not functional.)

HIV Replication and Host Cell Response

HIV replicates slowly in permissive cells, relative to many "fast" viruses (Fig. 13.4). If the infected lymphocyte is actively dividing, then viral replication proceeds at a maximal rate. If the infected T cell is resting, the provirus can enter the nucleus and integrate, but remains latent until the T cell begins to divide, when HIV replication proceeds. Recent data suggest that HIV may also replicate at low levels in some resting T cells.

From monocyte precursors in the bone marrow, various differentiated cell types are derived, including circulating monocytes, tissue macrophages, brain microglia, and dendritic cells of the skin and central lymphoid tissue. Each of these cells plays a role in the pathogenesis of infection in vivo, and some are permissive for HIV. HIV replicates in cultured monocyte-derived macrophages although they produce lower titers of infectious virus than do proliferating CD4$^+$ T lymphocytes (Fig. 13.4).

Dendritic cells are of particular importance because in vivo they appear to bind, sequester, and conserve infectious virions at their extracellular surface, although these cells are not permissive for virus replication. Recently, a lectin molecule that is expressed on the surface of dendritic cells, DC-SIGN, has been implicated in the binding and trapping of HIV particles that can then infect T lymphocytes. It is possible that virus captured by dendritic cells at mucosal sites is carried to draining lymph nodes where it can infect permissive T lymphocytes.

Cell Killing In Vitro HIV varies in its ability to cause cell killing in permissive cells, depending on both the cell type and the viral strain used for infection. The cytopathic properties of the virus cannot be predicted based only on its replication kinetics/properties. TCL-adapted viruses replicate well and cause cytopathic effects in many TCL (such as SUP-T1 cells), but these viruses also replicate in a monocyte-like cell line (U937) with no apparent killing or reduction in the rate of cell division (Fig. 13.4). Most HIV strains replicate and cause marked cellular destruction in primary lymphocyte cultures, although the amount of cell killing differs between strains. Only a subset of HIV strains replicate in macrophage cultures; in these cells, the level of replication is typically low relative to lymphocyte cultures, and the cultures remain viable and continue to produce virus for many weeks.

HIV or SIV can destroy CD4$^+$ lymphocytes by several mechanisms. CD4$^+$ lymphocytes vary in their susceptibility to infection and therefore in their vulnerability, depending on their physiologic status. Activated proliferating CD4 cells are much more susceptible that resting cells, and certain subpopulations of activated cells may be particularly susceptible, depending partly on the levels of X4 or R5 core-

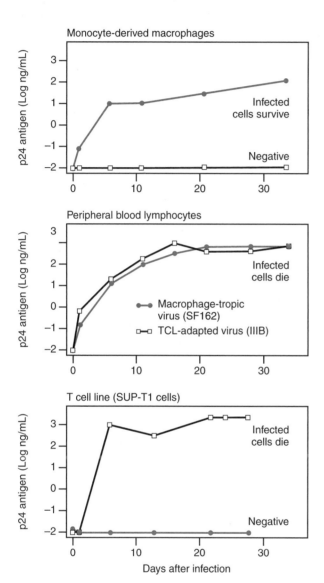

FIG. 13.4

The replication of prototypical HIV isolates, a "macrophage-tropic" virus (SF162) and a T-cell line (TCL)–adapted virus (IIIB) in different cell types. These graphs illustrate the relatively slow rate of virus replication, and the ability of both strains to replicate in stimulated primary blood mononuclear cells, containing dividing CD4$^+$ T lymphocytes. SF162 replicates well in primary macrophage cultures but not in a TCL (SUP-T1 cells), whereas the converse is true of strain IIIB. Also, the data demonstrate the ability of HIV to replicate in nondividing cells (most monocyte-derived macrophages do not divide in culture). HIV infection may or may not be cytopathic, since both primary T cells and transformed TCL are usually killed during infection while macrophages are not. (After Collman R, Hassan NF, Walker R, et al. Infection of monocyte-derived macrophages with human immunodeficiency virus type 1 (HIV-1). *J Exp Med* 1989;170:1149–1163.)

ceptors expressed, which can vary widely. Single-cell killing is due to apoptosis, initiated in part by the tat protein, and mediated through Fas signaling and the caspase cascade (see Chapter 4). In addition, infection of contiguous cells results in syncytium formation and cell death.

HIV strains vary in their cytopathic activity, which correlates with replicative capacity, syncytium induction, and coreceptor utilization. Viruses with a rapid/high, syncytium-inducing, X4 phenotype are more destructive than those with a slow/low, non–syncytium-inducing, R5 phenotype (viral variation is discussed below).

Cell Killing In Vivo   In addition to the direct virus-induced cytopathic effects seen in cell cultures, immune-mediated destruction of infected cells occurs in vivo. HIV-infected patients mount a cellular immune response in which CD8 lymphocytes through their T-cell receptors recognize and lyse infected (and perhaps uninfected) cells that present viral peptides in the context of class I major histocompatibility complex (MHC).

In vivo, only a small minority (less than 1%) of $CD4^+$ cells are infected at any time, which suggests that, in addition to destruction of infected $CD4^+$ cells, other mechanisms probably play a role in $CD4^+$ lymphocyte depletion. HIV or SIV infection causes a generalized activation of all lymphocyte populations (CD4, CD8, natural killer cells, and B cells), and a high proportion of activated cells undergo rapid apoptosis. Furthermore, HIV infection may interfere with the regeneration of CD4 cell populations either in the bone marrow or in the thymus (pathogenic mechanisms are discussed later in this chapter).

SEQUENCE OF EVENTS IN HIV INFECTION

Transmission, Portal of Entry, and Sequential Spread of Infection

HIV is transmitted by three major routes: via sexual contact (accounting for more than 90% of infections worldwide), from mother to child, or by blood or blood products. Mucosal secretions and blood contain both cell-free and cell-associated virus. Cell-free virions are infectious and can readily initiate experimental infections in primates. The relative importance of cell-associated virus as a vehicle for transmission is unclear because most inocula contain cell-free virus as well as infected cells. Under special experimental circumstances,

cell cultures can be infected with cell-associated but not with cell-free virus, suggesting that in some instances infected cells may be responsible for transmission. Epidemiologic studies show that the presence of sexually transmitted diseases increases the risk of HIV infection, which may be due both to ulcerations and to abrasions, and to increased number of inflammatory cells at the site of infection.

The details of transmission by sexual contact are not entirely clear, specifically how the virus breaches the epithelial or mucosal barrier in order to reach susceptible T lymphocytes or macrophages in the submucosal tissue. Undoubtedly, abrasions of the mucosa are important in some instances, but in nonhuman primates infection can also be initiated by atraumatic application of SIV to the vaginal mucosa or the tonsillar surface. Where transmission occurs via epithelium that overlies mucosa-associated lymphoid tissue, such as the tonsil or rectum, virus may transit M (microfold) cells to reach the underlying permissive mononuclear cells.

Some insights regarding the early target cells for infection have been gained in the SIV model, although it remains to be seen how experimental SIV transmission studies, which lead to a higher rate of infection than is seen in HIV transmission, model early events in HIV transmission. In the monkey experimentally infected by vaginal or tonsillar application of SIV, the virus can be first detected in the submucosal or lymphoid tissues, primarily in T lymphocytes (about 90%), with a few infected macrophages or dendritic cells (Fig. 13.5).

Within a few days, the local lymphoid tissue is heavily infected, with spread first to draining lymph nodes (about 1 week), and thence to distant nodes, spleen, and circulating mononuclear cells (1–3 weeks). In addition to the regional lymph nodes, there is a large population of T cells in the lamina propria and Peyer's patches of the gastrointestinal tract that may be heavily involved shortly after infection. Many of the infected T lymphocytes are dividing cells, which express high levels of SIV, but a substantial number of infected lymphocytes are resting cells that produce less virus.

Viremia, CD4 Counts, and Incubation Period

HIV produces a viremia that persists throughout the life of the infected individual and can be used to monitor the course of infection. In the blood, HIV is present both in association with infected cells (mainly $CD4^+$ T lymphocytes) and as free infectious virus in the plasma (Table 13.1). The level of

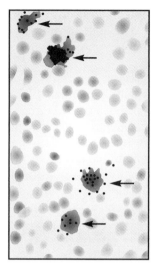
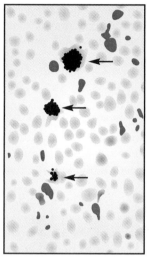

FIG. 13.5

Infected cells 7 days after atraumatic infection of the tonsillar mucosa with a dual-tropic strain SIVmac251, to show viral nucleic acid (black grains) radiolabeled with an antisense SIV probe. (*Left*) Infected cells are CD4-positive (*gray color, arrows*). (*Right*) Infected cells (*arrows*) are not macrophages (*gray color*) and (by inference) are T lymphocytes. (After Stahl-Hennig C, Steinman RM, Tenner-Racz K, et al. Rapid infection of oral mucosal-associated lymphoid tissue with simian immunodeficiency virus. *Science* 1999;285:1261–1265.)

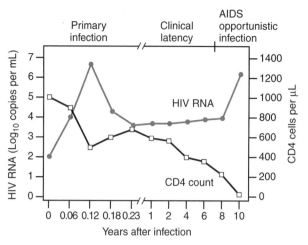

FIG. 13.6

Typical course of HIV infection as reflected in plasma viremia and concentration of CD4$^+$ cells in the blood. During primary infection there is widespread dissemination of virus to lymphoid tissues with or without an acute AIDS syndrome. The onset of AIDS is signaled by a drop in CD4 counts, a rise in viremia level, together with constitutional symptoms associated with opportunistic infections or neoplasms. (After Fauci AS, Desrosiers RC. Pathogenesis of HIV and SIV. In: Coffin JM, Hughes SH, Varmus HE. *Retroviruses*. Cold Spring Harbor, NY: Cold Spring Harbor Laboratory Press, 1997.)

viremia provides a window on the dynamics of infection. A useful surrogate for the course of disease is the concentration of CD4$^+$ T lymphocytes in the blood, which is inversely related to virus titer and is a harbinger of the functional loss of immune responses during clinical AIDS.

Absent antiretroviral therapy, there is an acute phase of infection (about 2 months duration) with high-titer viremia, followed by a subclinical phase with modest levels of viremia, that lasts for 1 to more

than 20 years, followed by a phase of clinical AIDS that lasts 1–4 years before death (Fig. 13.6). During acute infection, 3–6 weeks after transmission, a mononucleosis-like syndrome occurs in 50%–75% of patients, accompanied by a peak in viremia and an acute drop in the CD4 cell count in the blood. This is followed by induction of an immune response, at 1–3 months, which damps the infection and is associated with a dramatic drop in blood virus concentration of 10- to 1,000-fold below the acute peak level. How-

TABLE 13.1			
Variations in plasma and cell-associated viremia in HIV by stage of infection			
CD4$^+$ cells, count per 0.001 mL	Stage of infection	Plasma HIV, TCID per mL (median and range)	% CD4$^+$ cells infected (median and range)
>500	II	114 (1–500)	2.7 (0.2–11)
300–499	III	205 (1–500)	21 (2–35)
200–299		381 (25–500)	
<200	IV	1,466 (25–5,000)	30 (2–65)

HIV produces both a plasma and a cell-associated viremia that varies in intensity according to stage of infection. The % CD4$^+$ cells infected: based on in situ PCR. CDC infection stage: II, asymptomatic; III, persistent generalized lymphadenopathy; IV, clinical AIDS.

HIV, human immunodeficiency virus; TCID, titer of plasma virus expressed as tissue culture infectious doses; PCR, polymerase chain reaction; CDC, U.S. Centers for Disease Control and Prevention.

After Bagasra O, Seshamma T, Oakes JW, Pomerantz RJ. High percentages of CD4-positive lymphocytes harbor the HIV-1 provirus in the blood of certain infected individuals. *AIDS* 1993;7:1419–1425; and Pan L-Z, Werner A, Levy JA. Detection of plasma viremia in individuals at all clinical stages. *J Clin Microbiol* 1993;13:283–288.

ever, the infection is never completely cleared, and viremia usually stabilizes 3–9 months after infection at a level often called the virus "set point."

During the period of clinical latency, virus replication is occurring at a high rate, with concomitant rapid destruction and replacement of CD4 T lymphocytes (described below). The outcome of this dynamic process determines the next steps in infection and explains the diversity in the duration of subclinical infection. Rapid progressors develop AIDS in 1–3 years, whereas long-term nonprogressors remain well for more than 15 years or longer. Outcome is closely related to the virus set point (Fig. 13.7). In a cohort of infected patients, 90% of the quartile with the highest set points progress to AIDS in 5 years, while less than 10% of the quartile with the lowest set point has developed AIDS in that time. Patients with the slowest progression are often dubbed long-term nonprogressors (LTNP), defined as subjects who are AIDS free 10 years after infection. Of an initial cohort of infected persons, about 10% became LTNP, but at 20 years after infection less than 2% of the original cohort were AIDS-free LTNP. The slowest progression is seen in patients infected with HIV-2, more than 50% of whom remain AIDS free throughout their lives. The ability of

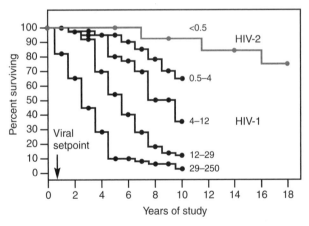

FIG. 13.7 _____

Progression to AIDS (percent surviving AIDS free) is associated with viremia set point. In this example, patients were divided into four quartiles, according to the set point determined 6–9 months after infection. At 10 years after infection, more than 95% of the quartile with the highest set points had progressed to AIDS whereas less than 40% of those in the lowest quartile had AIDS. For comparison, data are shown for HIV-2, which is less virulent than HIV-1 and has an even lower frequency of progression to AIDS. Set points are shown as RNA copies (×1,000) per milliliter plasma. (After Mellors JT, Rinaldo CR Jr, Gupta P, et al. Prognosis in HIV-1 infection predicted by the quantity of virus in plasma. *Science* 1996;272:1167–1170; and Whittle HC, Aryyoshi K, Rowland-Jones S. HIV-2 and T cell recognition. *Curr Opin Immunol* 1998;10:382–387.)

the host to contain an HIV infection indefinitely is central to understanding the dynamics of infection and is discussed further below.

Opportunistic Infections and Neoplasms

The drop in CD4 levels (normally more than 1,000 cells per microliter of blood) below a critical threshold (200–300 cells), often accompanied by a rise in virus set point, signals the advent of AIDS-defining illnesses. Constitutional symptoms include fever, fatigue, malaise, lymphadenopathy, gastrointestinal symptoms such as diarrhea, weight loss, and early evidence of immunodeficiency such as oral candidiasis (caused by *Candida albicans*) and hairy leukoplakia of the tongue (caused by Epstein–Barr virus [EBV] infection of epithelial cells).

Opportunistic infections are caused by a wide spectrum of parasites, including protozoa (such as *Toxoplasma gondii*), fungi (such as such as *C. albicans* and *Pneumocystis carinii*), bacteria (such as *Mycobacterium avium* complex and *M. tuberculosis*), and viruses (such as cytomegalovirus, herpes simplex virus, and varicella-zoster virus). In healthy persons, these agents produce clinically occult infections that usually manifest as overt illness only under conditions of immunodeficiency. Many of these organisms are intracellular parasites that are controlled mainly by the cellular immune response rather than by antibodies. The spectrum of AIDS-associated opportunistic infections differs geographically, reflecting the relative prevalence of different agents. Thus, tuberculosis is much more important as a manifestation of AIDS in developing than in industrialized countries.

Relative to the general population, AIDS patients are at increased risk for a selected number of neoplasms. Among these are polyclonal B-cell lymphomas, such as Burkitt's lymphoma (caused by EBV), cervical carcinoma (associated with human papillomavirus), and Kaposi's sarcoma (associated with HHV-8) (see Chapter 11). It is not clear why specific neoplasms are particularly associated with AIDS, but it probably reflects the compromise of certain immune surveillance mechanisms that are particularly important for control of these cancers rather than any direct effect of HIV.

IMMUNE RESPONSE TO HIV

HIV Proteins as Immunogens

HIV proteins are as immunogenic as the comparable proteins of other viruses, if purified and used as experimental immunogens. Some HIV proteins are synthesized in much higher copy numbers than others—for instance, p24 (CA)—and therefore elicit more robust antibody responses. The intensity of the immune response, whether antibody or cytotoxic T lymphocyte (CTL), does not predict its ability to prevent de novo infection or control an ongoing infection, both of which depend on the specific epitope that is targeted.

Antibody Responses

Most patients develop detectable antibodies against HIV-1 within 2 months of infection, with highest reactivity against the p24 (CA) and envelope (gp120, gp41, or gp160) proteins, quantified by enzyme-linked immunosorbent assay (ELISA) or Western blot methods. Compared to other viruses, the neutralizing response—as measured by current assays—in the infected but immunocompetent host is unusually low or absent, and less than 10% of patients have neutralizing titers greater than 1:100. Weak neutralizing responses are likely due to the structure of the envelope protein. As reconstructed from x-ray crystallography, the face of the SU protein that binds the CD4 receptor is heavily glycosylated (about 24 N-linked sites per SU molecule), and the fusion intermediates (Fig. 13.2) that bind to the coreceptors are transient structures. Both these phenomena reduce the probability of inducing epitope-relevant antibodies to block virus entry.

Neutralizing antibody, when it is induced, is capable of providing significant protection against experimental challenge with SIV. Thus, as shown in Fig. 13.8, a relatively low titer of passively acquired neutralizing antibody protects monkeys against subsequent challenge with a pathogenic SIV or SHIV (a chimeric virus with the HIV envelope on an SIV genetic background). Difficulty in inducing a neutralizing response probably plays a role in the failure of patients to contain the virus more effectively and has impeded the formulation of an effective preexposure vaccine (see Chapter 14).

Cellular Immune Responses

CD4 responses to HIV can be measured in a lymphocyte stimulation (LS) assay. Most patients show an HIV-specific LS response during acute infection, which is gradually lost over the next several years, even before severe depletion of circulating CD4 cells. Furthermore, patients who still have an active HIV-specific CD4 LS response control their infections better when treated with highly active antiretroviral therapy (HAART) than do patients lack-

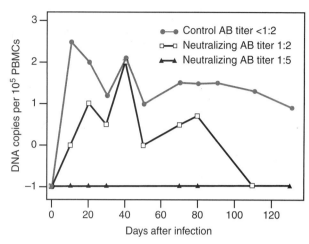

FIG. 13.8

Human immunodeficiency virus (HIV)–neutralizing antibodies will protect against a subsequent challenge with simian HIV (SHIV), showing that, once produced, they are as effective as similar antibodies against other viruses. In this experiment, monkeys were given intravenous injections of immunoglobulin that did or did not contain antibodies against HIV, and were challenged intravenously with a pathogenic dual-tropic SHIV DH12 a day later. The neutralizing titer in the plasma of passively immunized monkeys was determined just before challenge. Animals with a titer of approximately 1:5 were protected against challenge whereas those with a titer of approximately 1:2 were partially protected. Antibody titers are expressed as the highest dilution of plasma that neutralized 100% of 100 tissue culture infectious doses of the same virus used for infection. (After Shibata R, Igarashi T, Haigwood N, et al. Neutralizing antibody directed against the HIV-1 envelope glycoprotein can completely block HIV-1/SIV chimeric virus infections of macaque monkeys. *Nature Med* 1999;5:204–210.)

ing this response. This implies that virus-specific CD4 T cells play an important role in controlling infection, perhaps though their helper function, a point discussed further below.

Essentially all patients raise CD8-mediated CTL responses against HIV-1 that appear in 1–4 months after infection, concomitant with the decrease in peak viremia. Responses are against multiple epitopes, with the exact ones depending on the HLA haplotype of the infected person and the sequence of the infecting virus. Immunodominant epitopes have been detected in many of the viral proteins, most frequently in the gag, pol, and nef peptides. Virus-specific CTL are thought to be responsible for the drop in viremia at the end of the acute infection and play an important role in containing viremia levels during persistent infection. Thus, in experimental infection of monkeys with SIVmac, depletion of CD8 cells with an anti-CD8 antibody reduces the CD8 cell concentration in the blood and is accompanied by a significant transient increase in viremia levels (Fig. 13.9). CTL likely

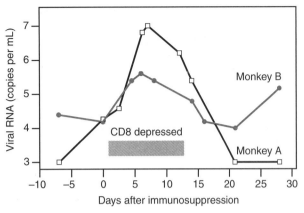

FIG. 13.9

The cellular immune response plays an important role in the control of simian immunodeficiency virus (SIV) infection. Monkeys infected with dual-tropic virulent SIVmac for more than 9 months had established stable virus set points. They were then treated with a potent antibody against CD8, which reduced the level of CD8 T lymphocytes in the blood by more than 99%. Data on two animals demonstrate the rise in viremia level during the period of immunosuppression and the reconstitution of immune control when immunosuppressive treatment was terminated. The effect of treatment is more pronounced in monkey A with initially lower viremia. (After Schmitz JE, Kuroda MJ, Santra S, et al. Control of viremia in simian immunodeficiency virus infection by CD8+ lymphocytes. *Science* 1999; 283: 857–860.)

work in conjunction with T-helper and B lymphocytes to control virus replication.

DYNAMIC ASPECTS OF INFECTION

Virus Turnover

During the long period of clinical latency (Fig. 13.6), HIV viremia remains quite stable, although the set point varies in different patients. As described in Chapter 2, virus in the blood is constantly removed and replaced (Fig. 13.10). Turnover in the plasma is similar to other viruses, with a half-life ($t_{1/2}$) of 10–30 minutes, based on studies of SIV in monkeys. In the presence of antiviral antibody, the $t_{1/2}$ is shortened by two- to fivefold. If the $t_{1/2}$ in plasma is approximately 15 minutes, then each day 10^8 to 10^{11} virions are shed into the blood in patients whose viral load ranges from 10^3 to 10^6 RNA copies per milliliter of plasma.

The main source of plasma virus is infected activated $CD4^+$ T lymphocytes in blood and lymphoid tissues (Fig. 13.10). Infected macrophages make a minor contribution because of their low levels of virus production. Each infected cell type has a characteristic half-life that can be estimated by following the effects of HAART when administered to patients

with long-term infection and stable CD4 levels (Fig. 13.11). After commencing HAART, there is a dramatic reduction in plasma viremia, which can be divided into three phases: a rapid drop of approximatley 100-fold over the first 10 days, due to the interruption of most cell to cell spread of virus and the die off of infected activated T cells; a slower decrease of about 10-fold over 2 months, reflecting the death of cells with a longer life, probably mainly macrophages; and a plateau that may be below the level of detection but reflects the indefinite persistence of residual virus.

Viral persistence has two main causes, latent infection of resting $CD4^+$ lymphocytes and persistent replication even in the presence of effective HAART. Latently infected resting T cells that carry either a provirus or a preintegration viral complex cannot be detected and eliminated by antiviral CTLs but may be activated intermittently to produce new virus. It is estimated that, after prolonged HAART has reduced residual virus to a plateau (Fig. 13.11), there remain 10^3 to 10^7 latently infected cells in individual patients. This reservoir of latently infected cells decays at a half-life estimated at about 6 months or longer, consistent with the life span of resting $CD4^+$ memory cells.

Even in the presence of HAART, active replication continues at a low level in both lymphocytes and monocytes in some patients, likely due to the failure of antiretroviral drugs to spread in effective concentration to every cell in solid tissues. Evidence for the persistence of a low level of active replication is (a) the presence of episomal cDNA intermediates, which are labile products indicative of active viral replication; and (b) mutations in virus sampled at intervals during prolonged HAART, which are seen in some patients. Organs such as the thymus, brain, lung, and kidney require further study as potential additional reservoirs of latent virus.

Cell Turnover

In uninfected individuals with stable CD4 cell levels in the blood, there is a dynamic equilibrium between the death (or removal) of cells and their replacement. This is altered in HIV-infected persons, resulting in a gradual reduction in CD4 T-cell concentration. Methods have been developed to measure the effect of HIV infection on cellular turnover, using labeling with either BrdU (in monkeys) or tritium-labeled glucose in humans. The results can be expressed as either the half-life ($t_{1/2}$) of circulating CD4 cells or the daily replacement of cells as a percentage of the total circulating population.

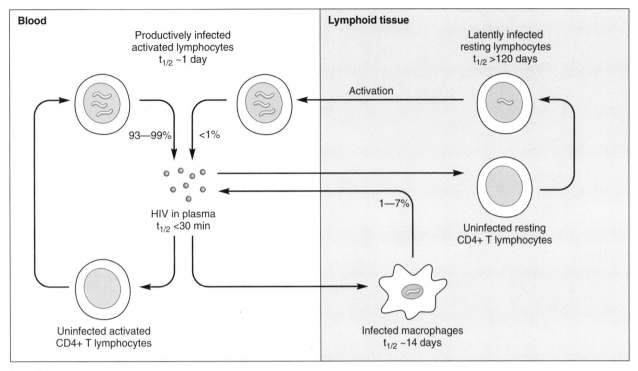

FIG. 13.10 _____

Kinetics of HIV or simian immunodeficiency virus in vivo. Plasma virus is produced mainly by infected activated lymphocytes with a modest contribution from infected macrophages. Latent infection of resting lymphocytes is unimportant during active infection but presents an impediment to eradication by highly active antiretroviral therapy. $t_{1/2}$, half-life; LN, lymph node; LPL, lamina propria lymphoid tissue. (After Igarashi T, Brown C, Azadega A, et al. Human immunodeficiency virus type 1 neutralizing antibodies accelerate clearance of cell-free virions form blood plasma. *Nature Med* 1999;5:211–216; Mittler JE, Markowitz M, Ho DD, Perelson AS. Improved estimates for HIV-1 clearance rate and intracellular delay. *AIDS* 1999;13:1415–1417; Ramratnam B, Mittler JE, Zhang L, et al. The decay of the latent reservoir of replication competent HIV-1 is inversely correlated with the extent of residual viral replication during prolonged anti-retroviral therapy. *Nature Med* 2000;6:82–86; Sharkey ME, Teo I, Greenough T, et al. Persistence of episomal HIV-1 infection intermediates in patients on highly active antiretroviral thealpy. *Nature Med* 2000;6:76–81; Zhang L, Dailey PJ, He T, et al. Rapid clearance of simian immunodeficiency virus particles from plasma of rhesus macaques. *J Virol* 1999;73:855–860.)

In uninfected healthy subjects, with a CD4 level of more than 1,000 cells per microliter, approximately 1% of the circulating CD4 population is replaced daily, and the $t_{1/2}$ is about 75 days. In HIV-infected persons, with moderately reduced but stable CD4 counts (approximately 350 cells per microliter), the daily replacement rises to about 3% and the $t_{1/2}$ is reduced to around 25 days. The reduced life of CD4 cells is due to direct cell killing by HIV, to indirect killing mediated by HIV-specific CTLs, and to activation and apoptosis of uninfected CD4$^+$ cells. The turnover data indicate that HIV-induced reduction in CD4 half-life is compensated by an increase in the appearance of new CD4 cells in the blood. The eventual reduction in CD4 count must reflect a failure of production to completely compensate for cell destruction, but this is a very subtle effect that is too small to measure. In other words, if 99.9% of CD4 cells are replaced daily (rather than the 100% in uninfected individuals), the CD4 level would decrease by threefold (from about 1,000 to about 350) over 3 years.

CD8 levels in the circulation are usually slightly increased during the subclinical phase of infection but may drop during the end stages of clinical AIDS. However, the turnover of CD8 lymphocytes or of B lymphocytes is increased from the onset of HIV infection. This implies that there is a generalized activation of all populations of lymphocytes, which reduces cellular half-life, since many activated cells undergo rapid apoptosis. In addition, generalized activation can lead to immune dysregulation and dysfunction of CD8 T cells (discussed below).

The turnover of CD4 cells in HIV-infected persons who have been successfully treated with HAART gradually returns to that seen in uninfected subjects, with a reduction in daily replacement and an increase in half-life, within about a year of therapy.

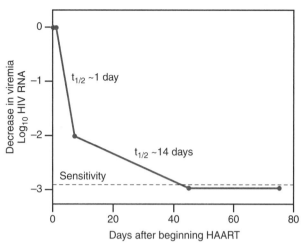

FIG. 13.11

The reduction in viremia following initiation of highly active antiretroviral therapy (three drugs). The curve can be divided into three phases: early rapid decrease of about 100-fold due to disappearance of infected activated CD4$^+$ T lymphocytes; a slower phase due to disappearance of longer lived infected macrophages; and a plateau of persistent infection that may be below the level of detection. (After Blankson JN, Finzi D, Pierson TC, et al. Biphasic decay of latently infected CD4$^+$cells in acute human immunodeficiency virus type 1 infection. *J Infect Dis* 2000;182:1636–1642; Grossman Z, Polis M, Feinberg MB, et al. Ongoing HIV dissemination during HAART. *Nature Med* 1999;5:1099–1104; Ramratnam B, Mittler JE, Zhang L, et al. The decay of the latent reservoir of replication competent HIV-1 is inversely correlated with the extent of residual viral replication during prolonged anti-retroviral therapy. *Nature Med* 2000;6:82–86; Mittler JE, Markowitz M, Ho DD, Perelson AS. Improved estimates for HIV-1 clearance rate and intracellular delay. *AIDS* 1999;13:1415–1417; Sharkey ME, Teo I, Greenough T, et al. Persistence of episomal HIV-1 infection intermediates in patients on highly active anti-retroviral therapy. *Nature Med* 2000;6:76–81.)

VIRAL VARIATION

Retroviruses exhibit a high rate of mutation due to the absence of proofreading during reverse transcription of RNA to DNA. The HIV genome is 10^4 bases, and there is approximately 1 base mismatch per 10^4 nucleotides or one mutation per virion replication. In an HIV-infected subject, approximately 10^{10} new virions are produced daily, so that each base in the genome undergoes mutation many times each day. Thus, although HIV infection is often initiated by only a few virus particles, the original viral genome quickly evolves into a quasi-species or swarm of genetically related viruses. During in vivo infection, the high mutation rate permits the selection of viral variants that have a selective advantage due to their replication potential or ability to escape host defenses. It has also led to the emergence of viral variants resistant to antiviral drugs.

Biological Phenotypes and Their Evolution During Persistent Infection

HIV primary isolates differ in their biological properties, such as growth in cell culture, syncytium induction (SI) and cytopathicity in cultured T cells, tropism for macrophages or TCL, and use of coreceptors (Table 13.2, Fig. 13.12). Primary isolates from the clinically asymptomatic phase of infection are frequently slow/low, non-syncytium-inducing (NSI), and macrophage-tropic, while AIDS isolates are frequently rapid/high, SI, and TCL-tropic. It is unclear whether the change in phenotype causes the development of immunodeficiency or whether immunodeficiency selects for viruses with the late phenotype, and both mechanisms may be operative. (Distinct from these variations in phenotype, laboratory-passaged viruses acquire properties rarely seen in wild-type isolates.)

SIV infection provides further information on the relationship between biological phenotype and the pathogenesis of AIDS. If a cloned SIV is used to infect macaques, isolates obtained during the progression to AIDS show increasing virulence as well as progression from a slow/low NSI to a rapid/high SI phenotype (Fig. 13.13). SHIV (simian human immunodeficiency viruses with an HIV envelope on an SIV genetic backbone) have been used to compare the effect of tropism on pathogenesis. A TCL-tropic SHIV (using CXCR4) spreads to lymph nodes with marked destruction of T cells and a reduction in circulating CD4 cells, whereas a macrophage-tropic SHIV (using CCR5) infects and depletes the gastrointestinal-associated lymphoid tissue (GALT) but does not cause a marked reduction in circulating CD4 cell count (Fig. 13.14).

The majority of viruses obtained early in HIV-1 infection are R5 (macrophage-tropic viruses that use the CCR5 coreceptor). The importance of R5 viruses in transmission is underlined by the finding that individuals who are homozygous for a genetic deletion of CCR5 are less susceptible to infection (discussed below and in Chapter 12). However, it is unclear exactly why R5 viruses are differentially transmitted because in some models of SIV infection most of the infected cells first detected after mucosal challenge are T lymphocytes that are susceptible to R5 or X4 viruses. It has been suggested that R5 viruses replicate better in resting T cells than do X4 viruses, which could give them an advantage in early infection.

Nonpathogenic HIV and SIV Infections

Untreated HIV-1 infections have a very high fatality rate, close to 100%, which is unusual among viruses

TABLE 13.2		
Correlation of biological phenotype of primary HIV isolates with stage of infection		
Biological phenotype	Asymptomatic infection	AIDS
Replication kinetics	Slow/low	High/rapid
SI	NSI	SI
Cellular host range	Macrophage-tropic	TCL-tropic
Coreceptor usage	CCR5	CXCR4 +/− CCR5

Although the biological phenotype of primary HIV isolates is correlated with the stage of infection, there are many exceptions. All viruses can replicate in primary cultures of CD4+ T lymphocytes. This table excludes laboratory passaged virus strains that have properties not seen in wild-type primary isolates.

SI, syncytium induction; NSI, no syncytium induction; TCL, T-cell line.

After Albert J, Koot M. In: Schuitemaker H, Miedema F, Schuitemaker H, eds. *AIDS pathogenesis. Immunology and medicine*, Vol. 28. New York: Kluwer Academic, 2000.

of animals or humans. However, lower virulence is seen with HIV-2 in humans, and HIV-1 in chimpanzees (the original host) rarely causes illness or death. Many SIV have been isolated from different species of monkeys, and most of them, such as SIVagm in African green monkeys or SIVsm in sooty mangabeys, appear to be relatively benign in their natural hosts. Pathogenic SIV infections have mainly resulted from passage into a monkey species other than the natural host. High virulence, as exhibited by HIV-1 in humans, is clearly not required for survival of lymphotropic lentiviruses and may be a phenomenon associated with recent transmission across a species barrier.

Studies of SIV strains in their natural hosts have indicated that there are at least two distinct mechanisms that explain the benign nature of nonpathogenic infections. In some instances, virus replication is contained at low levels, similar to those seen in long-term nonprogressors infected with HIV-1. Under such circumstances, homeostatic mechanisms are apparently able to replace lost CD4 lymphocytes indefinitely, leading to asymptomatic lifelong infections. However, in other examples, such as SIVsm infection of sooty mangabeys, virus replicates to high titers, but CD4 cells are not reduced and there is little evidence of generalized ac-

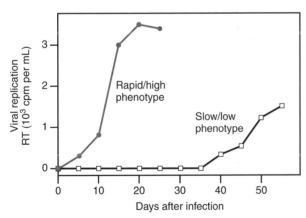

FIG. 13.12

Comparative replication of slow/low and rapid/high variants of wild-type primary isolates of HIV. A T-cell line (HUT 78) that is permissive for most primary isolates was infected with similar inoculums of two viruses (HIVSF2 and HIVSF13) recovered early and late from the same patient, and virus replication was followed by assay for reverse transcriptase in the culture supernate. A primary isolate is a virus obtained from a patient and passed a minimal number of times in primary blood mononuclear cells. (After Levy J. *HIV and the pathogenesis of AIDS*. Washington, DC: ASM Press, 1998.)

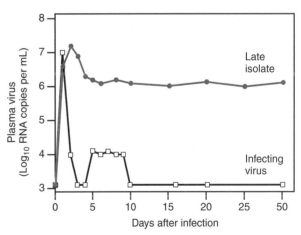

FIG. 13.13

The virulence of simian immunodeficiency virus (SIV) can increase during long-term infection. In this example, macaques (*M. nemestrina*) were infected with a cloned macrophage-tropic strain of SIV (Mne clone 8) and virus was isolated at intervals during infection. This figure compares the inoculated virus and a representative late isolate, which were used to infect a new group of macaques (one animal per group is shown), and indicates that the late isolate was more virulent than the original clone. (After Kimata JT, Kuller L, Anderson DB, et al. Emerging cytopathic and antigenic simian immunodeficiency virus variants influence AIDS progression. *Nature Med* 1999;5:535–541.)

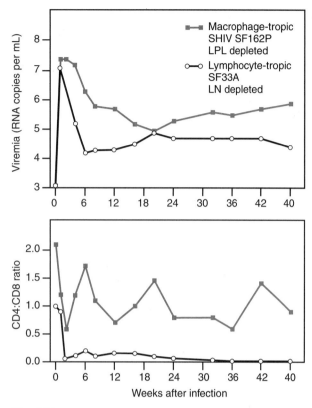

FIG. 13.14

Macrophage-tropic and T-lymphocyte-tropic viruses differ in their pathogenesis, as shown in this comparison of simian human immunodeficiency virus (SHIV) SFA33A (T-cell line tropic) and SHIV SF162P (macrophage tropic). *(Top)* Both viruses produced similar levels of viremia. *(Bottom)* T-cell tropic SHIV SFA33A reduced circulating CD4 T lymphocytes as indicated by the reduced CD4 to CD8 ratio, and also depleted CD4 cells in the lymph nodes but not CD4 cells in the lamina propria. SHIV SF162P, in contrast, depleted CD4 cells in the lamina propria of the gut but did not deplete CD4 cells in lymph nodes or reduce the circulating CD4 to CD8 ratio. (After Harouse JM, Gettle A, Tan RCH, et al. Distinct pathogenic sequelae in rhesus macaques infected with CCR5- or CXCR4-tropic strains of SHIV. *Science* 1999;284:816–819.)

tivation of lymphocytes. Thus, further studies of these models may provide insights into the mechanisms of HIV-induced immunodeficiency.

Immunologic Escape Mutants

HIV persists and replicates in the presence of an active humoral and cellular response, which might select for immunologic escape mutants. As CD8-mediated CTL activity appears to play a crucial role in the control of HIV infection (Fig. 13.9), variants with mutations in immunodominant CTL epitopes could have a selective advantage. In persistent SIV infection of macaques—a model for HIV in humans—mutations accumulate in epitopes to which the animals have developed CTL responses, sug-

gesting that viral variation may play a role in SIV persistence, and escape mutants have also been observed in HIV-1 infections. Thus, there is considerable evidence for the selection of viral variants that escape specific CTL responses, and this phenomenon may play a role both in persistence and in development of immunodeficiency.

Clades

Extensive genomic studies have identified ten major subtypes (clades) within the M group of HIV-1, which causes more than 95% of HIV-1 infections. Clades and recombinants among the clades appear to have arisen in Africa, where HIV-1 first emerged. It is not clear whether there are any consistent biological or epidemiologic differences among the clades, and all of them spread readily in humans. Outside of the African continent, certain clades predominate (such as clade B in North America) presumably due to a founder effect when the virus was originally introduced. Clades do not correspond to immunotypes, and there is considerable cross-antigenicity among clades, based on both neutralization (using the few sera with strong neutralizing capacity) and CTL assays.

MECHANISMS OF IMMUNE SUPPRESSION

When CD4 cell counts diminish, functional immunodeficiency begins to appear, signaled by the inability to contain a selected group of viruses, bacteria, and parasites. Most opportunistic infections represent activation of long-time persistent infections that the AIDS patient has been harboring in a latent or controlled form. It is likely that containment of these infections is mainly mediated by cellular immune mechanisms, particularly CTL. When treatment-naïve patients with opportunistic infections begin therapy with effective HAART, their opportunistic infections "melt away" as their circulating levels of CD4 cells rise. By inference, there must also be a reconstitution of effector CD8 lymphocytes that target and control specific opportunistic infections.

Paradox: CD4 Depletion vs. CD8 Dysfunction

Manifestations of acquired immunodeficiency are mainly due to a lack of effector CD8 cells that contain potential opportunistic infections. At first glance, this presents an apparent paradox because CD4 lymphocytes and macrophages—not CD8 cells—are targeted by HIV. The reduction of CD8 function is probably due to the concatenation of

	TABLE 13.3	
	Mutations that influence susceptibility to HIV infection or the rate of progression to AIDS	
Genetic locus	Mutation and its cellular effects	Influence on progression to AIDS
CCR5	D32: homozygous abrogates CCR5 expression	Protects vs. infection
	D32: heterozygous reduces CCR5 expression	Retards progression
CCR5	P1: homozygous CCR5 promoter mutation	Accelerates progression
CCR2	64I heterozygous mechanism?	Retards progression
SDF1	3A: homozygous ligand for coreceptor CXCR4	Retards progression (some studies)
CX3CR1	1249: homozygous reduced chemokine binding	Accelerates progression
B*35 or Cw*04	Class I HLA alleles: heterozygous ? alters immune response	Accelerates progression
B*5701	Class I HLA allele: heterozygous ? alters CTL epitope specificity	Retards progression

See chapter 12 for references.

several effects. First, there is a paucity of CD4 helper cells that are required to induce new antigen-specific CD8 effector cells from the post-thymus naïve T-cell pool. Also, HIV infection leads to a reduction in the replenishment of the general pool of naïve T cells emigrating from the thymus. Second, the state of general lymphocyte activation produced by HIV infection results in a high level of apoptosis that causes a nonspecific physiologic exhaustion of committed clones of CD8 cells. Third, continuous antigen-specific immune stimulation, produced by ongoing infections, activates pools of parasite-specific memory cells and eventually exhausts them. Together these effects erode cellular immune surveillance of opportunistic pathogens. Nonpathogenic SIV models, in which high levels of viral infection are well tolerated, strongly imply that generalized immune activation—and perhaps other secondary effects of infection—plays a key role in the pathogenesis of immunodeficiency.

GENETIC DETERMINANTS OF HOST SUSCEPTIBILITY

Epidemiologic studies have identified a number of genetic determinants that influence susceptibility to infection or the rate of progression to AIDS (see Chapter 12). Most prominent of these is the mutation in CCR5 (Δ32) that abrogates the expression of that gene. In its homozygous form Δ32 markedly reduces the risk of infection and, in its heterozygous form, reduces the rate of progression to AIDS (Table 12.6). A number of mutations have been reported

that slow or accelerate progression to AIDS (Table 13.3). Most of these mutations affect other coreceptors, such as CXCR4 or other ligands. In addition, a few specific HLA haplotypes have a marked influence on the rate of progression to AIDS, but their mechanism of action has not been elucidated (Table 12.7).

REPRISE

HIV is a persistent virus that has a complex pathogenesis with many unusual features. Characteristics that define the infection include (a) involvement of the helper subset of T lymphocytes and macrophages due to the use of CD4 as the primary cellular receptor; (b) escape from immune surveillance mainly due to the occurrence of latency in resting $CD4^+$ T lymphocytes in which an integrated provirus can be maintained for many years, complemented by the ability to escape neutralization by antibodies, and the development of variants that evade epitope-specific CTLs; (c) the ability to destroy activated T-helper lymphocytes at a high rate, which eventually exhausts homeostatic replacement by the host; and (d) the initiation of an acquired immunodeficiency due to the loss of $CD4^+$ T cells and the perturbation of the function of CD8 effector T lymphocytes. This unique concatenation of characteristics combines to produce an infection that is fatal in close to 100% of cases.

Another unusual characteristic of HIV is its extremely low transmission (risk less than 1:100

contacts) by sexual contact, the primary mode of host-to-host spread, reflecting the requirement for the transmitted virus to contact T lymphocytes and the inability to infect epithelial cells. Similarly, a minority of HIV-infected pregnant women transmit HIV to their infants (up to 35% of women who breast-feed). Low transmission rate and 100% fatality would ordinarily limit the ability of a virus to survive in its host population. Paradoxically, HIV has spread rampantly in spite of these limitations, due mainly to the long incubation period that provides many opportunities for infected but apparently healthy persons to transmit infection.

Furthermore, HIV has been peculiarly resistant to control measures. This can be attributed in part to the social stigma associated with an infection that is transmitted by sexual contact or the use of injected drugs. In addition, HIV often targets poor countries or socially disadvantaged subgroups in industrialized countries. The insidious nature of an infection that is widely seeded before impacting an infected population aggravates the problem. In summary, the features of HIV pathogenesis have combined to cause a global pandemic that presents the greatest infectious disease challenge ever known in the history of medical science.

FURTHER READING

Reviews and Chapters

Berger EA, Murphy PM, Farber JM. Chemokine receptors as HIV-1 coreceptors: roles in viral entry, tropism, and disease. *Annu Rev Immunol* 1999;17:657–700.

Burton DR, Parren PWHI. Vaccines and the induction of functional antibodies: time to look beyond the molecules of natural infection? *Nature Med* 2000;6:123–125.

Chan DC, Kim PS. HIV entry and its inhibition. *Cell* 1998;93:681–684.

Fauci AS, Desrosiers RC. Pathogenesis of HIV and SIV. In: Coffin JM, Hughes SH, Varmus HE. *Retroviruses*. Cold Spring Harbor, NY: Cold Spring Harbor Laboratory Press, 1997.

Levy J. *HIV and the pathogenesis of AIDS*. Washington, DC: ASM Press, 1998.

McCune JM. The dynamics of CD4 T-cell depletion in HIV disease. *Nature* 2001;410:974–979.

McMichael AJ, Rowland-Jones SL. Cellular immune responses to HIV. *Nature* 2001;410:980–987.

Schuitemaker H, Miedema F, Schuitemaker H, eds. *AIDS pathogenesis, Immunology and medicine series,* Vol. 28. New York: Kluwer Academic, 2000.

Wyatt R, Sodroski J. The HIV-1 envelope glycoproteins: fusogens, antiges, and immunogens. *Science* 1998;280:1884–1888.

Classic Contributions

Eckert DM, Malashkevich VN, Hong LH, et al. Inhibiting HIV-1 entry: discovery of D-peptide inhibitors that target the gp41 coiled-coil pocket. *Cell* 1999;99:103–115.

Mellors JT, Rinaldo CR Jr, Gupta P, et al. Prognosis in HIV-1 infection predicted by the quantity of virus in plasma. *Science* 1996;272:1167–1170.

Original Publications

Allen TM, O'Connor DH, Jing P, et al. Tat-specific cytotoxic T lymphocytes select for SIV escape variants during resolution of primary viraemia. *Nature* 2000;407:386–390.

Bagasra O, Seshamma T, Oakes JW, Pomerantz RJ. High percentages of CD4-positive lymphocytes harbor the HIV-1 provirus in the blood of certain infected individuals. *AIDS* 1993;7:1419–1425.

Bartz SR, Emerman M. Human immunodeficiency virus type 1 tat induces apoptosis and increases sensitivity to apoptotic signals by up-regulating FLICE/Caspase-8. *J Virol* 1999;73:1956–1963.

Blankson JN, Finzi D, Pierson TC, et al. Biphasic decay of latently infected CD4$^+$ cells in acute human immunodeficiency virus type 1 infection. *J Infect Dis* 2000;182:1636–1642.

Brooks DG, Kitchen SG, Kitchen CMR, et al. Generation of HIV latency during thymopoiesis. *Nature Med* 2001; 7:459–464.

Camerini D, Su H-P, Games-Torre G, et al. Human immunodeficiency virus type 1 pathogenesis in SCID-hu mice correlates with syncytium-inducing phenotype and viral replication. *J Virol* 2000;74:3196–3204.

Collman R, Hassan NF, Walker R, et al. Infection of monocyte-derived macrophages with human immunodeficiency virus type 1 (HIV-1). *J Exp Med* 1989;170:1149–1163.

Davey RT Jr, Bhat N, Yoder C, et al. HIV-1 and T cell dynamics after interruption of highly active antiretroviral therapy (HAART) in patients with a history of sustained viral suppression. *Proc Natl Acad Sci USA* 1999;96:15109–15114.

Evans DT, O'Connor DH, Jing P, et al. Virus-specific cytotoxic T-lymphocyte responses select for amino-acid variation in simian immunodeficiency virus Env and Nef. *Nature Med* 1999;5:1270–1276.

Grivel J-C, Margolis LB. CCR5- and CXCR4-tropic HIV-1 are equally cytopathic for their T-cell targets in human lymphoid tissue. *Nature Med* 1999;5:344–346.

Grossman Z, Polis M, Feinberg MB, et al. Ongoing HIV dissemination during HAART. *Nature Med* 1999;5:1099-1104.

Harouse JM, Gettle A, Tan RCH, et al. Distinct pathogenic sequelae in rhesus macaques infected with CCR5 or CXCR4-tropic strains of SHIV. *Science* 1999;284:816–819.

Huang Y, Paxton WA, Wolinsky SM, et al. The role of a mutant CCR5 allele in HIV-1 transmission and disease progression. *Nature Med* 1996;2:1240–1243.

Igarashi T, Brown C, Azadega A, et al. Human immunodeficiency virus type 1 neutralizing antibodies accelerate clearance of cell-free virions from blood plasma. *Nature Med* 1999; 5:211–216.

Kimata JT, Kuller L, Anderson DB, et al. Emerging cytopathic and antigenic simian immunodeficiency virus variants influence AIDS progression. *Nature Med* 1999;5:535–541.

Lapham CK, Zaitseva MB, Lee S, et al. Fusion of monocytes and macrophages with HIV-1 correlates with biochemical properties of CXCR4 and CCR5. *Nature Med* 1999;5:303–308.

Migueles SA, Sabbaghian MS, Shupert WL, et al. HLA B*5701 is highly associated with restriction of virus replication in a subgroup of HIV-infected long term nonprogressors. *Proc Natl Acad Sci USA* 2000;97:2709–2714.

Mittler JE, Markowitz M, Ho DD, Perelson AS. Improved estimates for HIV-1 clearance rate and intracellular delay. *AIDS* 1999;13:1415–1417.

Mohri H, Bonhoeffer S, Monard S, et al. Rapid turnover of T lymphocytes in SIV-infected rhesus macaques. *Science* 1998;279:1223–1227.

Mothes W, Boerger AL, Narayan S, et al. Retroviral entry mediated by receptor priming and low pH triggering of an envelope glycoprotein. *Cell* 2000;103:679–689.

Ogg GS, Jin X, Bonhoeffer S, Dunbar PR, et al. Quantitation of HIV-1-specific cytotoxic T lymphocytes and plasma load of viral RNA. *Science* 1998;279:2103–2106.

Pan L-Z, Werner A, Levy JA. Detection of plasma viremia in individuals at all clinical stages. *J Clin Microbiol* 1993;13: 283–288.

Ramratnam B, Mittler JE, Zhang L, et al. The decay of the latent reservoir of replication competent HIV-1 is inversely correlated with the extent of residual viral replication during prolonged anti-retroviral therapy. *Nature Med* 2000;6:82–86.

Rosenberg ES, Altfeld M, Poon SH, et al. Immune control of HIV-1 after early treatment of acute infection. *Nature* 2000;407:523–526.

Schmitz JE, Kuroda MJ, Santra S, et al. Control of viremia in simian immunodeficiency virus infection by CD8$^+$ lymphocytes. *Science* 1999;283:857–860.

Sharkey ME, Teo I, Greenough T, et al. Persistence of episomal HIV-1 infection intermediates in patients on highly active anti-retroviral therapy. *Nature Med* 2000;6:76–81.

Shibata R, Igarashi T, Haigwood N, et al. Neutralizing antibody directed against the HIV-1 envelope glycoprotein can completely block HIV-1/SIV chimeric virus infections of macaque monkeys. *Nature Med* 1999;5:204–210.

Stahl-Hennig C, Steinman RM, Tenner-Racz K, et al. Rapid infection of oral mucosal-associated lymphoid tissue with simian immunodeficiency virus. *Science* 1999;285:1261–1265.

Veazey RS, DeMaria M, Chalifoux LV, et al. Gastrointestinal tract as a major site of CD4$^+$ T cell depletion and viral replication in SIV infection. *Science* 1998;280:427–431.

Whittle HC, Ariyoshi K, Rowland-Jones S. HIV-2 and T cell recognition. *Curr Opin Immunol* 1998;10:382–387.

Zhang L, Dailey PJ, He T, et al. Rapid clearance of simian immunodeficiency virus particles from plasma of rhesus macaques. *J Virol* 1999;73:855–860.

Zhang ZQ, Schuler T, Zupancic M, et al. Sexual transmission and propagation of SIV and HIV in resting and activated CD4$^+$ T cells. *Science* 1999;286:1353–1357.

PART IV
Prevention of Viral Infections

14. Viral Vaccines

Viral pathogenesis and immune protection; comparison of vaccine modalities, including inactivated and attenuated virus vaccines, and newer vectors such as recombinant viruses and naked DNA; mechanisms of protection by established vaccines; roles of antibody and cellular immunity; the correlates of protection; sterilizing immunity vs partial protection; and challenges for an AIDS vaccine

Chapter 14
Viral Vaccines

VIRAL PATHOGENESIS AND VACCINE-INDUCED PROTECTION

The prevention of viral diseases by the use of vaccines has been one of the great successes of preventive medicine. Jenner's introduction in the late 18th century, of vaccinia immunization to ameliorate smallpox, is a landmark in public health and one of the earliest demonstrations of the basic principles of immunology. Vaccinia also provided a precedent for the use of live attenuated viruses to induce effective long-lasting protection, an example that even today inspires vaccinologists. During the last half of the 20th century a large number of safe and effective viral vaccines were developed for use in humans and animals. However, vaccine development has been largely an empirical science, and there is a paucity of data on the mechanisms of protection.

This chapter is based on the premise that vaccine-induced protection can best be understood in the context of viral pathogenesis, informed by analysis of humoral and cellular immunity. The major vaccine modalities are described, with their strengths and limitations, followed by an analysis of the mechanisms of vaccine-induced protection as exemplified by a few of the best studied vaccines. Finally, progress in development of an AIDS vaccine is summarized to illustrate one of the remaining unresolved challenges in vaccinology.

VACCINE MODALITIES

There are certain immunologic principles that govern the induction of protective responses by any vaccine modality (Sidebar 14.1). Delivery of an immunogen to professional antigen-presenting cells (APC) is the most effective way to initiate immune induction, which can be modulated to emphasize either cellular or humoral responses. There is a physiologic limit to the expansion of naïve T lymphocytes during the primary response but, once rested, committed memory lymphocytes can be restimulated to undergo further expansion (often called an anamnestic or booster response). Adjuvants can bring professional APC into contact with antigens through their proinflammatory action or exploit cytokines to increase proliferation of antigen-responsive lympho-

---

■ ■ ■
SIDEBAR 14.1

Principles of Immune Induction Relevant for Vaccine Efficacy

- Immune induction is much more efficient if an immunogen is presented by professional APCs, such as macrophages and dendritic cells.
- There is a relationship between the amount of antigen presented and the number of naïve lymphocytes that are induced to respond. The number of T lymphocytes induced during the active response determines the number of antigen-specific memory T lymphocytes that are generated.
- Following immune induction about 10–15 cell divisions occur in antigen-responsive T lymphocytes at which time there is no further proliferation. After a "rest" of weeks to months, antigen-committed T cells may then be induced to proliferate again, to produce an anamnestic immune response.
- For many viral infections, both immunoglobulin and cellular effector systems can participate in protective immunity, but their relative role varies for different viruses.
- Immune induction can be manipulated to favor either T_H1 (cellular) or T_H2 (antibody) responses, by formulation of immunogen, route of immunization, and the use of adjuvants.
- Adjuvants can enhance the immune response in a variety of ways, mediated by their proinflammatory action or by the induction of cytokines.
- Presentation of antigen to the mucosa-associated lymphoid system can induce local immunity, which may provide an effective barrier to viruses that invade via mucosal tissues. ■

---

cytes. Newer vaccine modalities attempt to exploit these immunologic principles to both enhance and focus the immune response to maximize protective efficacy.

Vaccine modalities fall into three broad categories: attenuated live viruses, nonreplicating preformed antigens, and vectors with limited replicative capacity. Each of these modalities has its advantages and disadvantages, and it is unpredictable as to which one will produce the most successful vaccine for a given viral disease (Table 14.1). The earliest vaccines were attenuated viruses that were derived using primitive methods, although in some instances molecular sequencing and virus cloning have been employed to produce improved versions. With the beginnings of experimental virology, technology was developed that led to the earli-

est nonreplicating viral vaccines, formulated by chemical or physical inactivation of virulent viruses. Further advances permitted the production of individual viral proteins that could be used as immunogens. Most recently, a variety of vector systems have been introduced to express viral proteins, and these are currently under active development as potential vaccine modalities.

Table 14.2 contains a list of the viral vaccines in common use for humans. Most of these are either live attenuated viruses or inactivated viruses since the newer vector systems are yet to be adopted as standard licensed vaccines.

Live Attenuated Viruses

Attenuated viruses produce infections that are milder than the illnesses produced by the counterpart virulent wild-type viruses from which they are derived. Attenuated variants may differ in several ways from wild-type isolates (see Chapter 8). They are often host range mutants, so that their replicative capacity—relative to their wild-type counterparts—is high in selected cell culture systems but much lower in vivo. Also, attenuated vaccine viruses are selected for differential tropism in vivo, in contrast to their virulent parents. For instance, cold-adapted influenza viruses replicate quite well at 33°C but poorly at 37°C (Table 8.5). In vivo, the cold-adapted virus replicates in the upper respiratory tract (nasal epithelium) but very little in the lower respiratory tract (alveolar epithelium), whereas the virulent virus replicates well in both sites. Attenuated oral poliovirus vaccine (OPV) exhibits a different pattern of tropism than does wild-type poliovirus because it replicates well in the gastrointestinal tract but poorly in the central nervous system (CNS), whereas wild-type virus replicates robustly in both sites (Table 14.3). Also, OPV causes little or no viremia so it rarely reaches the CNS. These properties reduce the pathogenicity of the attenuated virus while retaining its immunogenicity.

Efficacy In general, infection with a wild-type pathogenic virus, if the host survives, induces long-lasting—often lifelong—immunity that protects against illness upon reexposure to the same virus. Therefore, the immunity induced by "natural" infection has been considered the gold standard for vaccines. Because attenuated variant viruses can—under optimal circumstances—mimic the protection afforded by natural infection, they are often considered the vaccine modality of choice.

Attenuated viruses are usually effective im-

	TABLE 14.1
	Advantages and disadvantages of different vaccine modalities: attenuated viruses and nonreplicating antigens
Safety and efficacy	**Advantages and disadvantages**
	Live Attenuated Viruses
Safety	*Advantages* None *Disadvantages* Residual pathogenicity Reversion to increased pathogenicity Unrecognized adventitious agents Possible persistence
Efficacy	*Advantages* Local immunity at portal of entry Cellular and humoral immunity induction Long-lasting immune response Herd immunity Less expensive to manufacture *Disadvantages* Interference between serotypes Interference by adventitious viruses Loss of infectivity on storage Cold chain required to maintain infectivity
	Inactivated or Subunit Viruses, or Recombinant Proteins
Safety	*Advantages* Avoids dangers of attenuated viruses *Disadvantages* Potential residual infectious pathogenic virus Safety tests difficult and expensive Induction unbalanced immune response
Efficacy	*Advantages* No viral interference Avoids limitations of attenuated viruses *Disadvantages* No induction of local immunity Poor induction of cellular immunity May not mimic native epitopes for humoral immunity Short duration immunity More expensive to manufacture

munogens, assuming that they replicate sufficiently. The full cycle of replication in vivo, regardless of the specific cellular target, appears to provide sufficient protein substrate to professional APC to load both class I and class MHC (major histocompatibility complex) molecules with oligopeptides, in addition to binding immunoglobulin molecules to stimulate cognate B lymphocytes (see Chapter 5). Laboratory passage and cloning of viruses can produce variants with different degrees of attenuation as measured by the mildness of illness and the reduction in in vivo replication, and the degree of attenuation usually reflects the number of attenuating mutations. In general, immunogenicity decreases with reduction in in vivo replication, and "overat-tenuation" can lead to viruses that are poorly immunogenic.

The advantages and disadvantages of attenuated viruses as vaccines arise from the requirement for replication in the nonimmune host. For instance, OPV is subject to "interference" due to concomitant silent natural infection with other enteroviruses. As a result, when OPV is administered in mid-winter in areas with a temperate climate, a high proportion (more than 90%) of seronegative vaccinees undergo immunizing infections. In contrast, a single dose of trivalent OPV can produce a lower frequency of "takes" when administered to young children in the tropics (Table 14.4). OPV is a trivalent vaccine that includes types 1, 2, and 3 strains of virus. When OPV

	TABLE 14.2		
	Licensed viral vaccines for humans: a selected list		
Date of licensure United States	Virus Disease	Vaccine modality Route administration	Currently (2001) used in USA
Before 1900	Variola Smallpox	Attenuated Intradermal	Yes exposure only
~1939	Yellow fever	Attenuated Subcutaneous	Yes exposure only
1955	Polio Poliomyelitis	Inactivated Intramuscular	Yes all infants
1963	Polio Poliomyelitis	Attenuated Oral	Yes special circumstances
1963	Measles	Attenuated Subcutaneous	Yes all infants
1967	Mumps	Attenuated Subcutaneous	Yes all infants
1969	Rubella German measles	Attenuated Subcutaneous	Yes all infants
1971	Influenza	Inactivated Intramuscular	Yes high risk only
1980	Rabies	Inactivated Intramuscular	Yes high risk only
1981	Hepatitis B	Inactivated Intramuscular	No no longer made
1986	Hepatitis B	Recombinant HBs protein Intramuscular	Yes all infants
1995	Varicella Chicken pox	Attenuated Subcutaneous	Yes all infants
~1996	Hepatitis A	Inactivated virus Intramuscular	Yes high risk only
1999	Rotavirus Infant diarrhea	Attenuated Oral	No withdrawn

For definitive information, see Plotkin SA, Orenstein WA. *Vaccines*. Philadelphia; Saunders, 1999.

was under development, it was found that there was interference between the three types, and the relative titers had to be carefully balanced to achieve the max-

	TABLE 14.3		
	Differences in patterns of tropism between attenuated vaccine viruses and wild-type viruses		
Type 1 poliovirus strain	TCD_{50} per mL	Enterotropism TCD_{50} per PO ID_{50}	Neurotropism TCD_{50} per IC PD_{50}
Virulent Mahoney (CNS suspension)	10^6	$10^{3.3}$ (monkeys)	$10^{1.9}$
Attenuated LSc (tissue culture fluid)	$10^{7.6}$	$\sim10^4$ (humans)	$>10^{7.6}$

Attenuated vaccine viruses often exhibit a different pattern of tropism than the corresponding wild–type viruses. Virulent and an attenuated type 1 poliovirus are compared to show that they both replicate well in cell culture (primary monkey kidney cells) and are enterotropic (infectious after oral administration) but differ markedly in their neurovirulence after intracerebral injection in cynomolgus monkeys.

PD_{50}, 50% paralytic dose; ID_{50}, 50% infectious dose; TCD_{50}, 50% tissue culture dose; CNS, central nervous system; LSc, LiSchaeffer.

After Sabin AB, Hennessen WA, Winsser J. Studies on variants of poliomyelitis virus. *J Exp Med* 1954;99:551–576; Sabin AB. Properties and behavior of orally administered attenuated poliovirus vaccine. JAMA 1957;164:1216–1223.

imal number of conversions to all three types following oral administration to seronegative children.

Safety The search for an acceptable attenuated vaccine strain requires identification of variants that fall in a putative window of robust immunogenicity with minimal disease potential. In spite of diligent efforts to achieve complete safety, some attenuated vaccine viruses retain residual pathogenicity. For instance, OPV causes an occasional case of paralytic poliomyelitis, at a frequency of about 2 cases per 1 million primary immunizations (Table 14.5). In addition, some attenuated vaccine viruses may revert in virulence during passage in the primary vaccine recipient, which can be a problem if the virus is excreted. Thus, OPV often increases in virulence upon a single human passage due to revertant mutations, and in 2000 several small outbreaks of poliomyelitis were traced to reverted strains of vaccine virus.

Another problem with live virus vaccines is that they may be inadvertently contaminated with adventitious agents. For instance, yellow fever vaccine produced a massive epidemic of hepatitis B in the 1940s that was traced to a batch of vaccine that contained human serum obtained from an asymptomatic

		PERCENT SEROCONVERSION		
TABLE 14.4				
Interference of attenuated virus vaccines under various conditions				
Location	OPV	Type 1	Type 2	Type 3
Toluca, Mexico Tropics	Trivalent	68	82	43
Leningrad, USSR Winter	Trivalent	82	80	71
Leningrad, USSR Winter	Monovalent Order: 1,3,2	97	100	96

Some attenuated virus vaccines can be subject to interference under certain conditions. Frequency of seroconversion following administration of OPV to children in the tropics compared to seroconversion following OPV administration during the winter months in a temperate climate. Conversions are greater in Leningrad in the winter when other enteroviruses are relatively uncommon compared to Toluca where about 50% of children are excreting nonpolio enteroviruses at any one time. Also, type 2 interferes with the other types, and type 3 is most subject to interference. When the three types are fed individually and sequentially, the proportion immunized is greater than that elicited by trivalent vaccine.

OPV, oral poliovirus vaccine.

After Sabin AB, Alvarez MR, Amezquita JA, et al. Effects of rapid mass immunization of a population with live, oral poliovirus vaccine under conditions of massive enteric infection with other viruses. In: *Second international conference on live poliovirus vaccines*. Scientific Publication 50, Geneva: WHO, 1960.

individual who was later shown to be a carrier of hepatitis B virus (HBV) (Fig. 14.1). Another contaminant of yellow fever vaccine was avian leukosis virus, presumably acquired from the eggs used to prepare chick embryo cultures in which the vaccine virus was grown; this problem has now been eliminated by the use of leukosis-free eggs.

Inactivated Viruses, Subunit Vaccines, and Recombinant Proteins

Inactivated viruses are the other common modality among licensed viral vaccines. A number of chemical and physical methods can be used to inactivate viruses without destroying the integrity of the virus particle or much of its antigenicity. For instance, inactivated poliovirus vaccine (IPV) is manufactured by treating the virus with dilute formalin (formaldehyde gas dissolved in water) at 37°C for several weeks. The chemical treatment denatures the outer capsid protein that acts as the viral attachment protein sufficiently to prevent viral entry while retaining epitopes that induce neutralizing antibodies. β-Propriolactone is another chemical that acts in a manner similar to formalin and has been used to prepare inactivated rabies virus vaccines. An alternative is a so-called split product vaccine, produced by treatment of the virion with mild detergent or ethyl ether that dissociates the particle to yield a suspension of proteins and nucleic acids that are noninfectious but retain antigenicity. This method has been used to produce influenza virus vaccines.

Efficacy Inactivated virus products are usually injected intradermally, subcutaneously, or intramus-

		Paralytic rate per 10^6 primary infections	
TABLE 14.5			
Pathogenicity of attenuated virus variants used for vaccines			
Virus	Study period	or immunizations	Relative rates
Wild type	1931–1954	7,000	~3,000
OPV	1961–1978	2.3	1

Attenuated virus variants used for vaccines often retain some residual pathogenicity. In this example, the attenuated vaccine virus (OPV), which is highly attenuated but still causes a small number of cases of paralysis, is compared to wild-type poliovirus. In the United States, OPV is associated with about seven paralytic cases per 3.5 million primary immunizations, including cases in immunologically compromised and normal recipients, and cases in family contacts (usually parents) of immunized infants.

OPV, oral poliovirus vaccine.

After Nathanson N, McFadden G. Viral virulence. In: Nathanson N, et al., eds. *Viral pathogenesis*. Philadelphia: Lippincott–Raven Publishers, 1997.

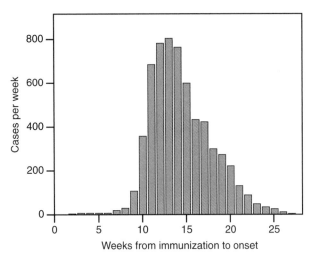

FIG. 14.1

Live attenuated vaccine viruses may be inadvertently contaminated with extraneous infectious agents capable of causing disease. In this example, a batch of yellow fever virus was contaminated with hepatitis B virus (HBV) introduced with human serum used to "stabilize" the yellow fever vaccine. Investigation showed that one of the serum donors was an asymptomatic carrier of HBV. The graph shows cases tabulated by weeks from immunization with 17D yellow fever vaccine to onset of jaundice. (After Sawyer WA, Meyer KF, Eaton MD, et al. Jaundice in army personnel in the western region of the United States and its relation to vaccination against yellow fever. *Am J Hyg* 1944;39:337–387.)

cularly. Such vaccines induce circulating antibodies but are usually poor at inducing mucosal (IgA) antibodies because the antigens are not presented to the mucosal-associated lymphoid tissue. Also, it is thought that inactivated virus vaccines are poor inducers of cellular immune reponses, although most have never been examined using modern methods to measure cellular immunity. Finally, it is often asserted that immune responses induced by inactivated vaccines are not long lasting, based on the view that continuous antigenic stimulus is required to maintain immune responses. However, in some experimental systems it appears that antigen-committed memory T cells can persist well in the absence of antigen (see Chapter 5). Furthermore, there are examples of antibody responses induced by inactivated vaccines that appear to be long lasting in comparison with the responses induced by the cognate wild-type virus.

Safety Inactivated virus vaccines are often formulated from pathogenic virus strains, and their safety is contingent on total inactivation. Successful inactivation requires a careful scientific analysis of the mechanism and kinetics of inactivation and the ability to test each vaccine batch for residual virus. On occasion, failures in inactivation have caused disease, such as occurred during the Cutter "incident," which confounded the introduction of IPV. A related problem is that "overinactivation" done to ensure safety can compromise the immunogenicity of inactivated vaccines.

On rare occasions, inactivated vaccines can induce an "imbalanced" immune response that leads to untoward effects. For instance, inactivated measles virus elicited an immune response to one of the two envelope proteins, the viral hemagglutinin, but not to the fusion protein. When children immunized in this manner were exposed to natural measles, they were not protected but developed "atypical" measles with unusual symptoms. Similarly, early attempts to test an inactivated vaccine against respiratory syncytial virus, an important respiratory virus of children, resulted in enhanced disease rather than protection.

Recombinant Proteins A modern alternative to inactivated viruses is the preparation of a recombinant viral protein for use as an immunogen. A successful example is the use of HBs, the envelope protein of HBV, as a vaccine. The original HBV vaccine was prepared by purifying the virus from the serum of human carriers because it was impossible to grow HBV in cell culture. The harvested virus was purified and inactivated with formalin. However, this process was fraught with dangers, such as the incomplete inactivation of HBV or the possibility of accidentally including extraneous agents present in donor serum. In this instance, the recombinant protein was clearly a safer product, and luckily it proved to be very effective as a vaccine. However, industrial scale production, purification, and stabilization of recombinant proteins is a daunting challenge, and such products are often expensive to manufacture.

Vectors: Recombinant Viruses, Replicons, and DNA Vaccines

In the last few years, there has been a burst of research activity dedicated to novel modes of antigen presentation, sometimes called vectors or "platforms." These new approaches include recombinant viruses, replicons, and purified DNA.

Recombinant Viruses DNA viruses are readily engineered to introduce new genetic sequences, which can also be done with RNA viruses whose genomes can be transcribed into infectious DNA clones. Although many virus genomes can be manipulated to express foreign antigens, the largest viruses, such as poxviruses and herpesviruses, are most suitable for this purpose. Poxviruses have been used more frequently than other viruses, and vaccinia virus is the basis for some licensed animal vaccines, such as a rabies virus vaccine that has been deployed for the successful immunization of wildlife.

There are distinct limits to the amount of genetic information that can be added to the genome of smaller viruses without compromising their replicative capacity. However, smaller genomes may be used successfully in specialized instances. For instance, the 17D attenuated vaccine variant of yellow fever virus can be chimerized to express the surface glycoprotein of other virulent flaviviruses, such as Japanese encephalitis or West Nile viruses, providing an immunogen with established safety that will induce neutralizing antibodies against a human pathogen. Likewise, dengue virus, another flavivirus, can be chimerized with Langat virus to produce an attenuated recombinant virus that induces protective immunity against tick-borne encephalitis viruses.

There are several considerations in selecting a replicating virus for use as a vaccine platform, including safety, immunogenicity, and prior immunity of the target population. Current safety standards make it much more acceptable to use a virus that has already had widespread use in the human populations, such as vaccinia virus. Even here there are safety problems because vaccinia causes serious complications albeit at the frequency of less than 1 case per 100,000 vaccinees. Thus, certain attenuated strains of vaccinia virus, such as modified virus Ankara (MVA) or NYVAC (an attenuated strain of vaccinia virus), are preferred to standard vaccinia virus.

The immunogenicity of a recombinant virus depends in part on the cells that it targets. Some viruses infect macrophages and dendritic cells, and this maximizes their ability to deliver proteins to professional APC, thereby enhancing the immunogenicity of the recombinant proteins that they encode. Since many recombinant constructs are based on human viruses, vaccinees may have been previously infected with the wild-type counterpart, and this pre-existing immunity can reduce the replication of the recombinant virus and compromise its immunogenicity. For instance, recombinant vaccinia viruses are somewhat less immunogenic in persons who were previously vaccinated than in vaccinia-naïve subjects.

Replicons Replicons are virus-like particles that enter a target cell, undergo limited transcription and translation to synthesize encoded proteins, but do not produce infectious progeny. Replicons consist of a virus genome that has been engineered to insert a new protein and to delete some of the genes of the parent virus. Such genomic constructs are often transfected into packaging cell lines that provide a viral envelope in trans, permitting the assembly of a virus-like particle with the cellular specificity associated with the envelope. In contrast to recombinant viruses, replicons cannot spread beyond the cells that they initially "infect." Replicons are a lower risk platform than recombinant viruses and can exploit the attributes of many wild-type viruses that would be unacceptable for use as an infectious recombinant virus.

One replicon now under test as a potential HIV immunogen is a modified Venezuelan equine encephalitis virus (VEEV) that has shown the ability to induce cellular immunity to simian immunodeficiency virus (SIV) or simian human immunodeficiency virus (SHIV) in monkeys. Immunized animals exhibit considerable protection compared with naïve control animals challenged with the same pathogenic SIV or SHIV.

The efficacy of replicons depends on their ability to reach a sufficient number of target cells, to produce enough novel immunogen, and to deliver the immunogen to professional APC in regional lymphoid tissues. In addition, it may be difficult to produce certain replicons on the industrial scale needed for vaccine deployment. Finally, replicons must pass safety tests based on the assurance that they will not recombine with cellular sequences to reconstitute the potentially pathogenic viruses from which they are derived. Only future investigation will determine whether replicons are a practical platform for vaccine formulation.

DNA-Based Immunogens It was first discovered in the early 1990s that a DNA plasmid, encoding a protein, could be used as an immunogen by simple injection of the "naked" DNA. This novel technology is currently under active investigation. DNA vaccine plasmids usually utilize a promoter such as the cytomegalovirus promoter, which is highly active in most eukaryotic cells, driving a genetic insert expressing the protein of interest, followed by a transcriptional terminator and a polyadenylation sequence. Modifications of the protein sequence, such as a signal sequence or a transmembrane domain, can be used to influence how the protein is processed in APC. If the protein is secreted then it may be targeted to the MHC class II exogenous pathway while if it is retained in the cytosol it can be targeted to the MHC class I endogenous pathway.

DNA constructs are usually administered to the skin or muscle. DNA bound to small gold beads can be administered intradermally by gene gun or injected into the epidermis as an aqueous suspension. The DNA enters epithelial cells, dendritic cells, and macrophages, traffics to the nucleus, and is transcribed, although the steps in this process are not well understood. To be immunogenic, the DNA-encoded protein must be presented by professional APC, based on experiments with chimeric mice. Proteins expressed in epithelial cells would be taken up by APC via the exogenous pathway, whereas proteins expressed in APC could enter the endogenous pathway. Gene gun injections induce responses with less DNA than is required for soluble DNA, but tends to induce T_H2 responses biased toward antibody and not to cellular immunity.

After intramuscular injection, the DNA is deposited in the extracellular space and may be transported directly to the draining lymph nodes where it is taken up by APC. In addition, injected DNA is taken up by muscle cells, but it is unclear whether myocyte expression is required for immune induction, which can occur even if the injected muscle is excised as early as 10 minutes after injection. Unmethylated CpG motifs in plasmid DNA provide an adjuvant potentiation of DNA.

DNA has been used for single or multiple immunizations or in concert with other vaccine platforms. DNA used to prime an immune response followed by boosting with a recombinant poxvirus expressing the same immunogen has been particularly successful in eliciting immune responses for several infectious agents (Fig. 14.19). Also, DNA can be adjuvanted with plasmids encoding cytokines, such as interleukin-2 (IL-2) (Fig. 14.18).

As a vaccine, DNA possesses several advantages. First, it represents a well-defined and stable immunogen that can be precisely characterized and controlled, as well as produced on a large scale at relatively low cost. It appears to be biologically safe assuming that it is adequately purified, and it avoids some of the dangers intrinsic in attenuated viruses, inactivated viruses, and certain vectors.

MECHANISMS OF PROTECTION BY ESTABLISHED VACCINES

A large number of viral vaccines have been developed, licensed, and are in use for the prevention of

disease in humans (Table 14.1). These successful established products provide a potential source of information about the ways in which a vaccine confers protection. To elucidate these mechanisms, information on several of these vaccines is reviewed below. Since most relevant research was conducted decades ago, prior to the introduction of modern methods for the measurement of cellular immune responses, the data focus mainly on the role of antibody in vaccine-induced protection. Keeping this caveat in mind, a few generalizations can be made about the mechanism of vaccine-induced protection (Sidebar 14.2).

Poliovirus

The pathogenesis of poliovirus is understood at an organ level, although many of the specific cellular details have never been elucidated. As shown in Figure 14.2, the virus is ingested, and invades via the tonsils and the lymphoid tissue of the small intestine, spreads to regional lymph nodes, and is transmitted through efferent lymphatics to the blood, where it circulates as a cell-free plasma viremia. Blood-borne virus invades the CNS either directly across the blood–brain barrier or indirectly by invading peripheral nerves or peripheral ganglia followed by neuronal spread to the CNS. When po-

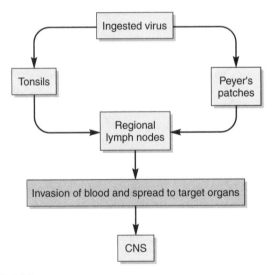

FIG. 14.2

Pathogenesis of poliovirus infection. This diagrammatic summary indicates that the virus invades via tonsils and Peyer's patches (lymphoid tissue accumulations in the walls of the small intestine), spreads to regional lymph nodes, produces a plasma viremia, and then invades the central nervous system. A variant of this scheme, not shown, suggests that the virus invades the peripheral nervous system from the blood and travels thence to the CNS. (After Bodian D. Emerging concept of poliomyelitis infection. *Science* 1955;122:105–108.)

> ■ ■ ■
> ### SIDEBAR 14.2
>
> #### Mechanisms of Vaccine-Induced Protection Against Viral Diseases: Some Tentative Principles
>
> - The mode of vaccine-induced protection can be best understood in the context of the pathogenesis of a specific viral infection.
> - Immune mechanisms involved in preexposure protection may be different from those involved in recovery from primary infection.
> - Vaccine-induced protection may be due to a combination of several protective mechanisms rather than a single component of the immune response. Different mechanisms may be involved in protection against different viruses.
> - Neutralizing antibody, if present at the portal of entry and at the time of infection, can act more rapidly than any other defense to inactivate the challenge virus.
> - Cellular immune responses, mediated by a variety of cell types and acting through diverse mechanisms, may contribute to protection, particularly if the challenge virus is not "sterilized" at the portal of entry. ■

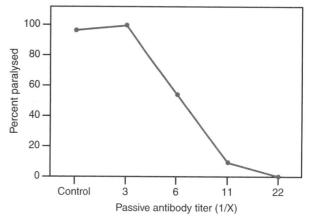

FIG. 14.3

Passive antibody protects against paralytic poliomyelitis. In this experiment cynomolgus macaques were given a single intramuscular injection of graded doses of poliovirus immune immunoglobulin. One day later the level of neutralizing antibody was determined and the animals were challenged with the virulent Mahoney strain of type 1 poliovirus by the intramuscular route. An antibody titer of approximately 1:10 conferred more than 90% protection. (After Nathanson N, Bodian D. Experimental poliomyelitis following intramuscular virus infection. III. The effect of passive antibody on paralysis and viremia. *Bull Johns Hopkins Hosp* 1962;111:198–220.)

liovirus infection is followed sequentially, after injecting a virulent wild-type virus into macaques (Fig. 2.5), viremia is observed for about a week, followed by the appearance of neutralizing antibody, simultaneous with the disappearance of infectious virus. These observations suggest that a potential weak link in the pathogenesis is the transit of virus through the blood and that antibody might be capable of blocking that step and preventing invasion of the target organ.

To test the hypothesis that neutralizing antibody might protect against the paralytic consequences of poliovirus infection, experiments can be performed with passive immunization. Monkeys are injected with graded doses of a pool of anti-poliovirus antiserum. One day after injection, the titer of passive antibody in the serum of the recipient animals is measured, and the animals are challenged with a dose of virulent poliovirus. The results provide clear evidence that antibody, present prior to virus challenge, protects against paralysis (Fig. 14.3). This strongly suggests that a vaccine that elicited neutralizing antibody might protect against paralytic poliomyelitis.

These considerations led to the formulation by Jonas Salk of an inactivated preparation of poliovirus (IPV) as a candidate immunogen. The 1954

Field trial of IPV provided an opportunity to test the hypothesis that neutralizing antibody could account for protection. The proportion of vaccinees, seronegative prior to immunization, who responded with different levels of antibody was compared with the estimated efficacy of the vaccine (Table 14.6). There was a good correlation between the proportion of vaccinees who responded at a titer of 1:4 or greater and the estimated efficacy of the vaccine (approximately 65%). This correlation suggested that a minimal level of neutralizing antibody could account for protection.

When attenuated strains of poliovirus were licensed as an OPV it became possible to compare the ability of IPV and OPV to prevent enteric infection, a different aspect of vaccine-induced protection. Such a comparison indicated that IPV conferred minimal protection against enteric infection but that OPV reduced fecal excretion significantly (Fig. 14.4). The ability of OPV to induce mucosal immunity was confirmed by demonstrating low levels of anti-poliovirus IgA in fecal samples from subjects immunized with OPV. It seems likely that OPV generates local immunity by inducing antibody production by B cells in the gastrointestinal-associated lymphoid tissue (GALT), although there is little direct evidence for this speculation.

The widespread use of poliovirus vaccines has led to the eradication of wild-type polioviruses from

TABLE 14.6

Correlation of serum neutralizing antibody with protection against paralytic poliomyelitis

Immunization status	Parameter	Number
Unvaccinated	Paralytic cases observed	40
Vaccinated	Percent <1:4	35.4%
	Paralytic cases expected at <1:4	14.2
	Paralytic cases observed	14

A low level of serum neutralizing antibody correlates with protection against paralytic poliomyelitis. In 1954, a large-scale clinical trial was done to test whether inactivated poliovirus vaccine would protect against poliomyelitis. To determine the level of antibody that correlated with protection, the proportion of children with seroconversions at different titers were compared with the proportion protected by the vaccine. At a level of 1:4 it appeared that about 35% failed to convert, leading to a prediction that the paralytic rate would be reduced by about 65% in vaccinated compared with placebo control children. Since the observed number of cases was similar to that predicted, it was concluded that a titer of about 1:4 probably was sufficient to protect against paralytic disease.

After Francis TJ Jr, Napier JA, Voight R, et al. Evaluation of the 1954 field trial of poliomyelitis vaccine. School of Public Health, University of Michigan, Ann Arbor, Michigan, 1957.

the United States (about 1980) and from the rest of the Western Hemisphere (about 1995). By any standard, this is an impressive success. Can this efficacy be attributed entirely to neutralizing antibody? Information relevant to this question has been derived from the cumulative experience with OPV over the last 40 years. It was noted above that OPV causes rare cases of poliomyelitis in vaccine recipients (about 1 per million in vaccine recipients). Some of these cases are in children with inherited immunodeficiencies that were unrecognized at the age (1–6 months) at which the child was immunized. Uni-

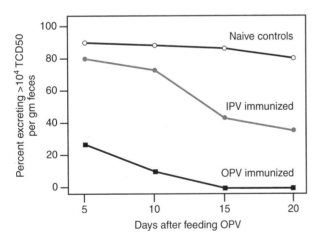

FIG. 14.4 _____

Oral poliovirus vaccine (OPV) provides greater protection against enteric infection than does IPV. Three groups of children (naïve unimmunized; immunized with OPV; immunized with IPV) were tested for fecal excretion after feeding of OPV. Since the serum antibody levels were similar in the OPV and the IPV immunized groups, it was inferred that OPV induced local mucosal immunity more effectively than IPV. (After Henry JL, Jaikara ES, Davies JR, et al. A study of poliovaccination in infancy: excretion following challenge with live virus by children given killed or living poliovaccine. *J Hyg* 1966;64:105–120.)

formly, these children have been diagnosed as hypo- or agammaglobulinemic; strikingly, children with inherited T-cell defects (such as the diGeorge syndrome) do not seem to be at risk of vaccine-associated poliomyelitis. Furthermore, some of the children with immunodeficiency-associated poliomyelitis continue to excrete the vaccine virus for months to years, in spite of the fact that many of them are treated with pooled normal immune globulin (by intramuscular injection).

These observations imply that clearance of enteric poliovirus replication is mediated by antibodies. Furthermore, the failure of systemic globulin treatment (which would provide circulating neutralizing poliovirus antibodies) to clear infection suggests that antibody produced by the GALT is important for the clearance of enteric infection. The absence of any data on the development of cellular immune responses to poliovirus vaccines precludes definitive conclusions, but there is little suggestion that CD8-mediated mechanisms play a role in protective immunity against poliovirus.

Rabiesvirus

Rabies presents a special challenge for immunization in part because of its unusual pathogenesis and in part because it is one of the few infections where postexposure vaccination is frequently used. Rabiesvirus is often acquired through the bite of a rabid animal (Fig. 1.1). Following injection into muscle or other peripheral site, the virus replicates locally, crosses the neuromuscular junction, and travels by the neural route to the CNS where it produces an invariably fatal encephalomyelitis (Fig. 2.2). Importantly, rabiesvirus never produces a viremia.

One peculiar aspect of rabies pathogenesis is the

variability in the incubation period. The virus may transit to the CNS within a few days or may be sequestered in an extraneural site for weeks to months before it invades the nervous system. The length of the rabies incubation period is determined by a variety of parameters, particularly the strain of virus. Thus, a neuroadapted rabiesvirus, challenge virus standard (CVS), produces rabies with a high frequency and a short incubation period, whereas a freshly isolated wild-type strain (so-called "street" virus) usually produces a lower frequency of "takes" and a much longer incubation period. Other parameters that influence incubation period are virus dose and experimental host.

The long incubation period following exposure to street rabiesvirus provides the opportunity for postexposure prophylaxis. In the United States, preexposure vaccination is limited to veterinarians or others who are at occupational risk. Because the general population is not routinely immunized, postexposure prophylaxis is the major mode of rabies prevention. The protective mechanisms of pre- and postexposure prophylaxis are probably somewhat different and are considered separately.

Preexposure Prophylaxis It appears that neutralizing antibody plays an important role in preexposure prophylaxis. Passive administration of antibody protects animals against subsequent challenge with rabiesvirus, the degree of protection being correlated with the titer of antibody, the timing of administration, and the strain and dose of rabiesvirus used for infection. If a group of animals are immunized with a rabies vaccine and tested for antibody just before challenge by injecting rabiesvirus at a peripheral site, there is a strong correlation between antibody

titer and the degree of protection (Table 14.7). Additional evidence for the protective role of antibodies is based on the protective efficacy of rabiesvirus monoclonal neutralizing antibodies. Also, vaccinia recombinant viruses or DNA constructs that express only the envelope glycoprotein provide excellent protection, which is proportional to the titer of neutralizing antibody.

It is likely that antibody acts at several different levels: at the site of virus injection, at the neuromuscular junction, and even within the CNS. Thus, if antiserum is applied at the site of rabiesvirus injection, it will reduce mortality. Specific depletion of antibody responses, by treatment with anti-μ antiserum, will potentiate intracerebral infection with an attenuated nonlethal rabiesvirus, implying that antibody can even reduce transsynaptic transmission within the CNS.

Postexposure Immunization Passive antibody, given shortly after infection with street rabiesvirus, does not reduce overall mortality but does prolong the incubation period (Fig. 14.5). Active immunization, begun just after infection with street rabiesvirus, will reduce overall mortality, and passive antibody will synergize this protective effect, reducing mortality even further. This synergistic effect is likely due to the ability of antibody to delay virus spread thereby providing the host an advantage in the "race" between the virus and induction of an active immune response.

Since active immunization elicits both antibody and cellular immune responses, which is responsible for the protective effect? Likely, both arms of the immune response play a role. If mice are immunized by intracerebral vaccine administration and chal-

TABLE 14.7			
Correlation of protection against rabies by preexposure immunization and neutralizing antibody levels at time of challenge			
Immunization status	Neutralizing antibody titer at challenge	Mortality Dead/total	Mortality (%)
Unimmunized	<2	14/17	82
Immunized	<2	8/10	80
	3–9	2/5	40
	10–99	4/18	22
	100–999	0/21	0
	>1,000	1/13	8

Protection against rabies conferred by preexposure immunization correlates with neutralizing antibody levels at the time of challenge. Monkeys were immunized with rabies vaccine and were then challenged intramuscularly with 10^5 mouse IC LD$_{50}$ of street rabies virus. Neutralizing antibody >1:100 was associated with about 100% protection, whereas lower titers were associated with partial protection.

After Sikes RK, Cleary WF, Koprowski H, et al. Effective protection of monkeys against death from street virus by post-exposure administration of tissue-culture rabies vaccine. *Bull WHO* 1971;45:10.

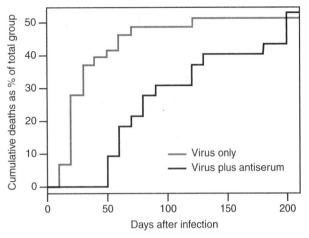

FIG. 14.5

Anti–rabies virus antiserum prolongs incubation period without reducing mortality. Two groups of mice were infected by foot-pad injection of a street rabies virus with a long incubation period (17–120 days), and one group was given an intramuscular dose of a high titer equine anti–rabies virus antiserum (calculated to provide a neutralizing titer in the recipient animal of about 1:1,000) 1 day after infection. The cumulative mortality in each group is plotted as a percentage of all the animals in the group. At 50 days, the treated group had experienced little mortality (compared to about 35% mortality in the virus-only group), but eventually the cumulative mortality rose to about 50% in both groups. (After Baer GM, Cleary WF. A model in mice for the pathogenesis and treatment of rabies. *J Infect Dis* 1972;125:520–527.)

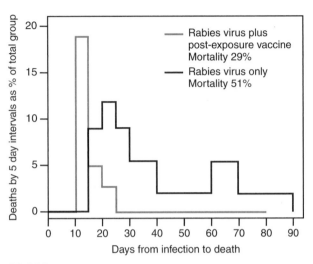

FIG. 14.6

Postexposure immunization against rabies can produce early death as well as partial protection. Mice were infected with 3,000 intracerebral LD_{50} of street rabies virus. One group (43 animals) was untreated (mortality 51%), whereas the other group was immunized with inactivated rabies virus vaccine 1 day after infection (mortality 29%). Although the vaccine clearly reduced mortality, it also accelerated the time to death in animals that were not protected. (After Baer GM, Cleary WF. A model in mice for the pathogenesis and treatment of rabies. *J Infect Dis* 1972;125:520–532.)

lenged by the same route, there is a correlation between the degree of protection and cytotoxic T lymphocyte (CTL) activity (Table 14.8). Although this is a somewhat contrived experimental system conducted with primitive CTL assays, it does suggest that, in addition to antibody, CD8 cells can play a role in the outcome of rabiesvirus infection of the CNS.

Finally, rabies immunization illustrates a much discussed but probably rare phenomenon, that is, immune-mediated disease enhancement by use of a vaccine (Fig. 14.6). In this example, not only

were the number of long incubation period cases markedly reduced, but there was an absolute increase in short incubation period cases. The excess of short incubation period cases implies immune enhancement, although the mechanism awaits definitive elucidation.

Hepatitis B Virus

The pathogenesis of HBV is characterized by a number of unusual features. The virus replicates mainly—perhaps exclusively—in the liver where it produces a large amount of viral envelope (HBs) protein ($>10^{10}$ filamentous and spherical particles per milliliter of plasma) together with a smaller

TABLE 14.8

Correlation of cytolytic T-cell response with postexposure protection against rabies.

Virus	Vaccine	^{51}Cr release (%) Day 7	Neutralizing antibody Day 7	Mortality (%)
Day 0	None	0	120	100
Day 0	Day −1	28	2,000	0
Day 0	Day 0	19	2,000	14
Day 0	Day +1	0	1,000	100

Mice were infected intracerebrally by 20 LD_{50} of street rabiesvirus and received a single intracerebral injection of inactivated rabies virus vaccine at different intervals. Protection correlated with the level of ^{51}Cr release among animals all of whom had similar levels of neutralizing antibody.

From Wiktor TJ. Cell-mediated immunity and postexposure protection from rabies by inactivated vaccines of tissue culture origin. *Dev Biol Stand* 1978;40:255–265.

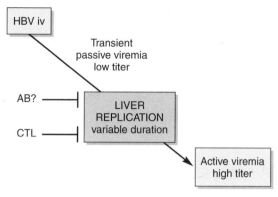

FIG. 14.7

Pathogenesis of hepatitis B, showing that virus enters the blood via transfusion or contaminated needle, and travels directly to the liver, where it replicates and produces an active viremia. In a proportion of infected persons, particularly infants, hepatitis B virus initiates a persistent infection with continuous release of high titers of virions into the blood. In adults, acute self-limited infections are common, with clearance of the virus often accompanied by acute hepatitis. It appears likely that clearance is mediated by cytotoxic T lymphocytes and not by antibody.

number (10^6 per milliliter of plasma) of infectious virions (Fig. 14.7). The course of acute hepatitis B is shown in Figure 14.8, with replication in the liver, rising levels of circulating HBs, induction of an immune response that leads to waning of HBs, and the appearance of anti-HBs antibodies. The resolution of infection is accompanied by acute hepatitis that ranges from subclinical to severe or even fatal. The timing of events suggests that HBV is not cytopathic

and that the acute hepatitis is caused by the cellular immune response (see Chapter 6). An alternative course of infection, the persistent carrier state, is seen frequently in infants infected during birth. Such persistent infections are not accompanied by acute hepatitis, strengthening the view that infection alone does not cause hepatitis. However, neonatal infection carries a high risk of cirrhosis and hepatocellular carcinoma, which develop decades later.

Evidence for the immune mediation of viral clearance is provided by the use of a transgenic mouse model, in which mice express one or several HBV proteins in the liver. When these animals are adoptively immunized with HBs-specific T lymphocytes, the viral protein is cleared from hepatocytes (Fig. 14.9) but treatment with anti-HBs antibody has no effect. CD8-initiated viral clearance is mediated by cytokines (interferon γ and tumor necrosis factor α) secreted by effector cells that inhibit HBs expression rather than by cytolysis, explaining how it occurs in the absence of overwhelming hepatitis (Fig. 5.11). By inference, it is likely that persistent infection represents a state of HBs immune tolerance due

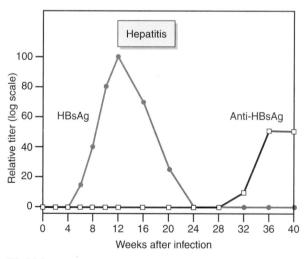

FIG. 14.8

Course of hepatitis B virus (HBV) infection. In this example, HBV produces an acute infection, with clinical hepatitis, following which the virus is cleared and the illness is resolved, leaving antibodies against HBV surface protein (anti-HBsAg) as a permanent footprint of infection.

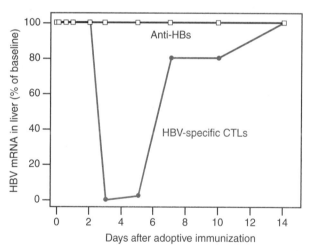

FIG. 14.9

Clearance of hepatitis B virus (HBV) infection is mediated by CD8 T lymphocytes and not by anti-HBs antibodies. In this transgenic model, HBs is expressed in hepatocytes and is shed into the blood. Congenic nontransgenic mice are immunized with HBs and used as a source of T-cell clones specific for HBs as well as for anti-HBs antisera. T cells clear the liver although only transiently, presumably due to the exhaustion of the transferred immunocytes. Other experiments indicate that clearance is mediated by cytokines (interferon γ and tumor necrosis factor α) secreted by the CD8 cells and not by cytotoxic T-lymphocyte–mediated cytolysis, explaining how the mice survive the clearance process (see Fig. 5.11). (After Guidotti LG, Ando K, Hobbs MV, et al. Cytotoxic T lymphocytes inhibit hepatitis B virus gene expression by a noncytolytic mechanism in transgenic mice. *Proc Natl Acad Sci USA* 1994;91; 3764–3768.)

TABLE 14.9			
Protection of infants against perinatal hepatitis B virus carriage by passive antibody (hepatitis B immune globulin) treatment at 0, 3, and 6 months			
HBs antigen *Anti-HBs antibody*	HBIG treated (57) (%)	Placebo (61) (%)	Protection
No HBs *No anti-HBs*	11	—	Total
HBs transient *Anti-HBs*	58	—	Partial
HBs persistent *No anti-HBs*	26	95	None
No HBs *No anti-HBs*	5	5	Not exposed

HBIG conferred varying degrees of protection, ranging from complete ("sterilizing" immunity) to none, with the majority of subjects being partially protected.

HBs, hepatitis B surface antigen; HBIG, hepatitis B immune globulin.

After Beasley RP, Hwang LY, Stevens CE, et al. Efficacy of hepatitis B immune globulin for prevention of perinatal transmission of the hepatitis B virus carrier state: final report of a randomized double-blind placebo-controlled trial. *Hepatology* 1983;3:135–141.

to "exhaustion" or "deletion" of HBs-reactive CD4 and/or CD8 clones.

There is an effective HBV vaccine that consists of a recombinant form of the HBs expressed in a eukaryotic cell system. It is often asserted that this vaccine protects by inducing anti-HBs antibodies that neutralize the virus. This is based in part on the protective effect of hepatitis B immune globulin (HBIG). In some developing countries, a high proportion of women of childbearing age are chronic carriers of HBV and frequently transmit infection to their infants during delivery, with the subsequent development of persistent infections at a high frequency. If infants are given repeated doses of HBIG beginning at birth, several effects are observed: a small proportion (about 10%) are totally protected from infection, and a large proportion (about 60%) undergo a short-term self-limited immunizing infection (Table 14.9). When this latter group is followed sequentially, circulating levels of passive antibody are seen for the first 6 months; when antibodies wane, there is a transient appearance of HBs antigenemia at 6–9 months, followed by the development of an active anti-HBs antibody response (Fig. 14.10).

Since HBV infection is probably confined to the liver, it appears unlikely that passive antibody alone could clear the established liver infection that is signaled by antigenemia at 6–9 months. More likely, passive antibody reduces the initial infecting inoculum in the early postpartum period, and this tips the balance so that the newly infected host is now able to mount an active response that clears the infection. From studies on transgenic mice—quoted above—it

appears likely that clearance requires participation of the host's effector T lymphocytes. This scheme of immunization resembles an outmoded form of protection called "passive–active" immunization, that employed passive antibody to reduce the dangers of active immunization with a pathogenic virus.

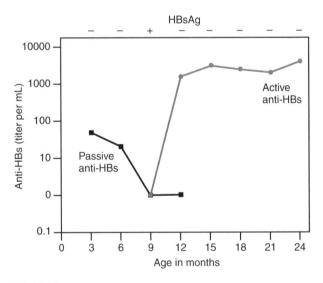

FIG. 14.10 _____

Protection of infants against perinatal hepatitis B virus carriage by passive antibody (hepatitis B immune globulin) at 0, 3, and 6 months. This figure shows a subgroup of infants who developed a transient antigemia and then an active antibody response after the termination of passive immunization. (After Beasley RP, Hwang LY, Stevens CE, et al. Efficacy of hepatitis B immune globulin for prevention of perinatal transmission of the hepatitis B virus carrier state: final report of a randomized double-blind placebo-controlled trial. *Hepatology* 1983;3: 135–141.)

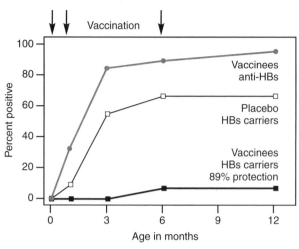

FIG. 14.11

Prevention of the hepatitis B virus carrier state in vaccinated infants. Babies born to mothers who were hepatitis B surface antigen (HBs) carriers were divided into placebo and vaccinated groups, and the vaccinees were given a whole virus inactivated vaccine at 0, 1, and 6 months. Immunization induced anti-HBs antibodies that appeared at 1–3 months postpartum. In spite of the delayed appearance of antibodies, immunization markedly reduced the proportion of infants who became carriers. (After Xu ZY, Liu CB, Francis DP, et al. Prevention of perinatal acquisition of hepatitis B carriage using vaccine: preliminary report of a randomized double-blind placebo-controlled and comparative trial. *Pediatrics* 1985;76:713–718.)

With this reconstruction in hand, we turn to data on the active immunization of infants against HBV (Fig. 14.11). When infants born of mothers who are HBV carriers are immunized with an inactivated whole virus (or recombinant HBs) vaccine at birth and thereafter, a remarkable result is seen. Once again, a high proportion (about 90%) of infections are "converted" from persistent to acute and self-limited, which is surprising because the active response to the vaccine only appears 1–3 months after birth (i.e., 1–3 months after infection). This strongly implies that HBV infection is established in the liver and is subsequently cleared. Again, it seems likely that a host cellular immune response—elicited by either the vaccine or the active infection—plays a role in vaccine-induced protection. In summary, a speculative reconstruction invokes the synergistic cooperation of both humoral and cellular immunity to explain the efficacy of HBV vaccine.

AIDS VACCINE

History and Challenges

The effort to develop an AIDS vaccine represents a landmark chapter in the history of viral vaccines. On the one hand, AIDS is arguably the greatest public health challenge of the first part of the 21st century and it is broadly held that any effective strategy to control this global pandemic will depend on an effective vaccine. On the other hand, development of a safe and effective vaccine presents a much more difficult scientific challenge than any prior viral vaccine. This account focuses on the immunobiological issues that must be resolved in order to formulate a successful AIDS vaccine.

When HIV was identified as the cause of AIDS in 1983/84 it was immediately recognized that vaccine development was an important priority. Yet 15 years later we lack a vaccine despite the greatest investment that has ever been made in vaccine research. How did we blunder into this impasse? There are several reasons for the present dilemma. Initial efforts were premised on the assumption that it would be possible to formulate an AIDS vaccine based on past successes. Almost all successful viral vaccines have been formulated as either a live attenuated variant virus, or an inactivated virus or viral protein.

HBV vaccine, introduced in 1986, is formulated as a recombinant form of the major glycoprotein of HBV. Coming to fruition concomitant with the isolation of HIV, the outstandingly successful HBV vaccine suggested that a similar approach could be applied to HIV, a virus that also has a single viral attachment glycoprotein. Implicit in this approach is the assumption that immunization with the protein will induce neutralizing antibodies. Unfortunately, HIV did not follow the pattern of HBV, for several reasons. First, both natural infection with HIV or immunization with recombinant gp120 induces antibodies, but these have little or no ability to neutralize the virus. Second, the recombinant HIV envelope protein is a monomer that does not readily form trimers, which are the native macromolecular form found on the mature virion. The monomer lacks some of the conformational neutralizing epitopes found on the trimer and induces antibodies that will bind the monomer but not neutralize infectious HIV. Although the consensus opinion in the scientific community is pessimistic, there is an ongoing phase III efficacy trial of monomeric gp120 that should provide a more definitive test of its possible efficacy.

The other factor, attenuated variant virus, has also been explored in some depth but the results have been disappointing. A naturally attenuated strain of HIV was discovered in Sydney, Australia, to have infected a cohort of about ten persons who had been infected via blood transfusions. This variant virus had a major deletion in the *nef* gene, which is known to attenuate SIV. For about 10 years the "Sydney cohort" followed a benign course similar to that seen in nonprogressors infected with HIV, producing optimism about the outlook for a safe attenuated HIV variant. However, between years 10–15 after infection most members of the Sydney cohort began to lose their CD4 cells and showed other evidence of incipient albeit long-incubation-period AIDS. This experience has had a chilling effect on the outlook for a safe attenuated HIV strain because it would appear impossible to conduct a safety test of any attenuated HIV vaccine candidate in human volunteers.

Thus, more than 10 years after isolation of HIV, it has become clear that an empirical approach to the development of an AIDS vaccine has failed. Furthermore, there is a paucity of information about the mechanisms by which established effective viral vaccines confer their protection. The remainder of this section focuses on the immunobiological issues important for the development of an AIDS vaccine.

The Macaque Model

SIV denotes a group of viruses isolated from various monkey species. Some of the SIVs are pathogenic for macaques, in which they cause an infection and disease that closely resembles AIDS in humans. Furthermore, it is possible to construct recombinants between SIV and HIV that have the HIV *env* gene inserted into an SIV genetic backbone (SHIV), and some of these viruses also cause AIDS. There are a number of variables that influence the degree of pathogenicity of SIV or SHIV, including viral variation, dose and route of infection, and animal-to-animal variation. Thus, it is possible to simulate the variety of courses of infection that are seen in humans. Data from the macaque model has become critical for the development of an AIDS vaccine and for analysis of mechanisms of protection.

Will Partial Immunity Protect Against AIDS?

Natural infection with wild-type virus provides excellent long-lasting protection against rechallenge with the same virus and constitutes the "gold standard" for immune-mediated prophylaxis. However, even natural infection does not provide "sterilizing" immunity, which is rarely seen with effective vaccines (for an exception, see one subgroup in Table 14.9). It may be questioned whether an AIDS vaccine that produces only partial protection will prevent the occurrence of AIDS following exposure to

wild-type HIV. Figure 14.12 illustrates the problem. Vaccines that protect against acute virus diseases need only modulate the degree of spread or replication for a limited period of days to weeks to prevent disease. An AIDS vaccine of similar efficacy might reduce the viral set point but would not prevent the persistence of infection. Would this down-modulated but persistent infection eventually cause AIDS?

The issue is illustrated with an experimental vaccine tested in the macaque model (Fig. 14.13). The immunized animals showed a clear-cut reduction of about 100-fold in their viremia titers relative to control monkeys, but SIV persisted. This is analogous to differences seen in human cohort studies such as illustrated in Figure 14.14, in which HIV-infected subjects were divided into four quartiles according to their viral set point at 6–9 months after infection, and their AIDS-free survival plotted. In the quartile with the highest titer 90% had developed AIDS in 10 years whereas only 10% of the quartile with the lowest set points had developed AIDS in 10 years. The differences in set points between these two quartiles were 100-fold, similar to what was seen in vaccinated macaques. HIV-2 is also illustrated in Figure 14.13 to show that with an even lower set point a considerable proportion of infected persons do not develop AIDS within their lifetimes. At present, we do not know which paradigm might apply to an AIDS vaccine.

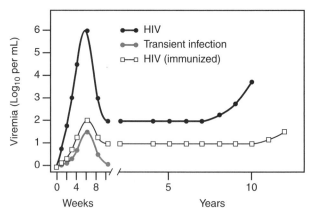

FIG. 14.12

The dynamics of HIV infection, illustrating the challenges to a prophylactic vaccine. Most viral vaccines are directed against acute infections, and an immunized subject is successfully protected if the extent of replication is reduced for a period of days or a few weeks. However, reduction of acute HIV infection will not prevent the establishment of a persistent infection. In contrast to many other persistent infections that can be innocuous, HIV may eventually erode the CD4 lymphocyte population sufficiently to produce immunodeficiency, albeit with an extended incubation period.

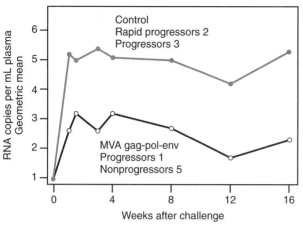

FIG. 14.13 _____

Partial protection by an experimental simian immunodeficiency virus (SIV) vaccine. Macaques were immunized with a recombinant vaccinia virus (modified virus Ankara strain) expressing the major proteins of SIV, and were challenged intravenously with a virulent SIV strain (E660), together with unvaccinated control animals. Immunization did not protect the animals against infection but did reduce the level of viremia and ameliorate the course of disease. (After Ourmanov I, Brown CR, Moss B, et al. Comparative efficacy of recombinant modified vaccinia virus Ankara expressing simian immunodeficiency virus (SIV) Gag-pol and or Env in macaques challenged with pathogenic SIV. *J Virol* 2000;74:2740–2751.)

Correlates of Protection

In the development of a vaccine, it is very useful to have an immunologic correlate of protection, which can be used to evaluate different vaccine formulations and immunization schedules. A vigorous search for such a correlate has been made in the macaque model, and a large literature is replete with conflicting claims. Perhaps the assumption that there will be a single correlate of protection is mistaken, explaining the inconsistency in claims. Instead, is it possible that several different immune modalities may synergize to provide protection (the "barrier" hypothesis)?

Synergy between CD4, CD8, and B lymphocytes has been documented in some animal models, particularly one involving Friend leukemia virus. The SHIV macaque model also provides data to support the barrier hypothesis because it is possible to protect in at least two different ways, either with preinfection with an attenuated strain of SIV or by immunization with HIV envelope. SIV must protect by cellular immunity directed against the non-env proteins, whereas HIV presumably protects by neutralizing antibody directed against the env proteins, perhaps supplemented with anti-env CD8 lymphocytes. Clearly, these are distinctly different modes of protection against a single virus challenge.

Evidence that both cellular immunity and antibody could contribute to an effective AIDS vaccine comes from the macaque model. If monkeys persistently infected with SIV are depleted of their CD8 cells, their viremia levels rise rapidly only to drop when CD8 cells are reconstituted (Fig. 14.15), indicating that cellular immunity will probably play a critical role in vaccine-induced protection. Neutralizing antibodies also can down-modulate an experi-

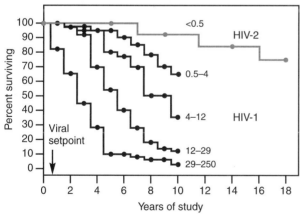

FIG. 14.14 _____

Survival of a cohort of patients infected with HIV-1 to show the influence of viral set point on the course of illness. In addition, a cohort of patients infected with HIV-2 is shown for comparison. The numbers at the end of the curves represent the set points at 6–9 months, expressed as RNA copies ($\times$ 1,000) per milliliter plasma. (After Mellors JW, Rinaldo CR, Gupta P, et al. Prognosis in HIV-1 infection predicted by the quantity of virus in plasma. *Science* 1996;272:1167–1170; and Whittle HC, Ariyoshi K, Rowland-Jones S. HIV-2 and T cell recognition. *Curr Opin Immunol* 1998;10:382–387.)

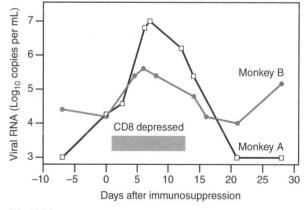

FIG. 14.15 _____

Cellular immunity plays an important role in controlling the viral set point (viremia titer) in simian immunodeficiency virus (SIV) infection of macaques. Viral set points are shown for two animals persistently infected with SIV that were treated with a monoclonal antibody directed against the CD8 receptor to temporarily deplete CD8 cells (during the period shown). Monkey A showed a marked rise in viremia level and monkey B a lesser one during the period of CD8 depletion. (After Schmitz J, Kuroda M, Santra S, et al. Control of viremia in simian immunodeficiency virus infection by CD8+ lymphocytes. *Science* 1999;283:857–860.)

mental infection as shown by the effect of pretreatment of macaques with immune globulin prior to challenge with SHIV (Fig. 14.16). Thus, there is persuasive evidence that several immune modalities could contribute to an AIDS vaccine.

Induction of Neutralizing Antibody and Cellular Immunity

HIV differs from most viruses in that infected persons, with some exceptions, fail to develop neutralizing antibody. Sera from infected patients exhibit antibodies against env and other proteins, but there is little neutralization even when tested against an isolate from the same patient. Also, although many anti-gp120 monoclonal antibodies have been isolated from HIV-infected persons, only a few of them have broadly neutralizing activity. It has been shown that SIV incorporates MHC class II proteins into the virion envelope (in addition to the viral envelope proteins) and that antibody against the MHC molecule has potent neutralizing activity. Taken together, these observations suggest that the difficulty in inducing neutralizing antibodies is due the properties of the gp120 molecule.

Resistance to neutralization is explained, at least in part, by the structure of gp120, which has a large number of glycosylation sites (typically about 24),

so that the surface of the molecule is shielded from antibodies. In addition, there are several loops (particularly the variable regions 1 and 2) that partly cover the domain on gp120 that binds to CD4, the viral receptor. These structural features presumably impede both the induction and the action of neutralizing antibodies. This speculation is supported by experiments with SIV that has been engineered to reduce either glycosylation sites or variable loops (Table 14.10).

By contrast, it appears relatively easy to induce cellular immunity to HIV. Infected patients have CD8 cells that are responsive in various assays for cellular immunity such as the Elispot and intracellular cytokine assays. Epitopes can be mapped to most of the viral structural genes (about 15 epitopes in total per patient), with the highest proportion of responses to the gag proteins. Also, a variety of candidate immunogens are capable of inducing CD8 responses in macaques, with partial protection against AIDS.

Cross-clade Immunity

HIV-1 viruses can be classified into about ten genotypes, usually called *clades*, all of which are present in Africa but which have quite different distributions in the Western Hemisphere, Europe, and Asia. Within a given clade, there is less than 10% nucleotide diversity. There is up to 25% diversity between clades, which is greatest in the *env* gene and less in the *gag* and *pol* genes. Originally, it was postulated that clades might represent immunotypes and that it would be necessary to formulate a multivalent vaccine for use in different geographic areas. However, it now appears that clades do not represent immunotypes and that there is a considerable degree of cross-clade immune responses. If HIV-infected subjects are tested and those few sera are selected that

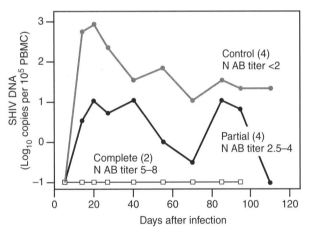

FIG. 14.16

Circulating neutralizing antibody present at the time of exposure can down-modulate simian immunodeficiency virus (SIV) infection of macaques. In this figure, an antiserum against SIV was administered to monkeys, and the level of neutralizing antibody in the recipients was determined prior to challenge with simian human immunodeficiency virus (SHIV) strain DH12 (100 TCID$_{50}$) by the intravenous route. A minimal level of antibody was sufficient to down modulate the subsequent course of infection as reflected in the viremia titer, compared with untreated infected control animals. (After Shibata R,R, Igarashi T, Haigwood N, et al. Neutralizing antibody directed against the HIV-1 envelope glycoprotein can completely block HIV-1/SIV chimeric virus infection of macaque monkeys. *Nature Med* 1999;5:204–210.)

TABLE 14.10		
Effects of removal of glycosylation sites on SIV mutants		
	NEUTRALIZING ANTIBODY TITERS AGAINST WILD-TYPE AND MUTANT VIRUS (MEDIAN AND RANGE)	
Immunizing virus	Wild type	Mutant
Wild type	~60 (40–80)	450 (1–1,500)
Mutant	450 (100–1,000)	1,500 (50–5,500)

Mutants of SIV with selected glycosylation sites removed are more effective at inducing neutralizing antibody and are more susceptible to neutralization than is wild-type SIV. Monkeys were infected with either SIV 239 (wild type) or a deglycosylated mutant and their postinfection sera were tested against both viruses.

After Reitter J, Means R, Desrosiers RC. A role for carbohydrates in immune evasion in AIDS. *Nature Med* 1998;4:679–684.

TABLE 14.11									

Cross-clade neutralization titers from a panel of 10 pairs of sera and isolates from patients infected with HIV-1

	LOG 2 NEUTRALIZATION TITER VS. VIRUS OF CLADE								
Sera	A	B	C	F1	E	F2	G	H	O2	O3
A	5	0	0	0	0	0	0	0	0	0
B	11	8	9	7	11	11	8	7	11	11
C	7	6	7	7	8	9	7	3	8	8
E	7	5	7	6	7	7	6	4	6	7
F1	7	6	8	5	7	8	5	5	6	6
F2	7	6	7	3	8	8	4	3	8	7
G	6	5	7	6	7	8	6	4	5	4
H	8	6	7	4	8	8	6	5	5	8
O2	6	6	7	0	7	5	0	0	7	8
O3	0	0	0	0	0	0	0	0	0	6

A panel of sera that showed significant neutralization of a clade-matched wild-type isolate were tested against isolates representing most other clades. Most sera showed cross-clade neutralization although there were some exceptions.

After Nyambi PN, Nkengasong J, Lewi P, et al. Multivariate analysis of human imunodeficiency virus type 1 neutralization data. *J Virol* 1996;70:6235–6243.

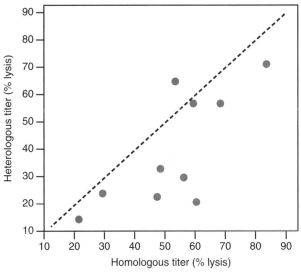

FIG. 14.17 _____

HIV-infected patients exhibit cross-clade cytotoxic T-lymphocyte responses. Lymphocytes from ten patients, half infected with clade B and half with clade E strains of HIV, were tested for their ability to lyse target cells expressing the homologous or the heterologous virus strain. There was a general correlation between the level of cytolysis against homologous and heterologous targets, although the heterologous effect was about 10% lower. (After Lynch JA, deSouza M, Robb MD, et al. Cross-clade cytotoxic T cell response to human immunodeficiency virus type 1 proteins among HLA disparate North Americans and Thais. *J Infect Dis* 1998;178: 1040–1046.)

exhibit some degree of homologous neutralization, most also show cross-clade activity (Table 14.11). Likewise, CTL from infected subjects show both homologous and heterologous activity (Fig. 14.17).

Candidate AIDS Vaccines

There are now several AIDS vaccine candidates in advanced human trials, although the recombinant gp120 is the only vaccine in phase III efficacy trials. As mentioned, although there is little optimism about the efficacy of this formulation, there should be some definitive results by the year 2002. A second candidate vaccine, formulated as a "priming" immunization with a recombinant poxvirus and a "boost" with the recombinant envelope protein, is currently in phase II immunogenicity trials. This vaccine has induced CTL activity but only in 30%–50% of vaccinees, so it is unclear whether it will be advanced to a phase III trial.

Much more promising are a variety of experimental immunogens that exploit different vector systems. Figure 14.18 shows one example that uses two DNA plasmids expressing viral proteins and IL-2 (as a biological adjuvant). This experimental vaccine induced considerable CD8 responses and provided impressive protection when immunized monkeys were challenged with a pathogenic SHIV. An equally impressive degree of protection was conferred by a DNA vaccine prime followed by a recombinant vaccinia virus boost (Fig. 14.19).

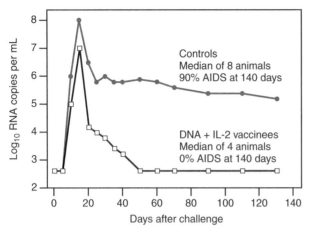

FIG. 14.18 _____

Protection against simian human immunodeficiency virus (SHIV) in macaques by a DNA vaccine enhanced with adjuvant. The animals were immunized with DNA plasmids encoding HIV *env*, SIV *gag*, and *interleukin*-2 at 0, 4, 8, and 40 weeks, and challenged by an intravenous injection of R5X4 pathogenic SHIV 89.6P at 46 weeks. (After Barouch DH, Santra S, Schmitz JE, et al. Control of viremia and prevention of clinical AIDS in rhesus monkeys by cytokine-augmented DNA vaccination. *Science* 2000;290:486–492.)

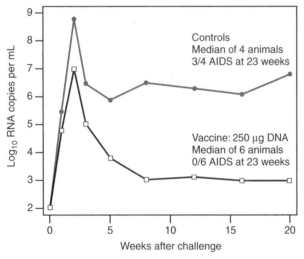

FIG. 14.19 _____

Protection against simian human immunodeficiency virus (SHIV) in macaques by a prime-boost mixed-modality vaccine. The animals were immunized intradermally at 0 and 2 months with 250 μg of a DNA construct expressing most of the proteins of SHIV 89.6, and were boosted at 6 months with a recombinant modified virus Ankara (attenuated vaccinia) virus expressing many of the proteins of SHIV 89.6. At 13 months, they were challenged by intrarectal infection with pathogenic SHIV 89.6P. The vaccinated monkeys contained their viremia (median is shown) whereas the control infected animals did not. Not shown is the loss of CD4 lymphocytes in all of the control infected animals, whereas five of six vaccinated animals maintained their CD4 counts. Since there is no cross-neutralization between SHIV 89.6 and SHIV 89.6P, the considerable protection observed must be mainly attributed to cellular immune mechanisms. (After Amara RR, Villinger F, Altman JD, et al. Control of a mucosal challenge and prevention of AIDS in rhesus monkeys by a multiprotein DNA/MVA vaccine. *Science* 2001;292:69–74.)

REPRISE

The mechanisms whereby immunization protects against virus disease depend on the pathogenesis of the specific infection. In some instances, preformed neutralizing antibody intercepts invading virus at the portal of entry and partially or (rarely) totally inactivates the viral inoculum. In other instances, circulating antibodies neutralize virus entering the blood, preventing dissemination to key target organs or tissues. Antibody may even block transit of viruses by the neural route. It is likely that cellular immune responses also play a role in protection induced by some vaccines. This speculation is based on the observation that, in vaccine-protected individuals, exposure to wild-type virus initiates a mild infection that is rapidly cleared, and this clearance likely involves CD8 effector lymphocytes. In some instances, CD4, CD8, B lymphocytes, and perhaps other lymphoreticular elements cooperate to provide vaccine-induced protection that is more effective than that mediated by any single component of the immune response.

There are many methods to produce a viral vaccine. Licensed human vaccines fall mainly into two groups: live attenuated variant viruses, and nonreplicating antigens manufactured from inactivated virions or recombinant viral proteins. A number of new vectors are now under development as vaccine platforms. These fall into three classes: (a) attenuated viruses—such as vaccinia or other poxviruses—that have been engineered to express additional viral proteins; (b) replicons, which are nonreplicating constructs that will introduce either RNA or DNA encoding viral proteins into host cells; and (c) naked DNA plasmids that encode a protein immunogen under the control of specific promoter sequences.

Different vaccine modalities may protect through a different spectrum of immune defenses. Each vaccine modality has its advantages and disadvantages, and it is unpredictable which approach will be most effective to protect against a specific virus. The newer vector systems now under development may provide improved approaches to vaccines, particularly if combined with the use of cytokines or other biological and chemical adjuvants.

FURTHER READING

Reviews and Chapters

Ellis RW, ed. *New vaccine technologies*. Austin, TX: RG Landes, 2001.

Hasenkrug KJ, Dittmer U. The role of CD4 and CD8 T cells in recovery and protection from retroviral infection: lessons from the Friend virus model. *Virology* 2000;272:244–249.

Lowrie DB, Whalen RG, eds. *DNA vaccines*. Totowa, NJ: Humana Press, 2000.

Nathanson N, Mathieson BJ. Biological considerations in the development of a human immunodeficiency virus vaccine. *J Infect Dis* 2000;182:579–589.

Nathanson N, McFadden G. Viral virulence. In: Nathanson N, et al., eds. *Viral pathogenesis.* Philadelphia: Lippincott–Raven Publishers, 1997.

Paterson Y, ed. *Intracellular bacterial vaccine vectors.* New York: Wiley-Liss, 1999.

Perlmann P, Wigzell H, eds. Vaccines. *Handbook of experimental pharmacology*, Vol. 133. Berlin: Springer-Verlag, 1999.

Plotkin SA, Orenstein WA, eds. *Vaccines.* Philadelphia: WB Saunders, 1999.

Classic Papers

Arthur LO, Bess JW, Urban RG, et al. Macaques immunized with HLA-DR are protected from challenge with simian immunodeficiency virus. *J Virol* 1995;69:3117–3124.

Bodian D. Emerging concept of poliomyelitis infection. *Science* 1955;122:105–108.

Francis TJ Jr, Napier JA, Voight R, et al. Evaluation of the 1954 field trial of poliomyelitis vaccine. School of Public Health, University of Michigan, Ann Arbor, 1957.

Fynan EF, Webster RG, Fuller DH, et al. DNA vaccines: protective immunizations by parenteral, mucosal, and gene gun inoculations. *Proc Natl Acad Sci USA* 1993;90:11478–11492.

Guidotti LG, Ando K, Hobbs MV, et al. Cytotoxic T lymphocytes inhibit hepatitis B virus gene expression by a noncytolytic mechanism in transgenic mice. *Proc Natl Acad Sci USA* 1994;91:3764–3768.

Reitter J, Means R, Desrosiers RC. A role for carbohydrates in immune evasion in AIDS. *Nature Med* 1998;4:679–684.

Ulmer JB, Donnelly JJ, Parker SE, et al. Heterologous protection against influenza by injection of DNA encoding a viral protein. *Science* 1993;259:1745–1749.

Original References

Amara RR, Villinger F, Altman JD, et al. Control of a mucosal challenge and prevention of AIDS in rhesus monkeys by a multiprotein DNA/MVA vaccine. *Science* 2001;292:69–74.

Baer GM, Cleary WF. A model in mice for the pathogenesis and treatment of rabies. *J Infect Dis* 1972;125:520–532.

Barouch DH, Santra S, Schmitz JE, et al. Control of viremia and prevention of clinical AIDS in rhesus monkeys by cytokine-augmented DNA vaccination. *Science* 2000;290:486–492.

Beasley RP, Hwang LY, Stevens CE, et al. Efficacy of hepatitis B immune globulin for prevention of perinatal transmission of the hepatitis B virus carrier state: final report of a randomized double-blind placebo-controlled trial. *Hepatology* 1983;3: 135–141.

CDC. Outbreak of poliomyelitis—Dominican Republic and Haiti, 2000. *MMWR* 2000;49:1094–1103.

Henry JL, Jaikara ES, Davies JR, et al. A study of polio vaccination in infancy: excretion following challenge with live virus by children given killed or living poliovaccine. *J Hyg* 1966;64:105–120.

Lodmell DL, Ray NB, Ulrich JT, et al. DNA vaccination of mice against rabies virus: effects of the route of vaccination and the adjuvant monophosphoryl lipid A (MPL). *Vaccine* 2000;6: 1059–1066.

Lynch JA, deSouza M, Robb MD, et al. Cross-clade cytotoxic T cell response to human immunodeficiency virus type 1 proteins among HLA disparate North Americans and Thais. *J Infect Dis* 1998;178:1040–1046.

Mellors JW, Rinaldo CR, Gupta P, et al. Prognosis in HIV-1 infection predicted by the quantity of virus in plasma. *Science* 1996;272:1167–1170.

Monath TP, Levenbook I, Soike K, et al. Chimeric yellow fever virus 17D-Japanese encephalitis virus vaccine: dose-response effectiveness and extended safety testing in rhesus monkeys. *J Virol* 2000;74:1742–1751.

Nathanson N. Epidemiologic aspects of poliomyelitis eradication. *Rev Infect Dis* 1984;6(Suppl 2):S308–S312.

Nathanson N, Bodian D. Experimental poliomyelitis following intramuscular virus infection. III. The effect of passive antibody on paralysis and viremia. *Bull Johns Hopkins Hosp* 1962;111:198–220.

Nathanson N, Hirsch VM, Mathieson BJ. The role of nonhuman primates in the development of an AIDS vaccine. *AIDS* 1999;13(Suppl A):S113–S120.

Nyambi PN, Nkengasong J, Lewi P, et al. Multivariate analysis of human imunodeficiency virus type 1 neutralization data. *J Virol* 1996;70:6235–6243.

Ogra PL, Karzon DT, Righthand F, MacGillivray M. Immunogobulin response in serum and secretions after immunization with live and inactivated poliovaccine and natural infection. *N Engl J Med* 1968;279:893–900.

Ourmanov I, Brown CR, Moss B, et al. Comparative efficacy of recombinant modified vaccinia virus Ankara expressing simian immunodeficiency virus (SIV) Gag-pol and or Env in macaques challenged with pathogenic SIV. *J Virol* 2000;74:2740–2751.

Sabin AB. Properties and behavior of orally administered attenuated poliovirus vaccine. *JAMA* 1957;164:1216–1223.

Sabin AB, Alvarez MR, Amezquita JA, et al. Effects of rapid mass immunization of a population with live, oral poliovirus vaccine under conditions of massive enteric infection with other viruses. In: *Second international conference on live poliovirus vaccines*. Scientific Publication 50. Geneva: WHO, 1960.

Sabin AB, Hennessen WA, Winsser J. Studies on variants of poliomyelitis virus. *J Exp Med* 1954;99:551–576.

Sawyer WA, Meyer KF, Eaton MD, et al. Jaundice in army personnel in the western region of the United States and its relation to vaccination against yellow fever. *Am J Hyg* 1944;39:337–387.

Schmitz J, Kuroda M, Santra S, et al. Control of viremia in simian immunodeficiency virus infection by CD8[+] lymphocytes. *Science* 1999;283:857–860.

Shibata R, Igarashi T, Haigwood N, et al. Neutralizing antibody directed against the HIV-1 envelope glycoprotein can completely block HIV-1/SIV chimeric virus infection of macaque monkeys. *Nature Med* 1999;5:204–210.

Sikes RK, Cleary WF, Koprowski H, et al. Effective protection of monkeys against death from street virus by post-exposure administration of tissue-culture rabies vaccine. *Bull WHO* 1971;45:10.

Valtanen S, Roivainen M, Piirainen L, et al. Poliovirus-specific intestinal antibody responses coincide with decline of poliovirus excretion. *J Infect Dis* 2000;182:1–5.

Watanabe M. Polio outbreak threatens eradication program. *Nature Med* 2001;7:135.

Whittle HC, Ariyoshi K, Rowland-Jones S. HIV-2 and T cell recognition. *Curr Opin Immunol* 1998;10:382–387.

Wiktor TJ. Cell-mediated immunity and postexposure protection from rabies by inactivated vaccines of tissue culture origin. *Dev Biol Stand* 1978;40:255–265.

Xu ZY, Liu CB, Francis DP, et al. Prevention of perinatal acquisition of hepatitis B carriage using vaccine: preliminary report of a randomized double-blind placebo-controlled and comparative trial. *Pediatrics* 1985;76:713–718.

Glossary

A2G an inbred mouse strain

AAV adeno-associated virus, a parvovirus that depends on adenovirus as a "helper"

ADCC antibody-dependent cell-mediated cytolysis

ADV Aleutian disease virus

AIDS acquired immunodeficiency syndrome

A/J an inbred mouse strain

Aleutian disease a disease caused by a parvovirus that is particularly pronounced in the Aleutian strain of mink

alphaherpesviruses herpesviruses that infect neurons

ALSV avian leukosis-sarcoma virus

ALV avian leukosis virus

amphotropic a class of murine leukemia viruses that infect cells from mice and other species

anti-HBs antibody against HBs antigen

AP-2 adapter complex that recruits transmembrane proteins to clathrin-coated pits

APC antigen-presenting cell

arboviruses arthropod-borne viruses

ASC antibody-secreting cell

ASV avian sarcoma virus

ATL acute T-cell leukemia

B cell a lymphocyte that matures in bone marrow and is the progenitor of antibody-secreting plasma cell

B19 designation of a human parvovirus that infects erythrocytes

B8R a poxvirus protein homologous to the IFN-γ receptor

BALB/c an inbred mouse strain

BCR B-cell receptor

betaherpesviruses herpesviruses that infect lymphoid cells

Bgp1 biliary glycoprotein 1, a cellular membrane protein that also serves as a receptor for mouse hepatitis virus

BK polyoma virus a human polyoma virus, named after a patient from whom it was isolated

BLV bovine leukemia virus

BrdU bromodeoxyuridine, a nucleotide used to label cells

C3, C1q, C4b complement proteins

C4b-BP plasma protein that binds C4b

CA capsid protein

cAMP cyclic adenosine monophosphate

carbolic acid dilute phenol, an antiseptic

CC chemokines chemokines with a cysteine-cysteine motif

CCR5 5th receptor for CC chemokines

CD cluster of differentiation, a cell marker identified by a cluster of monoclonal antibodies

CD4 cluster of differentiation 4, a cell surface marker used to define helper T lymphocytes

CD8 cluster of differentiation 8, a cell surface marker used to define effector T lymphocytes

CDK cyclin-dependent kinase

CDV canine distemper virus, a morbillivirus of dogs

CMV cytomegalovirus, a gammaherpesvirus

c-myc cellular myelocytomatosis protooncogene, a transcription factor

CNS central nervous system

colchicine a drug that can block fast axoplasmic transport in neuronal processes

complement-fixing antibody antibody that is capable of binding complement through the Fc domain on the antibody molecule

condyloma wart

Coronaviridae a family of positive-stranded RNA viruses

CpG cytosine phosphate guanosine, an oligonucleotide

CPV canine parvovirus

CR1, CR2 conserved region 1 and 2, in adenovirus E1A and HPV E7 proteins

CRE cAMP-responsive element

CREB CRE binding

CRPV cottontail rabbit papillomavirus

CTL cytotoxic T lymphocytes, cells capable of lysis of antigen-specific target cells

cupping therapeutic bleeding

CVS challenge virus standard, a rabies virus strain

CXC chemokines chemokines with a cysteine-X-cysteine motif

CXCR4 4th receptor for CXC chemokines

Δ32 mutation a mutation in the *CCR5* gene that results in failure to express the CCR5 protein

dendritic cells specialized macrophages found in skin and lymphoid tissues

DHF/DSS dengue hemorrhagic fever/dengue shock syndrome

DTH delayed-type hypersensitivity

dysplasia precancerous change in cellular phenotype

E1A, E3 early proteins of adenovirus

E2 one of the envelope proteins of Sindbis virus, an alphavirus

E2F family of cellular transcription regulators

EBNA Epstein–Barr nuclear antigen

EBV Epstein–Barr virus

ecotropic a class of murine leukemia viruses that will only infect mouse cells

EGF epidermal growth factor

EIAV equine infectious anemia virus, a lentivirus

EIF eukaryotic initiation factor

ELISA enzyme-linked immunosorbent assay

Elispot an assay for functional CD8 cells that measures the secretion of cytokines such as IFN-α

endogenous applied to retroviral sequences that exist as sequences within germline DNA

enhancer a DNA sequence that enhances transcription by binding cellular transcription factor(s)

env envelope gene of retroviruses

ER endoplasmic reticulum

ev endogenous virus, retroviral sequences in the host genome

exogenous applied to retroviruses that circulate as replication-competent viruses and are transmitted from host to host

F1 first cross between two parental strains of an organism

FACS fluorescence-activated cell sorter

FcR receptor for the Fc domain of immunoglobulin molecules

FeSV feline sarcoma virus, a retrovirus

FIV feline immunodeficiency virus, a lentivirus

flv gene that influences susceptibility to flaviviruses

FMR Friend, Maloney, Rauscher group of murine leukemia viruses

F-MuLV Friend murine leukemia virus

fomites microbiologically contaminated materials that transmit infection

FPV feline panleukopenia virus, a parvovirus

G protein guanine nucleotide-binding protein

gag gene that encodes the major internal structural proteins of retroviruses

Gag group antigen of retroviruses, one of the virus structural polypeptides

GALT gastrointestinal-associated lymphoid tissue

gammaherpesviruses herpeviruses that infect lymphoid tissue and are associated with neoplasms

gE, gI, gC glycoproteins of herpes simplex virus

GH growth hormone

G-MuLV gross murine leukemia virus

gp120 the outer envelope glycoprotein of HIV, also known as the SU or surface protein

gp160 HIV envelope glycoproteins prior to cleavage into gp41 and gp120

gp41 the inner envelope protein of HIV, also known as the TM or transmembrane protein

GPCR G-protein-coupled receptor

Gross/AKR gross subgroup of murine leukemia viruses

H-2 major histocompatibility complex locus in mice

H5N2 designation for the hemagglutinin 5 and neuraminidase 2 of influenza virus

HA hemagglutinin of influenza virus, consisting of two peptides, HA1 and HA2

HAART highly active antiretroviral therapy

HBcAg the core antigen of HBV

HBIG hepatitis B immune globulin

HBs hepatitis B surface antigen, the viral envelope protein

HBsAg hepatitis B surface antigen

HBV hepatitis B virus

HCV hepatitis C virus

herd immunity protection conferred on unimmunized members of a partially vaccinated population because virus transmission is reduced in the immunized members of the group

HHV-8 human herpesvirus 8, also called Kaposi's sarcoma herpesvirus

HIV human immunodeficiency virus

HLA human leukocyte antigen

HPV human papillomavirus

HSV herpes simplex virus

HTLV-1 human T-cell leukemia virus type 1

ICE interleukin-1–converting enzyme

ICP infected cell protein, a term used to designate individual proteins of HSV

IE immediate early genes of HSV

IFN interferon

IFN-γ interferon-γ, or immune interferon

IFN-γR cellular receptor for IFN-γ

Ig immunoglobulin

IgA immunoglobulin A

IgG immunoglobulin G

IL-1 interleukin-1

IL-2 interleukin-2, also called T-cell growth factor

IL-4 interleukin-4

immortalized cell line a cell line that can be maintained indefinitely in culture

IN integrase, a viral enzyme essential for the integration of viral DNA into host DNA

in situ PCR a method for the histochemical identification of specific nucleic acids using sequence amplification followed by in situ hybridization

INOS inducible nitric oxide synthase

IPV inactivated poliovirus vaccine

IRES internal ribosomal entry site

ISVP infectious subvirion particle of reovirus

JAK Janus tyrosine kinase

JC polyoma virus a human polyoma virus, named after the patient from whom it was isolated

KIR killer inhibiting receptor, receptors in NK cells that inhibit perforin-mediated killing

knockout mice mice in which a specific gene has been inactivated using a method that involves homologous DNA recombination in embryonic stem cells

KS Kaposi's sarcoma, a skin cancer

KSHV Kaposi's sarcoma herpesvirus, also called HHV-8

L929 cells a murine cell line

LAT latency-associated transcript, a term used for RNA transcripts of the HSV genome that are produced during latency

LCMV lymphocytic choriomeningitis virus

LD$_{50}$ 50% lethal dose

LDA limiting dilution assay, used to quantify CTL precursor or memory cells

LMP lymphoblast membrane protein, a protein of EBV

LTR long terminal repeat, a noncoding region at the termini of retroviruses

M cells microfold cells, specialized cells in the epithelium of the intestine that are involved in antigen uptake and viral entry

M, N, O main, new, outlier, subgroups of HIV-1

MA matrix protein, a structural protein of many viruses

macrophages the principal phagocytic cells of the body

MAIDS murine AIDS

MALT mucosal-associated lymphoid tissue

MBP myelin basic protein

MCFV mink cell focus-forming virus

MCMV mouse cytomegalovirus, a herpesvirus

MDV Marek's disease virus, a herpesvirus

MHC major histocompatibility complex, a set of contiguous genes encoding the major determinants of graft rejection

MHV mouse hepatitis virus, a coronavirus

MHVR mouse hepatitis virus receptor, also biliary glycoprotein or Bgp

mos Maloney mouse sarcoma, a viral oncogene

MP mousepox, caused by ectromelia virus, a poxvirus

MPV mouse polyoma virus

MSV murine sarcoma virus

MT-2 a continuous cell line of human T lymphocytes

MuLV murine leukemia virus

Mx gene a genetic locus that influences host susceptibility to myxoviruses such as type A influenza virus

myc myelocytomatosis, a viral oncogene

NA a continuous mouse neuroblastoma cell line

NC nucleocapsid protein of retroviruses

Nef negative factor, an accessory protein of HIV

NK cell natural killer cell, a class of lymphocytes

NYVAC an attenuated strain of vaccinia virus

OPV oral poliovirus vaccine, also known as Sabin vaccine

p12 a gag protein of retroviruses

p15e 15-kd envelope protein of some retroviruses

p53 a tumor suppressor protein

papilloma wart

PBMC primary blood mononuclear cell

PCR polymerase chain reaction

pCTL precursor cytotoxic T lymphocyte

Peyer's patches lymphoid patches in the lining of the small intestine

PFU plaque-forming unit

pol polymerase gene of retroviruses that encodes protease, reverse transcriptase, and integrase

polytropic a class of murine leukemia viruses that will infect vertebrate cells other than mouse cells, also called MCF viruses

PR protease enzyme of retroviruses

Pr60^{gag} the gag protein encoded by the MAIDS virus, a variant of the normal gag protein

pRb retinoblastoma protein, a tumor suppressor

PRI Princeton Rockefeller Institute, designation of a strain of mice

promoter a DNA sequence that can bind RNA polymerases and initiate transcription of downstream exons

PVR poliovirus receptor

ras rat sarcoma, a viral oncogene

RDA representational difference analysis

rev regulator of expression of viral proteins, an accessory protein of HIV

RFC/B.5 an attenuated bunyavirus mutant selected by passage in cell culture

Rfv-1 a gene that affects recovery from Friend MuLV

Rfv-2 a gene that affects recovery from Friend MuLV

RIF Rous interfering factor, an ALV that interferes with superinfection by an RSV of the same subgroup

RPV rabbit papilloma virus

RSV Rous sarcoma virus

RT reverse transcriptase

RV194-2 an attenuated variant rabies

sarcoma tumor of transformed muscle cells

Schwann cells cells that form the myelin sheath around neuronal processes of peripheral nerves

SHIV simian human immunodeficiency virus, a chimeric virus that usually has the *env* gene of HIV inserted into a SIV backbone

SI stimulation index

Sindbis virus an alphavirus

SIV simian immunodeficiency virus

Skp1p proteosome targeting factor

src sarcoma, a viral oncogene

SSPE subacute sclerosing panencephalitis, a chronic progressive fatal infection of humans caused by measles virus

STAT signal transducer and activator of transcription

stratum corneum the outermost layer of the epidermis

stratum granulosum an intermediate layer of the epidermis

stratum Malphighii the innermost layer of the epidermis

SU surface protein one of the two envelope proteins of HIV

SV40 simian virus 40, a polyoma virus

$t_{1/2}$ half-life

T cell lymphocyte that has matured in the thymus

T_H1 CD4 helper cells that induce cellular immune responses

T_H2 CD4 helper cells that induce B lymphocyte proliferation and differentiation

tat transactivator of transcription, an accessory gene of HIV

TATA a frequently occurring DNA sequence in enkargotic cell promoters

tax trans-acting protein of the HTLV-1 group of retroviruses

TCD tissue culture dose

TCL T-cell line, a cell line derived from transformed T lymphocytes

TCR T-cell receptor

thoracic duct the final conduit that carries lymph into the vena cava

TM transmembrane, a protein that crosses the viral envelope

TNF tumor necrosis factor

TNF-α tumor necrosis factor α, a cytokine with cytopathic properties

Tolerance a state of immunologic unresponsiveness

TRAF TNF receptor association factor

transformed cell line an immortalized cell line that shows an oncogenic phenotype

βTrCP protein that binds to Skp1p

trigeminal ganglion the organ containing the cell bodies of the trigeminal nerve fibers

trigeminal nerve a cranial nerve

ts temperature sensitive

TUNEL assay terminal deoxynucleotidyltransferase-mediated dUTP nick end labeling

U3 unique 3′ noncoding sequence in retrovirus genomes

U5 unique 5′ noncoding sequence in retrovirus genomes

VAP viral attachment protein

VCP vaccine complement controlling protein

VEEV Venezuelan equine encephalitis virus, an alphavirus

VEGF vascular endothelial growth factor

vif viral infectivity factor, an accessory protein of HIV

VMV visna maedi virus, an ovine lentivirus

VP1 virus protein 1, a structural protein of picornaviruses

vpr virion protein r, an accessory protein of HIV

vpu virion protein u, an accessory protein of HIV

VZV varicella-zoster virus, a herpesvirus

WHV woodchuck hepatitis virus, a hepadnavirus

WNV West Nile virus, a flavivirus

xenotropic a class of murine leukemia viruses that will infect vertebrate cells other than mouse cells

Subject Index

Page numbers followed by f, t, or s refer to figures, tables, or sidebars, respectively.

A

Acquired immunodeficiency syndrome (AIDS). *See also* Human immunodeficiency virus
  murine AIDS features, 90
  neoplasms, 176
  opportunistic infections, 92, 176
  susceptibility
    AIDS resistance, 164–165
    HLA-B57 in AIDS susceptibility, 165, 165t
Adenovirus
  apoptosis effects, 46, 47
  downregulation of major histocompatibility complex class I expression, 44–45, 87–88
  infection course, 148–149
  oncogenesis
    E1A role, 149–150, 149f
    E1B role, 149–150, 149f
    rodent models, 149
    serotype activity, 45t
    tumor suppressor protein inactivation, 143
Age, host
  virus immunosuppression determinant, 94
  virus susceptibility factors
    infancy, 165–167
    old age, 167–168
AIDS. *See* Acquired immunodeficiency syndrome
ALV. *See* Avian leukosis virus
Antibody. *See* Immunoglobulin
Antigen, features, 57
Antigen-presenting cell (APC). *See also* Dendritic cell
  function, 58
  vaccination role, 189
APC. *See* Antigen-presenting cell
Apoptosis
  adenovirus modulation, 46, 47
  assays, 46
  features, 46, 47s
  human immunodeficiency virus induction, 40, 46
  reovirus induction, 39–40
  Sindbis virus induction, 47
  virus proteins in induction, 46–47, 47f
ASV. *See* Avian sarcoma virus
Attenuated virus
  comparative pathogenesis of virulent and attenuated viruses
    host immune response, 108
    neural spread, 105–106
    portal of entry, 105
    target organ, 106
    tissue specificity, 107–108
    viremia, 105
  selection
    choice of viruses for virulence studies, 104

(Attenuated virus continued)
    cold-adapted variants, 103–104
    monoclonal antibody-resistant viruses, 104
    mutagenized viruses, 104
    temperature-sensitive mutants, 103
  vaccines
    efficacy, 190–192
    poliovirus vaccine attenuation, 9–10, 100, 193t
    safety, 192–193, 193t
    tropism, 190, 192t
Autoimmune disease, molecular mimicry by viruses, 78–80
Avian influenza virus
  blocking of infection, 43f
  susceptibility factors, 162
  tropism variability, 35t, 36
Avian leukosis virus (ALV)
  blocking of infection, 43f
  oncogenesis, 137–138
  receptor saturation, 42
  susceptibility factors, 162
Avian sarcoma virus (ASV)
  oncogenesis, 137–138
  receptor saturation, 42

B

B cell
  Epstein—Barr virus infection, 150
  immune response overview, 48f, 57, 59f
  immunoglobulin production pathway, 58
  kinetics of humoral response, 62–63
  maturation, 56
  mucosal immune response, 67–68
  protection against reinfection, 67
  specificity of response, 57–58
Bcl-2
  apoptosis inhibition, 47
  Sindbis virus susceptibility role, 166
Bgp1, mouse hepatitis virus susceptibility role, 162
Blood, shedding of viruses, 22
Bunyavirus, virulence, genetic determinants, 107f, 109
Burkitt's lymphoma. *See* Epstein—Barr virus

C

Canine distemper virus (CDV), immunosuppression induction, 90, 90f
Carbohydrate, viral receptors, 29f
CD4 lymphocyte. *See also* T cell
  cell-mediated immunopathology effector function, 71–72, 73
  cytokine production, 65
  effector functions, 65
  function, 58
  helper cell subsets, 58, 60
  human immunodeficiency virus targeting, 27–28, 44f, 83, 92–93, 93f, 121, 173–174

(CD4 lymphocyte continued)
    immunopathology role, 75t
    SU, CD4 binding, 171, 172f, 177
    virus effects on helper cell balance, 60
CD8 lymphocyte. *See also* T cell
  assays, 64t
  cell-mediated immunopathology
    clinical disease, 73–76
    experimental examples, 72–73
    mechanisms, 70–72
  cytokine production, 71
  effector functions, 65
  function, 58
  hepatitis B virus clearance, 201–202
  human immunodeficiency virus effects, 182, 183
  kinetics of virus response, 65–66
  lymphocytic choriomeningitis virus induction of tolerance, 85–86
  rabies virus response, 200
CDV. *See* Canine distemper virus
Cell cycle, virus regulation, 45
Cervical cancer. *See* Human papillomavirus
Chemokine receptors
  CCR5 mutations
    AIDS resistance, 164–165
    human immunodeficiency virus infection resistance, 163t, 164, 182
  human immunodeficiency virus utilization
    apoptosis induction, 46
    receptor types in binding, 171
    signaling, 40
    strain variations, 171
  mutations and human immunodeficiency virus infection resistance, 164, 182, 183
Chromium-51 release assay, cellular immunity, 63f, 64
Clearance, virus, 18, 19
CMV. *See* Cytomegalovirus
Complement
  C1q binding by antibodies, 62, 87
  cascade, 57f
  host defense, 55–56
  viral interference of cascade, 87
Conjunctiva, virus entry, 14
CREB, transcription regulation by viruses, 41
Cytomegalovirus (CMV)
  latency, 119–120
  transcription regulation, 41
Cytotoxic T lymphocyte. *See* CD8 lymphocyte

D

Dendritic cell
  function, 56–57, 58
  human immunodeficiency virus infection, 173
  lymphocytic choriomeningitis virus targeting, 83–84